Transmission Electron Microscopy of Metals

WILEY SERIES ON THE SCIENCE AND TECHNOLOGY OF MATERIALS

Advisory Editors: J. H. Hollomon, J. E. Burke, B. Chalmers, R. L. Sproull, A. V. Tobolsky

---

PRINCIPLES OF SOLIDIFICATION
Bruce Chalmers

APPLIED SUPERCONDUCTIVITY
Vernon L. Newhouse

THE MECHANICAL PROPERTIES OF MATTER
A. H. Cottrell

THE ART AND SCIENCE OF GROWING CRYSTALS
J. J. Gilman, editor

COLUMBIUM AND TANTALUM
Frank T. Sisco and Edward Epremian, editors

MECHANICAL PROPERTIES OF METALS
D. McLean

THE METALLURGY OF WELDING
D. Séférian—Translated by E. Bishop

THERMODYNAMICS OF SOLIDS
Richard A. Swalin

TRANSMISSION ELECTRON MICROSCOPY OF METALS
Gareth Thomas

PLASTICITY AND CREEP OF METALS
J. D. Lubahn and R. P. Felgar

INTRODUCTION TO CERAMICS
W. D. Kingery

PROPERTIES AND STRUCTURE OF POLYMERS
Arthur V. Tobolsky

MAGNESIUM AND ITS ALLOYS
C. Sheldon Roberts

PHYSICAL METALLURGY
Bruce Chalmers

FERRITES
J. Smit and H. P. J. Wijn

ZONE MELTING
William G. Pfann

THE METALLURGY OF VANADIUM
William Rostoker

INTRODUCTION TO SOLID STATE PLASTICS, SECOND EDITION
Charles Kittel

Transmission Electron Microscopy of Metals

by

Gareth Thomas

ASSISTANT PROFESSOR OF METALLURGY,
UNIVERSITY OF CALIFORNIA,
BERKELEY

John Wiley & Sons, Inc.

New York · London · Sydney

THIRD PRINTING, AUGUST, 1966

Library of Congress Catalog Card Number: 62-8790

Printed in the United States of America

Preface

Over the last twenty-five years electron microscopy has become an increasingly popular technique for examining materials, but it is only recently that microscopes have been made available which are capable of resolving details of the order of atomic dimensions. It is fortunate that specimen preparation techniques have progressed to the stage where it is now possible to see directly through solid materials so that a wealth of new experimental evidence concerning their inner structures has been obtained. One of the most striking advances in electron microscopy has been the direct observation of defects in crystals because hitherto these could only be studied by indirect means. The tremendous advantage of the transmission technique is, of course, that the results obtained are visual and therefore convincing.

As a result of the rapid progress which has recently been made, it is now possible to interpret electron micrographs of crystals by using the theory of electron diffraction. However, although this theory and its application to electron microscopy is available in many journals and conference reports, as yet there is no single textbook which deals with the subject from first principles and in a form which is readily understood by the nonphysicist. Consequently, I have written this book as an introduction to the field; it is based on the lecture course given in the Mineral Technology Department, University of California, Berkeley.

In this book the emphasis has been placed upon the principles and applications of the transmission technique; thus I have not discussed in any detail the theory of the instrument or electron optics. Instead I have attempted to give a comprehensive account of the fundamental principles which are involved in the study of metals using transmission electron microscopy. Because of the rapid developments which have occurred since 1956, a demand has arisen for trained microscopists to work in the field of metallurgy and metal physics. There is no doubt that this demand will increase in the near future. Unfortunately, at the present time there are not enough experienced teachers generally available who can impart their knowledge to students and to those who already have some acquaintance with the electron microscope. It is to be expected that electron microscopy and electron diffraction will soon be universally taught at the universities so that every student is made aware of the potentialities of the technique. At the moment there are few universities where a facility of this kind is available.

Because of the specialized nature of the subject at the present time, students who have taken the course at Berkeley have all been seniors or graduates who have previously taken courses in crystallography and X-ray diffraction, although these are not necessarily prerequisites. In the book it is assumed that the reader has a working knowledge of crystallography, so that this subject has not been included, but I have outlined an *introductory treatment* of electron diffraction. Thus in the first part of the book I have discussed electron scattering, the theory of electron diffraction, and its relation to image formation and contrast. This is followed by an account of electron microscope and specimen preparation techniques (including replication), and in the final chapter I have summarized the major application of transmission electron microscopy to the study of the structure of metals. References have been given wherever possible so that the reader may pursue the topics more deeply.

It is essential that practical use of the microscope and sample preparation complement the lecture course, and so students are encouraged to attempt short research problems in addition to laboratory demonstrations in order to gain practical experience in as short a time as possible. For this to be of maximum efficiency, the laboratory must be adequately equipped to provide preparation space and darkroom facilities. Since many people are unfamiliar with the requirements for establishing a laboratory for electron microscopy, an appendix describing this has been included. I am greatly indebted to the Siemens and Halske Company for supplying most of this information.

Acknowledgments to authors and publishers who have very kindly allowed many diagrams and plates to be reproduced in this book are listed separately. Here it is a pleasure to record the invaluable assistance and experience I gained while at the Metallurgy Department of the University of Cambridge, England, in particular to Dr. Jack Nutting who first introduced me to this subject, and to many former colleagues for their help and encouragement. I am indebted to Drs. P. B. Hirsch, A. Howie, and M. J. Whelan of the Cavendish laboratory, for allowing me to quote at length from their published results concerning the theory of contrast discussed in Chapters 2 and 5. I also wish to thank the members of the staff of the Department of Mineral Technology at Berkeley who have assisted directly and indirectly in the preparation of this book. I am particularly grateful to Professor Earl R. Parker for encouraging publication of the material and to Drs. V. E. Cosslett, R. D. Heidenreich, J. A. Ibers, P. R. Swann, R. de Wit, and J. Washburn for making valuable suggestions and criticisms after reading parts of the manuscript. Finally, I wish to express my gratitude to my wife for her unfailing patience and help at all times.

Berkeley, California
September 1961

GARETH THOMAS

Acknowledgments

The subject matter presented in this book has been derived from various sources, and I am grateful to many authors and publishers for allowing me to use their material. I also wish to thank the following publishers and companies for giving permission to print their illustrations:

Acta Metallurgica
Butterworths Scientific Publications
Clarendon Press, Oxford
Hitachi, Ltd.
The Institute of Metals and the Iron and Steel Institute
McGraw-Hill Publishing Company
Phillips Electron Optics
Philosophical Magazine
Review of Scientific Instruments
The Royal Society
Siemens and Halske
Springer Verlag
John Wiley and Sons
Zeitschrift für Naturforschung
Zeitschrift für Physik

I am also indebted to many colleagues for supplying me with photographs. The acknowledgments and references for these are given in the corresponding Figure captions.

Contents

1

Electron Scattering

1.1 Introduction

Although electron microscopy has developed with astonishing rapidity, its allied technique, namely electron diffraction, has not been nearly as widely used. Probably the chief reason for the less rapid development of electron diffraction is that the interpretation of diffraction patterns, and the information which they provide, are sufficiently highly specialized to be out of the usual experience of many electron microscopists. In the electron microscopy of noncrystalline solids this situation is not of great seriousness, but when crystals are being investigated, electron diffraction and microscopy must assume equal importance. The reasons for making this statement will become obvious when a little attention is paid to the interactions which occur when an electron beam passes through a crystal. These interactions are not simple processes, and it is the purpose of this opening chapter to introduce and develop the concepts which we require to understand them. This approach is necessary in order to explain electron diffraction patterns; it also forms the basis for our discussions in Chapters 2 and 5 of the mechanisms of image formation and contrast. It is hoped that this procedure will place electron diffraction in its proper perspective to the microscopist.

To account for the diffraction of electrons by crystals the electrons must be treated as waves. Thus the initial problem is to consider the wavelike character of electrons and the scattering which is produced when they pass through a crystal.

1.2 Wave Nature of Electrons

When a crystal is bombarded with a beam of electrons (also known as β-particles), we find that one or more strong and well-defined beams of electrons are emitted from the bombarded areas of the sample. Such a beam is analagous to the diffracted light rays from a grating or X-ray diffraction from crystals. Trying to account for the strong diffracted beams on the supposition that electrons have only the nature of particles invokes complete failure. As a result of the theory of de Broglie,[1] however, a satisfactory explanation of the motion of material particles in terms of waves (de Broglie waves) has been obtained, and his theory, proposed in 1924, has been amply demonstrated by experiment. As with light waves, the wave fronts of an electron beam are normal to the rays, the rays being the directions of propagation, i.e., the motion of the electrons. For the purpose of describing electron paths in a microscope we normally treat the electron beam in terms of ray paths rather than as a wave motion. For a particle in uniform motion the wave fronts can be regarded as plane, and the beam is homogenous if all the particles have the same velocity, i.e., a monochromatic beam. Before dealing specifically with electron waves let us consider the general theory of waves.

It is difficult to describe a wave motion with generality and precision. Even the mathematical definition of it as the *solution* of a differential equation is open to question, for there are several modifications of the usual form which can also be considered wave equations.[2] A wave is an effect that has a definite value at a definite place and time, but in general varies continuously from position to position and from time to time. In the special case of simple harmonic motion, the variation with time is the same at every position (the monochromatic wave). At a single point this motion is described by

$$\psi = a \sin (\nu t + c) \tag{1a}$$

where a is the amplitude, t the time, $\nu/2\pi$ the frequency, $2\pi/\nu$ the period, and $(\nu t + c)$ the phase. The simplest wave of this type is the plane wave of constant amplitude traveling along the x axis

$$\psi = a \sin \left(\nu t \pm \frac{2\pi x}{\lambda}\right) \tag{1b}$$

where λ is called the wavelength. We can define the wave function ψ by stating that the square of the absolute value of ψ represents the probability of finding an electron at any distance x. Another form of this type of wave is the spherical wave, namely

$$\psi = \frac{a}{r} \sin \left(\nu t \pm \frac{2\pi r}{\lambda}\right) \tag{2}$$

Equation (1) can be expressed in the general form

$$\psi = a_{xyz} \sin (\nu t - \phi_{xyz}) \tag{3}$$

where a and ϕ are functions of the space coordinates, but not of time. A wave front is defined as a surface over which ϕ is constant, i.e., it is a surface of constant phase at a given time. As an analogy to Huygens' principle in geometrical optics, each point on the wave front can be considered as a source of secondary wavelets, and successive wave fronts can be determined by Huygens' construction.[3,4] This is only true when the velocity is always constant. When the velocity varies, it is necessary to apply dynamical arguments.[2]

For the types of waves given by equations (1) and (2), the phase velocity (v) is constant and has the value $\nu\lambda/2\pi$.

If we reconsider equation (1b), by differentiating we obtain

$$\frac{\partial\psi}{\partial x} = \pm\frac{2\pi}{\lambda}\cos\left(\nu t \pm \frac{2\pi x}{\lambda}\right) \tag{4}$$

and the second differential gives

$$\frac{\partial^2\psi}{\partial x^2} = \frac{4\pi^2}{\lambda^2}\sin\left(\nu t \pm \frac{2\pi x}{\lambda}\right) \tag{5}$$

which leads to the general form of the one-dimensional wave equation

$$\frac{\partial^2\psi}{\partial x^2} + \frac{4\pi^2\psi}{\lambda^2} = 0 \tag{6}$$

Solution of this equation are $\psi = \sin 2\pi(x/\lambda)$ or $\psi = \exp(-2\pi ix/\lambda)$ etc. In the general three-dimensional case, from equation (6) we have

$$\frac{\partial^2\psi}{\partial x^2} + \frac{\partial^2\psi}{\partial y^2} + \frac{\partial^2\psi}{\partial z^2} = \frac{1}{V^2}\frac{\partial^2\psi}{\partial t^2} \tag{7}$$

where the wave velocity V is written for $\nu\lambda/2\pi$. When V is constant, all plane waves of constant amplitude are propagated with equal speed; thus $v = V$. If the wave is periodic, it can be represented by a series of sines and cosines, i.e., a Fourier series. If the wave is not periodic, the representation becomes a Fourier integral and this gives an expression for a wave of any form, and many examples may be found in the book by Brillouin.[5]

De Broglie[1] postulated that material particles have a wavelike character and that the wave length of the associated wave is given by

$$\lambda = \frac{h}{mu} \quad \text{i.e.,} \quad \frac{h}{\text{momentum}} \tag{8}$$

where h is Planck's constant, m the mass of the particle, and u its velocity. The wave velocity V (equation (7)) is then related to the velocity of electrons u by $V = c^2/u$, where c is the velocity of light. Schrödinger developed the ideas of de Broglie further and applied them to the problem of an electron moving in an arbitrary potential. The procedure adopted is outlined as follows.

If the electrons, having a total energy W, are moving in a potential field E, its kinetic energy is given by the difference $(W - E) = \frac{1}{2}mu^2$. Combined with (8) this gives

$$\frac{1}{\lambda^2} = \frac{2m}{h^2}(W - E) \tag{9a}$$

which leads to the one-dimensional wave equation for electrons, namely,

$$\frac{d^2\psi}{dx^2} + \frac{8\pi^2 m}{h^2}(W - E)\psi = 0 \tag{9b}$$

satisfied by the wave function ψ at a distance x.

In three-dimensions, equation (9*b*) becomes

$$\frac{\partial^2\psi}{\partial x^2} + \frac{\partial^2\psi}{\partial y^2} + \frac{\partial^2\psi}{\partial z^2} + \frac{8\pi^2 m}{h^2}(W - E)\psi = 0 \tag{10}$$

or

$$\nabla^2\psi + \frac{8\pi^2 m}{h^2}(W - E)\psi = 0$$

which is the well-known Schrödinger equation referred to a set of Cartesian coordinates x, y, z. The solution of this equation for zero potential E is de Broglie's solution for a free electron.[1]

1.3 Electron Wavelengths

Let us first derive the simple equation for the wavelength of electrons. If a beam of electrons is accelerated by a potential E, then the energy of the electrons Ee is

$$Ee = \tfrac{1}{2}mu^2 \tag{11}$$

where e is the electronic charge and m is the mass of the moving electron. Substituting in the de Broglie relation (8), we obtain

$$Ee = \tfrac{1}{2}h^2/\lambda^2 m \qquad \text{or} \qquad \lambda = h/\sqrt{2mEe} \tag{12a}$$

By putting in the values of h, m, and e, this becomes

$$\lambda = \sqrt{\frac{150}{E_{\text{in volts}}}} \quad (\text{Å}) \tag{12b}$$

However, the mass m varies with the velocity u, and it is necessary to introduce the relativistic corrections, particularly for large values of E (and consequently u). We can see that this must be carried out because of difficulties with respect to the de Broglie wavelength (equation (8)) since when $u = 0$, $\lambda = \infty$, i.e., the position of electrons at rest appears to be indeterminate; however λu always remains finite. We must now take into account the wavelength related to the rest mass. This wavelength may be obtained from relativity mechanics as first proposed by de Broglie.[1] In his treatment of this problem, by using the Einstein relativity equation, the energy of the electrons with respect to the rest mass m_0 is expressed by the relation

$$Ee = mc^2 - m_0c^2 \tag{13a}$$

where the relativistic mass

$$m = \frac{m_0}{\sqrt{1 - u^2/c^2}} \tag{13b}$$

is used. We see that

$$(m + m_0)Ee = m^2u^2 \tag{13c}$$

Substituting for mu from equation (8) we get

$$\lambda = \frac{h}{\sqrt{(m + m_0)Ee}} \tag{14a}$$

$$= \frac{h}{\sqrt{2m_0Ee\left(1 + \dfrac{Ee}{2m_0c^2}\right)}} \tag{14b}$$

where the factor $[1 + (Ee/2m_0c^2)]^{1/2}$ is the relativistic correction factor and amounts to 2% for $E = 50$ kv and 10% for $E = 200$ kv. Thus, for low energy beams it is sufficient to calculate the wavelengths from equation (12). However, for the 50 to 100 kv electrons normally used in electron microscopes we need to calculate wavelengths from equation (14), and some of these are listed in Table I. It is to be noted that the wavelengths of electrons are considerably smaller than those of X-rays and light; the chief significance of this is the greater resolving power of electron beams compared to other forms of radiation.

From the preceding derivations we can see that the form of the Schrödinger equation (equation (10)) is inadequate since it only describes the electron in terms of the linear momentum mu, and neglects the existence of angular momentum (i.e., spin). The Schrödinger equation is thus only invariant when the coordinates are transformed to a set of moving axes. The fact that electrons have angular momenta implies that electron beams

Table I

E in volts	λ in Å	Velocity $u \times 10^{-10}$ (cm/sec)
1	12.26	0.00593
10	3.88	0.01876
100	1.23	0.05932
1,000	0.388	0.1873
10,000	0.122	0.5846
30,000	0.0698	0.9846
40,000	0.0601	1.1216
50,000	0.0536	1.237
60,000	0.0487	1.338
70,000	0.0449	1.427
80,000	0.0418	1.506
100,000	0.0370	1.644
300,000	0.0197	2.329
1,000,000	0.0087	2.822

should exhibit polarization phenomena since the spins might be aligned in a common direction. Although many attempts have been made to prove this, so far the experimental evidence is uncertain.

1.4 Interaction of Electron Beams with Matter

1.4.1 *Scattering Phenomena*

Because of the negative electrostatic charge on the electron, an electron beam passing through a crystalline specimen will suffer deflections owing to the fields of force exerted by the electronic fields within the atom. The beam leaving the specimen will thus be modified. Inside the specimen electrons may lose energy as a result of absorption, and this energy may appear in the form of heat, excitation, ionization, secondary emission, or X-rays. The effect of the positively charged nucleus is to produce a scattering of electrons with a change in direction only, i.e., with no loss in velocity. These electrons are then said to be elastically or coherently scattered. Some authors separate the elastic part into incoherent (wide-angle) scattering when considering amorphous films. Actually the scattering is still coherent and it is the atoms themselves which are incoherently arranged as compared to those in crystals. When the electron beam interacts with the atomic electrons inside the specimen, the scattered beam suffers a change in velocity, and they are then said to be inelastically scattered. These interactions are illustrated schematically in Fig. 1.

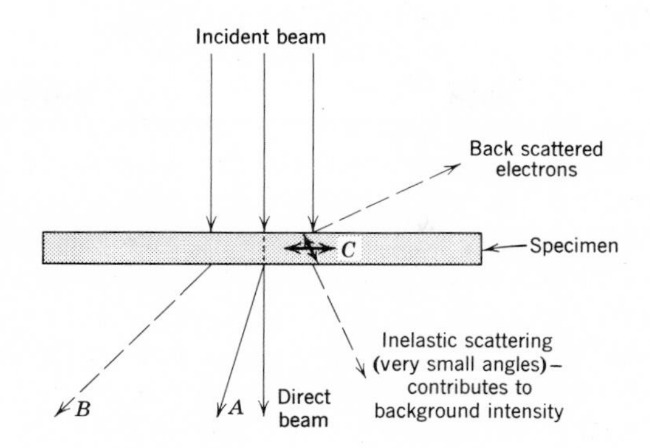

FIG. 1. Schematic illustration of the interaction of an electron beam with matter. *A*. Coherent elastic scattering (Bragg diffraction) from crystalline films. *B*. Incoherent elastic scattering from amorphous films. *C*. Absorption-energy lost inside specimen (heat, ionization, etc.)

From Fig. 1 we see that the beams leaving the specimen provide information which can be broadly classified into three groups: (1) diffraction, (2) microscopy, and (3) emission or spectrography—with the use of velocity analyzers. In this book we are concerned with the first two groups and thus we need to take particular notice of the coherently scattered electrons, since in crystals these give rise to the diffraction pattern and play the most important part in image formation and contrast.

1.4.2 *Rutherford Scattering*

A model[6] of the interactions which occur between incident electrons and the atom is illustrated in Fig. 2. This type of encounter was first envisaged by Rutherford to account for the scattering of α-particles in his experiments to establish the nuclear structure of the atom. Because the

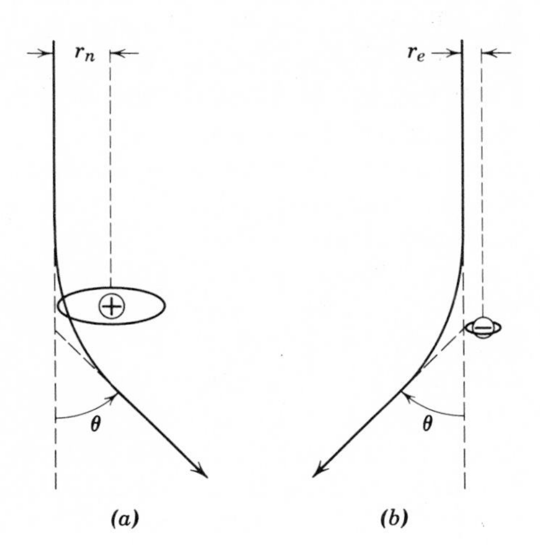

FIG. 2. Schematic representation of the deflection of an electron beam by an atom. (*a*) Elastic scattering from positive nucleus. (*b*) Inelastic scattering from negative atomic electron. (After Hall.[6] By permission from *Introduction to Electron Microscopy*, by C. E. Hall. Copyright 1953, McGraw-Hill Book Co., Inc.)

mass of the electron is small compared to that of the nucleus the latter can be considered to be fixed.

From Fig. 2 we see that an electron beam passing close to an isolated nucleus (i.e., stripped of its atomic electrons), will be attracted towards it, whereas electrons in the specimen would exert a repulsive force on the beam. The magnitude of these forces is given by the coulomb inverse-square law, and the paths are hyperbolic. The magnitude of θ, the angle of deflection, depends on the distance of approach. If the electron beam meets the field of force of the nucleus over a circle of radius r_n it is elastically scattered over an angle θ at the periphery, whereas inside the area of the circle the scattering is greater than θ. The area πr_n^2 is thus the elastic cross section for scattering through angles $>\theta$. The corresponding inelastic cross section for scattering is πr_e^2, where r_e is the effective radius of the force-field of an electron. Since in both cases the path of the electron beam will be hyperbolic, we see that for elastic scattering $r_n = Ze/V\theta$ and for inelastic scattering $r_e = e/V\theta$ where e is the electronic charge, V is the beam potential (esu), Z is the atomic number, and assuming $\sin\theta = \theta$. From this we see that the ratio of the inelastic and elastic scattering depends on Z as shown in Fig. 3, i.e., the inelastic scattering is smaller than the elastic scattering by the factor $1/Z$. The higher scattering power of the nucleus is due to the greater concentration of charge.

It should be noticed that at small angles the electron beam will pass at a relatively large distance from the nucleus so that the electron cloud of the atom will then exert a pronounced shielding effect. As we shall see in section 1.8, it is necessary to include this screening effect when considering the important small angle (Bragg) scattering of electron beams, since the oversimplified inverse-square law will not hold at all points along the path of the beam. In order to account for the shielding effect of atomic electrons, their potential distribution about the nucleus must be calculated by either employing quantum mechanical methods (Hartree

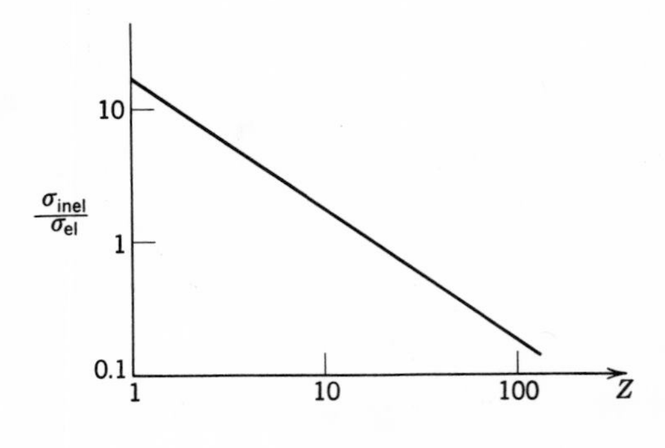

FIG. 3. Ratio of inelastic and elastic scattering of electrons as a function of atomic number. (After Lenz.[8])

model), or statistical methods (Thomas-Fermi model). This latter approach is generally adopted for heavier atoms containing a large number of electrons † ($Z > 18$).

If the screening effect is neglected i.e., assuming that all electrons are free, we can obtain the simple Rutherford scattering law by calculating the fraction of incident electrons scattered through angles $>\theta$ in a thin film of thickness t, atomic number Z, atomic weight A, and density ρ.[6] The number of atoms per cubic centimeter exposed to the beam is $\rho N_A/A$, where N_A is Avogadro's number. The probability of scattering of one electron through angles $>\theta$ is equal to the total number of scattering cross sections per unit area, i.e.,

$$\text{probability} = \frac{\rho N_A}{A}(\pi r_n^2 + Z\pi r_e^2)t \tag{15}$$

If N electrons are incident and dN are scattered over angles greater than θ, by substituting for r_n $(=Ze/V\theta)$ and r_e $(=e/V\theta)$, we obtain

$$\frac{dN}{N} = \frac{\rho N_A}{A}\frac{Z^2e^2}{V^2\theta^2}\left(1 + \frac{1}{Z}\right)t \tag{16}$$

Although this expression is inaccurate, it does show qualitatively the relationship between the number of scattered electrons, beam potential, specimen thickness, and atomic number. As is well known by experiment, better transmission is obtained when light atoms, thin specimens, and high beam potentials are employed.

1.4.3 *Concept of Transparency Thickness*

Since we are particularly concerned with the transmission of electrons through materials, it is important to consider the factors affecting the transparency thickness of specimens. This term is defined as the object mass thickness which can be pictured with tolerable transparency, where mass thickness is the physical thickness multiplied by density.[7] This concept is particularly important for noncrystalline films. In this case, the beam leaving the specimen is spread out over a broad angle, and its shape depends chiefly on the thickness of the specimen. However, the scattering may not be uniform from a uniformly thin specimen when more than one kind of atom is present (e.g., in carbon extraction replicas). (See section 4.1.4.) Thus in amorphous films contrast in the image depends mainly on the deficiency of electrons from regions of large scattering cross section (i.e., regions of low transparency thickness). This form of contrast

† See reference 10 for a discussion of these treatments.

may also be present in thin crystalline films containing different kinds of atoms (e.g., Fig. 134).

Figure 4, due to Lenz,[8] shows both the effect of the accelerating voltage and atomic number on the transparency thickness and is in agreement with expression (16). These results were calculated for total scattering from carbon, chromium, and gold. From this we see that specimens of the order of 1000 Å thick or less are required for work with electron transmission. Because of relativistic effects, the curves are much flatter for high beam voltages than for low ones. This means that the use of accelerating voltages much greater than a few hundred kv is not very promising as far as specimen transparency is concerned. The concept of transparency thickness in thin crystalline films is complicated, however, because of diffraction and the fact that interference may occur between

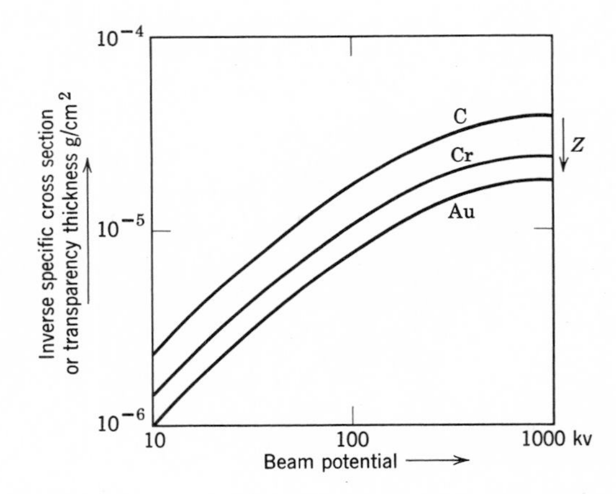

FIG. 4. Transparency thickness as a function of beam voltage. (After Lenz.[8])

the diffracted and transmitted beams. As we shall see in Chapter 2, this interference may produce total extinction. In microscopy a crystal perfectly oriented for diffraction would appear opaque, even though from Fig. 4 we would expect transparency. Because of this factor we must now take into account the effect of crystal structure.

1.5 Effect of Crystal Structure

1.5.1 *Bragg's Law*

Before it was suspected that electrons might possess wave properties, it was known that X-rays could be diffracted by crystals, and Bragg[9] showed that the position of the diffracted beams could be determined by a simple model as shown in Fig. 5. We can use the same model for electrons.

In Fig. 5 let us consider the effect of an incident plane wave of electrons upon rows of planes in a crystalline lattice, thereby treating the crystal as a diffraction grating. We need to find the condition that all waves scattered from all the lattice points shall reinforce, using the assumption that any point will give rise to a secondary wave, and neglecting all other factors. In Fig. 5, a and b represent the normals to the wave fronts incident at an angle θ to planes PQ, RS separated by distance d $(=AC)$. These waves will be diffracted, i.e., reinforce, provided that the path difference equals a whole number of wavelengths. The extra path length of the rays aCa' is $BC + CD$, i.e., $2d \sin \theta$; hence the condition for diffraction is

$$2d \sin \theta = n\lambda \tag{17a}$$

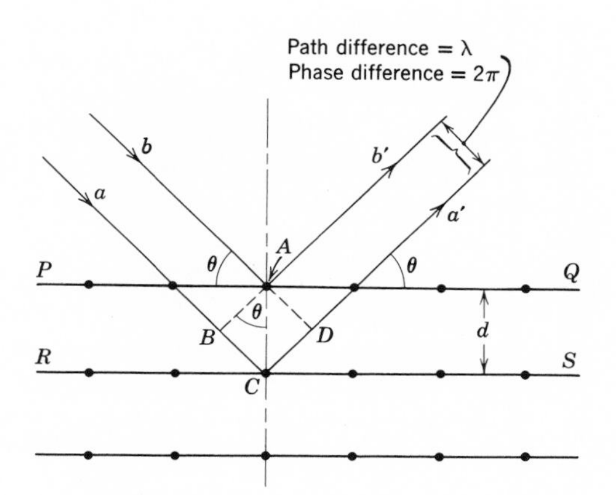

FIG. 5. Illustrating the Bragg law of diffraction.

where n is the order of the reflection and λ is the wavelength. This expression is the well-known Bragg law of diffraction.

If the reflecting planes have Miller indices hkl, then

$$2d_{hkl} \sin \theta = n\lambda \tag{17b}$$

or

$$2d_{nh+nk+nl} \sin \theta = \lambda \tag{17c}$$

Since the phase difference is $2\pi/\lambda$ times the path difference between waves scattered from the two lattice points, the condition for Bragg reflection can also be stated that the phase difference between successively scattered waves must be a multiple of 2π. Since electron wavelengths are very much smaller than the wavelengths of X-rays, the Bragg angles are consequently much smaller. Thus, for low order reflections $\sin \theta$ is approximately equal to θ (e.g., for a cubic metal, with $2d = 4$ Å and $\lambda = 0.04$ Å, for the first-order reflection, θ, is only 10^{-2} radian).

1.5.2 *Potential Energy of Electrons*

For a free electron (i.e., unbound to the nucleus) its kinetic energy is $\frac{1}{2}mu^2$ where u is its velocity, and by quantum theory its wavelength $\lambda = h/mu$ (equation 8); hence $\frac{1}{2}mu^2 = h^2/2m\lambda^2$, i.e., the energy of motion of the electron is inversely proportional to the square of the associated wavelength. It is more convenient to express this in terms of the wave number k, the magnitude of which is given by the reciprocal of the wavelength. The wave vector **k** is then the propagation vector of the wave in reciprocal space. (See sections 1.11 and 1.12.) The relationship between the wave number and energy of an electron is shown in Fig. 6.

Since the potential energy of a free electron at a distance r from the nucleus is given by the coulomb inverse-square law, viz., Ze^2/r, where Z is the atomic number and e is the electronic charge, the potential energy of an electron along a row of atoms in a one-dimensional crystal would be as shown in Fig. 7. It can be seen from this that the potential at the end of the row rises in a manner similar to that near an isolated atom, so that the electron here is under a considerable attractive force from the lattice as a whole. The surface can therefore be considered to be at a position $x_0/2$ from the end atom in the row.

In a real crystal, however, when the electrons fall upon certain families of reflecting planes such that Bragg's law is satisfied, they are diffracted and cannot continue to proceed undeviated through the crystal. This is true for electrons already present in the crystal as well as for those used to bombard it. For transmission, therefore, the electrons must move in directions away from the Bragg condition. Since an impinging beam on a specimen may contain electrons falling at different Bragg angles θ, there is an angular range about θ for which there is a mixture of transmission

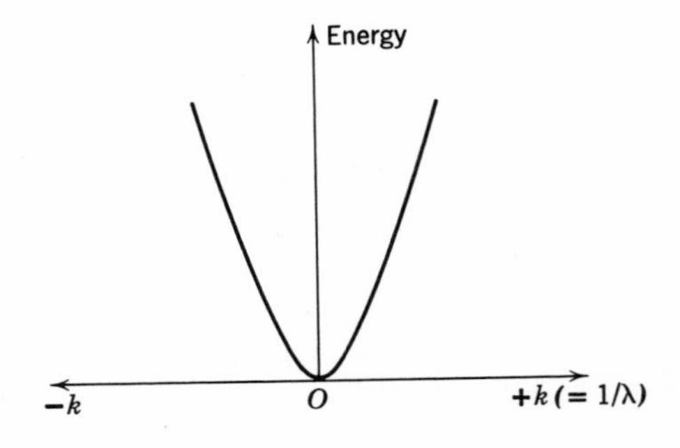

FIG. 6. One-dimensional representation of the wave-number energy relationship in crystals (schematic), for electrons in a uniform force-field.

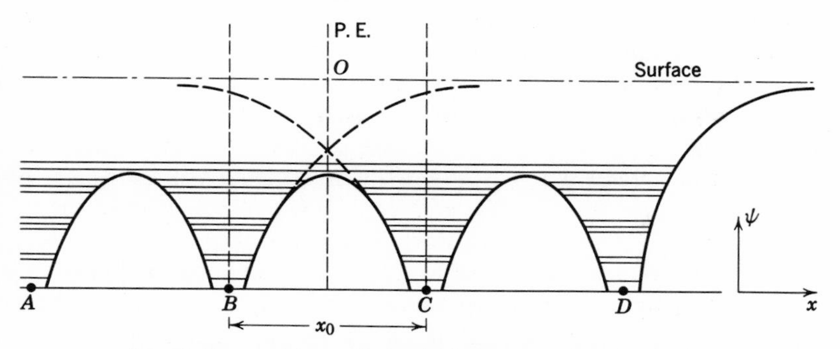

FIG. 7. Schematic representation of the potential energy of an electron in a lattice composed of a single row of atoms.

and reflection. This effect is analagous to the resolving power of a simple diffraction grating (section 1.13). The influence of this behavior is to transform the simple curve of Fig. 6 into that shown in Fig. 8, and consequently to modify the form of potential field shown schematically in Fig. 7. For a first order reflection $2dk_c \sin \theta = 1$, so that the first discontinuity shown in Fig. 8 corresponds to a critical value of the wave number equal to k_c. If we now construct the k-energy diagrams for a three-dimensional crystal taking into account all the possible reflections, we find that a series of zones exists around the origin in k-space (i.e., reciprocal space), the boundaries of which strongly diffract electrons. The shape of these zones is governed by the crystal structure being considered, and the zones themselves are known as Brillouin zones.†

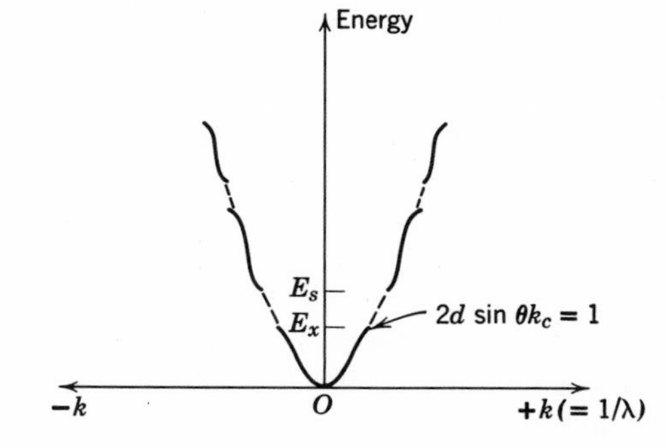

FIG. 8. One-dimensional representation of the energy-wave number relationship for an electron moving in the periodic field of a crystal (schematic).

† For more details of the zone theory of solids, see e.g., Kittel's book (bibliography section at the end of this chapter).

We see from Fig. 8 that there is an energy gap $E_s - E_x$ at the first Brillouin zone (and for all other such zones) which is forbidden to electrons. For electrons to be able to escape across the zone or escape from the surface, they must acquire an energy $>E_x$. If we assume that the potential barrier at the surface is E_s, then the energy input to release electrons must be greater than $(E_s - E_x)$. If we apply a voltage V_0 in order to accomplish this, V_0e must be greater than $(E_s - E_x)$, e being the electronic charge. V_0 is termed the work function of the surface at 0° K and is of the order of several electron volts. To obtain electron emission we can lower V_0e by lowering the potential barrier E_s, e.g., by absorption of radiant energy or fast particles as well as by thermal excitation. However, we will have more to say about the generation of electron beams in Chapter 3.

1.6 Solutions of the Wave Equation

To account completely for the interaction of an electron beam with a crystal, Schrödinger's equation (expression 10) must be solved rigorously, allowing for the boundary conditions set at the crystal surfaces as well as inside the crystal (i.e., at the Brillouin zones). The mathematical solutions of this equation which take into account the effect of the lattice potential upon the wave functions are complex, in fact, complete solutions are unknown. However, a simplified analysis is possible if we assume that in equation (10), $W > E$ and that E is constant; in other words, the crystal is regarded as a box with abrupt boundaries on all sides and with a uniform potential everywhere inside the box. Solutions of the Schrödinger equation will then lie along the curve of Fig. 6. This is the basic assumption used in the kinematical theory of diffraction. The amplitude can then be expressed as a complex exponential function:

$$\text{amplitude} \sim \cos\phi + i\sin\phi \sim e^{i\phi} \tag{18}$$

where ϕ represents the phase of the wave and depends upon its position in the crystal. Solutions of the wave equation allowing for the periodic variation in E, however, must follow the curve shown in Fig. 8, and thus have the discontinuities corresponding to the critical values of k_c, i.e., for electrons exactly in the Bragg condition for reflection. However, close to the Bragg condition, i.e., for values of k near k_c, the use of equation (18) may be justified. Thus the path of a wave through a crystal can be represented by an amplitude-phase diagram—an approach we shall use later in this chapter and when we consider contrast effects in images.

The parts of the crystal corresponding to the discontinuities of Fig. 8, i.e., the surfaces of the Brillouin zones, are known as the dynamical regions, and solutions of Schrödinger's equation are then given by the dynamical theory.[2] In these solutions the lattice potential E is usually represented as a three-dimensional Fourier series so that expression (10) becomes

$$\left(\frac{8\pi^2 me}{h^2}\right)E = \sum_h \sum_k \sum_l V_{hkl} e^{i\phi} \tag{19}$$

where V_{hkl} are the Fourier coefficients of the lattice potential.

Solutions of this equation are usually developed to obtain the amplitude of one strongly diffracted wave. Even for this case, however, the procedure is extremely complicated. However, most of the effects we are interested in here can be explained in terms of the kinematical theory, at least qualitatively.

Having discussed the general concepts of electron scattering, in the remaining part of this chapter we shall consider the coherent scattering of electrons by crystals, i.e., electron diffraction. In the first instance, by choosing a static model of a crystal structure, we need to find the geometrical conditions for the formation of a diffracted beam.

1.7 The Laue Equations

In the Bragg analysis of diffraction considered in section 1.5 we dealt with the crystal as a set of planes consisting of a number of scattering centers. In the derivation of the Laue equations we consider the nature of the diffraction of electrons produced by identical scattering centers located at the lattice points of a space lattice. For our model consider the unit cell shown in Fig. 9 in which the lattice points are at the cell corners, i.e., we assume that there is only one point per unit cell, and we allow a plane wave of electrons to fall upon it as shown. The incident wave fronts have normals AO' and BO, etc., making angles $\alpha_1, \beta_1, \gamma_1$, with respect to the axes a, b, c, and emerge as reflected waves at angles $\alpha_2, \beta_2, \gamma_2$ to these axes.

The path difference between rays scattered at the origin O and from the lattice point O' will be $O'P - OQ$, i.e., $c(\cos\gamma_1 - \cos\gamma_2)$ with respect to the c axis. Similarly, the path differences with respect to scattering points along the b and a axes are $b(\cos\beta_1 - \cos\beta_2)$ and $a(\cos\alpha_1 - \cos\alpha_2)$. In each case the condition for reinforcement is that each of these path differences must equal a whole number of wavelengths, i.e.,

$$a(\cos\alpha_1 - \cos\alpha_2) = h\lambda \tag{20}$$

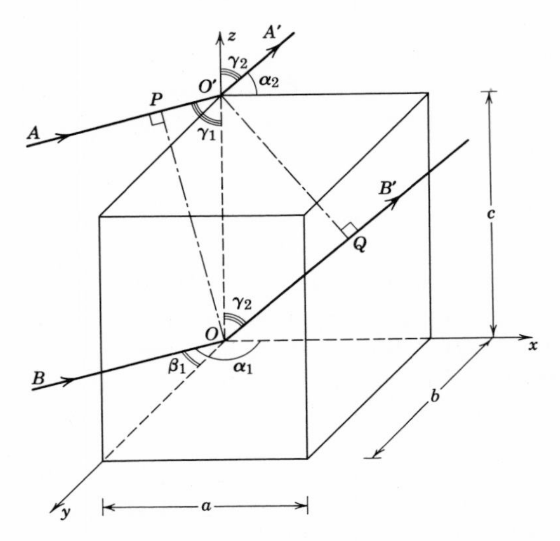

FIG. 9. Scattering of an electron wave by an atom at O' with respect to the origin O, illustrating the derivation of the Laue equations.

$$b\,(\cos\beta_1 - \cos\beta_2) = k\lambda \tag{21}$$

$$c\,(\cos\gamma_1 - \cos\gamma_2) = l\lambda \tag{22}$$

where h, k, l are integers. These equations illustrate effectively the geometrical conditions for the formation of a diffracted beam.

If the axes are orthogonal, $\cos^2\alpha_1 + \cos^2\beta_1 + \cos^2\gamma_1 = 1$ and for a cubic crystal $a = b = c$. Thus by squaring (20), (21), (22) and adding we get

$$a^2\{(\cos\alpha_1 - \cos\alpha_2)^2 + (\cos\beta_1 - \cos\beta_2)^2 + (\cos\gamma_1 - \cos\gamma_2)^2\} = \lambda^2(h^2 + k^2 + l^2) \tag{23}$$

By expanding, this reduces to

$$2a^2\{1 - (\cos\alpha_1\cos\alpha_2 + \cos\beta_1\cos\beta_2 + \cos\gamma_1\cos\gamma_2)\} = \lambda^2(h^2 + k^2 + l^2) \tag{24}$$

If 2θ is the angle between the incident and diffracted waves, equation (24) reduces still further to

$$2a^2(1 - \cos 2\theta) = \lambda^2(h^2 + k^2 + l^2) \tag{25}$$

i.e.,

$$\frac{4a^2\sin^2\theta}{h^2 + k^2 + l^2} = \lambda^2 \tag{26}$$

In the cubic lattice the spacing d between hkl planes is given by

$$d^2 = a^2/(h^2 + k^2 + l^2) \tag{27}$$

Thus equation (26) may be written

$$2d_{hkl} \sin \theta = \lambda$$

which is the Bragg law. This can similarly be shown for a more general lattice. The use of the Laue equations becomes necessary when we are dealing with crystals of dissimilar dimensions, and they are also used to determine conditions for diffraction in crystals where there is a continuous distribution of scattering points. This analysis alone, however, will only give the condition for reinforcement of waves from *corresponding* points in each cell.

1.8 Atomic Scattering

The intensity of a given diffracted wave depends on two factors: one corresponds to the fulfillment of the Laue and Bragg equations and the other is the atomic scattering factor. The latter describes the result of the interference effects which take place within the structure of the atom and this represents the efficiency of scattering by the atom. In X-ray scattering, the atomic scattering factor f_x is defined as the ratio of the amplitude scattered by the charge distribution in the atom to that scattered by a point electron. Since the intensity of a wave is the square of its amplitude, f_x^2 is the ratio of the intensity of the scattering of an atom to that from an electron. For X-rays, where scattering is produced from interactions with electrons, f_x depends only on the number of electrons in the scattering atom Z. In electron radiation, however, the chief contribution to scattering results from interactions with the potential field of the nucleus, so that the scattering factor for electrons depends on $(Z - f_x)$. In both cases the scattering factor varies with angle and is greatest at very small angles. Our implied assumption in section 1.7 that the amplitude of scattering is independent of the scattering angle is thus only an approximation. Let us now find out what the scattered amplitude is when we account for atomic scattering.

The amplitude of the coherently scattered electron wave may be calculated in terms of the simple coulomb field, i.e., assuming a uniform potential field about the nucleus (Fig. 6). This analysis yields the Rutherford scattering law discussed in section 1.4. However, this treatment neglects the screening effect of the atomic electrons. To take into account the effect of the electron cloud, the Schrödinger equation must be solved for an incident beam of electrons traveling in a lattice where the potential is always varying. In this case the dynamical theory of scattering must be employed, choosing a suitable dynamical model for the structure of the atom. Such a solution is extremely difficult to carry out. However, an approximation due to Born gives an equation for coherently scattered fast electrons assuming a spherical cloud of atomic electrons.[2,10] For unit incident amplitude the Born approximation is usually expressed in the form

$$f(\theta) = \frac{K(Z - f_x)}{\sin^2 \theta/\lambda^2} \tag{28}$$

where $K = me^2/2h^2$, $f(\theta)$ is the atomic scattering factor for electrons, and f_x is the atomic scattering factor for X-rays.† The scattering factor $f(\theta)$ is seen to depend on the ratio $\sin \theta/\lambda$ where θ is the Bragg angle; when $f_x \to 0$ and the scattering is produced only by the nucleus, this equation is transformed to Rutherford's equation for the scattering of high speed α-particles and through large angles.[10]

By inserting the value of K, equation (28) is written

$$f(\theta) = (Z - f_x)\left(\frac{\lambda \times 10^8}{\sin \theta}\right)^2 2.38 \times 10^{-10} \tag{29}$$

where λ, the wavelength of the incident beam, is expressed in centimeters. Unlike the atom factor for X-rays, which for unit incident amplitude is given by[11]

$$\frac{e^2}{mc^2} f_x \qquad \text{i.e.,} \qquad 2.82 \times 10^{-13} f_x \tag{30}$$

$f(\theta)$ for electrons is not dimensionless. As written in (29) $f(\theta)$ has dimensions of centimeters; $\{f(\theta)\}^2$ is the scattering cross section, i.e., the intensity of electrons scattered by the atom. For small angles we see that the value of $f(\theta)$ for electrons is about 10^4 times greater than the scattering factor for X-rays (for the same reflection). This result agrees with the well-known fact that a very thin sheet of material will strongly scatter electrons although it would produce very little effect on X-rays.

It should be pointed out that the accuracy of equation (29) decreases with increasing Z. It is also only true for fast electrons which are coherently scattered, so that when calculated values of scattered intensities are compared with experimental results, the background intensity due to inelastic scattering must be excluded. For accelerating voltages >50 kv, the value of K in expression (29) should be corrected for relativistic effects (section 1.3). Accurately calculated values of $f(\theta)$ have been obtained by Ibers,[12] for elements of atomic number 1 to 80, and these are listed in

† Tables of f_x can be found in most textbooks dealing with X-ray diffraction and in *Internationale Tabellen zür Bestimmung von Kristallstrukturen*, 1935, volumes 1 and 2, Berlin.

Appendix E. These data are the most reliable for use in crystal structure determination by electron diffraction.

From the preceding discussions it becomes clear that crystal structure determination is much less reliable by electron diffraction than by X-rays because of the complexities of the scattering processes. Since electrons are scattered from very thin films, there is also the danger of obtaining extra reflections from any contamination on the specimen, e.g., grease or carbon deposits from the vacuum system, which will certainly cause confusion in the interpretation of the diffraction pattern. Anomalous intensities are also obtained in diffraction patterns of very small crystals and when double diffraction occurs (see also section 1.14). Consequently, the interpretation of diffraction effects resulting from electron scattering is often complicated, particularly when it is necessary to apply dynamical theory so as to explain the experimentally observed intensities. Fortunately, in electron microscopy we are usually not concerned with structural analyses, and a good qualitative understanding of diffraction effects in transmission may be obtained from the kinematical theory of diffraction.

1.9 The Structure Factor

Until now we have only considered scattering from a cell containing one scattering point. When there are many points in the cell, the Laue equations give only the condition for the reinforcement of waves scattered by corresponding points in each cell. There will also be interference between waves scattered from the different points in each cell, so that in some cases there are orders of reflections given by the Bragg and Laue equations which are completely suppressed by this effect. The factor causing the disappearance or reinforcement of the diffraction maxima is known as the structure factor or structure amplitude which is usually denoted by F. F is the sum of all the scattered amplitudes from all the atoms in the unit cell. These amplitudes can be obtained using the kinematical expression (equation 18) discussed in section 1.6, where we saw that the amplitude is proportional to $e^{i\phi}$, where ϕ is the phase difference between waves from successive scattering points. The total amplitude scattered by an atom is then $fe^{i\phi}$, where f is the atomic scattering factor for electrons, and is thus the proportionality factor for expression 18. The phase difference between waves scattered from different lattice points can be obtained from the Laue equations (20–22):

$$a\,(\cos\alpha_1 - \cos\alpha_2) = h\lambda$$

$$b\,(\cos\beta_1 - \cos\beta_2) = k\lambda$$

and

$$c\,(\cos\gamma_1 - \cos\gamma_2) = l\lambda$$

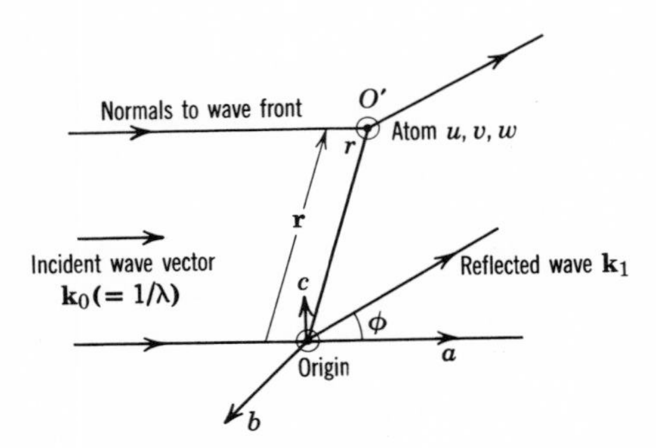

FIG. 10. The scattering of an incident electron wave by an atom at (u, v, w). The path difference is $(\mathbf{k}_1 - \mathbf{k}_0)\cdot\mathbf{r}$.

as represented in Fig. 9. The sum of these equations gives the total phase difference for the atom at (a,b,c) with respect to the origin, i.e., $2\pi(h + k + l)$, (phase difference $= 2\pi/\lambda$ times path difference). For an atom of coordinates $(u,0,0)$, the phase difference is $2\pi hu$ with respect to an atom at the origin, and so for an atom situated at u, v, w (Fig. 10) the phase difference ϕ is $2\pi(hu + kv + lw)$ or $2\pi/\lambda\, \mathbf{r}\cdot(\mathbf{k}_1 - \mathbf{k}_0)$. The resultant scattered amplitude of waves from the atom at u, v, w, can be obtained by plotting an amplitude-phase diagram where the phase angle $\phi = 2\pi(hu + kv + lw)$. This is shown in Fig. 11*a*. The resultant ampli–

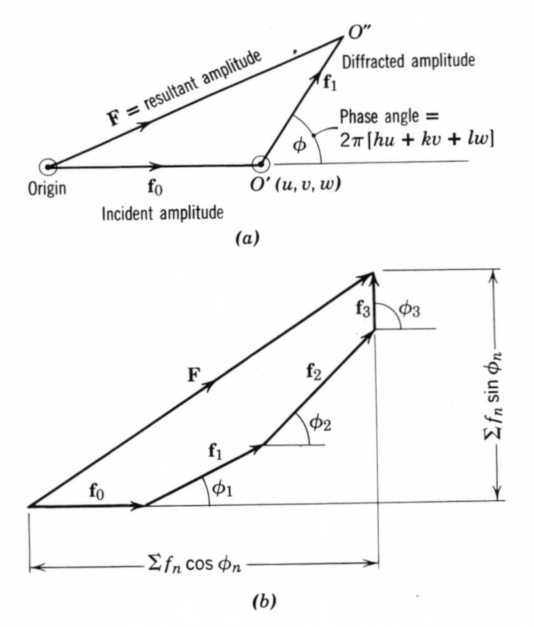

FIG. 11(*a*) Schematic amplitude-phase diagram for scattering from a unit cell containing a single lattice point at O'. (*b*) Amplitude-phase diagram for scattering from N atoms in the unit cell. The vector addition of the amplitudes of the diffracted rays gives the resultant amplitude F, i.e., the structure amplitude.

tude is thus the structure factor F and is given by the length of the vector drawn from the origin intersecting the diffracted amplitude vector at a distance equal to $|\mathbf{f}_1|$, i.e., at O'', from the scattering point (u, v, w). For a unit cell containing only one lattice point (Fig. 11*a*) $\mathbf{F}_{hkl} = \mathbf{f}_1 + \mathbf{f}_0$. Hence, for scattering from a unit cell containing N lattice points one must take the sum of all the scattered amplitudes to obtain $\mathbf{F}$, as shown in Fig. 11*b*. If we first resolve $\mathbf{f}_1$ into its components $\mathrm{f}_1 \cos \phi_1$ and $\mathrm{f}_1 \sin \phi_1$, we then obtain $|\mathbf{F}_{hkl}|^2$ for N different scattering points as follows:

$$|\mathbf{F}_{hkl}|^2 = (\mathrm{f}_1 \cos \phi_1 + \mathrm{f}_2 \cos \phi_2 + \ldots + \mathrm{f}_n \cos \phi_n)^2 + (\mathrm{f}_1 \sin \phi_1 + \mathrm{f}_2 \sin \phi_2 + \ldots + \mathrm{f}_n \sin \phi_n)^2 \quad (31a)$$

where $\mathrm{f}_1\, \mathrm{f}_2 \ldots \mathrm{f}_n$ are the scattering factors of the different atoms in the unit cell.

It is more convenient to write (31a) in the same form as (18), thus:

$$\left|\mathbf{F}_{hkl}\right| = \sum_{n=1}^{N} f_n \exp\left[2\pi i(hu_n + kv_n + lw_n)\right] \quad (31b)$$

If all the atoms are the same we can take f_n outside the summation; hence

$$\left|\mathbf{F}_{hkl}\right| = f_n \sum_{n=1}^{N} \exp\left[2\pi i(hu_n + kv_n + lw_n)\right] \quad (31c)$$

The intensity resulting from all the scattering points in the unit cell is given by $|\mathbf{F}_{hkl}|^2$. For each single scattering point the intensity depends on the scattering angle and indices h, k, l. Experimentally it is possible to measure the intensity and so establish F, but from this it is not possible to separate the sum of the cosine and sine terms. However, if the lattice has a center of symmetry, the sum of the sine terms is zero. Values of F^2 equal to zero correspond to extinction, and so it is possible to determine which values of h, k, l lead to reflection. If we consider a body-centered cubic lattice (Fig. 12), the atom coordinates are 0,0,0, at the corners and $\frac{1}{2}, \frac{1}{2}, \frac{1}{2}$, at the center of each unit cell. Because of the symmetry we can leave out the sine terms in the structure amplitude expression, and since all the atoms are the same, $f_1 = f_2$ etc., thus equation (31*c*) reduces to

$$|\mathbf{F}_{hkl}| = fe^{i\phi_1} + fe^{i\phi_2} \quad (32)$$

Since ϕ_1 for the atom at the origin is zero and ϕ_2 for the body-centered atom is $2\pi[h/2 + k/2 + l/2]$, we get

$$|\mathbf{F}_{hkl}| = fe^{2\pi i(0)} + f \exp\{2\pi i[h/2 + k/2 + l/2]\} = f[1 + e^{\pi i(h+k+l)}] \quad (33)$$

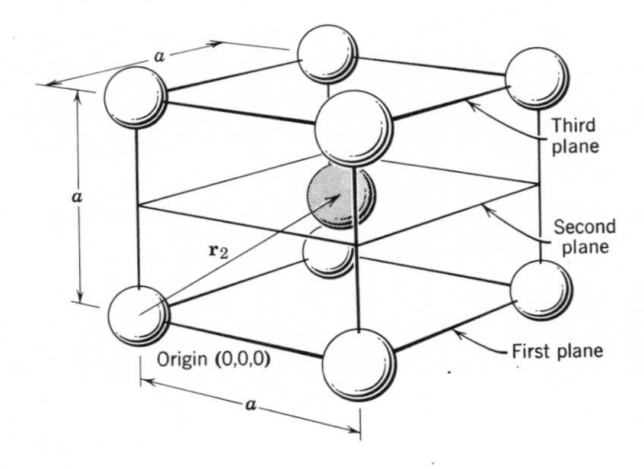

FIG. 12. The body-centered cubic lattice. Coordinates of body-centered atom (shaded) are $\frac{1}{2}, \frac{1}{2}, \frac{1}{2}$.

From this we see that

$$|\mathbf{F}_{hkl}|^2 = 4f^2 \qquad \text{if } h + k + l \text{ is even,}$$

and

$$|\mathbf{F}_{hkl}|^2 = 0 \qquad \text{if } h + k + l \text{ is odd.}$$

Thus in BCC lattices reflections occur only from planes having the indices h, k, l in which $h + k + l$ is even. Hence the (100) reflection vanishes for the body-centered cubic lattice. A (100) reflection normally occurs when the reflections from the first and third plane in Fig. 12 are out of phase by 2π. However, the second plane is equivalent to the first and third planes, and since it is situated halfway between them, it gives a reflection out of phase by π, thereby canceling the contribution from the other plane. This only occurs if these planes are identical in composition, whereas in the CsCl structure {100} reflections are obtained, since the Cl and Cs atoms are situated in alternate {100} planes.

For face-centered cubic lattices {scattering points at (0,0,0,), ($\frac{1}{2}$,0,$\frac{1}{2}$,), (0,$\frac{1}{2}$,$\frac{1}{2}$,), ($\frac{1}{2}$,$\frac{1}{2}$,0)}, reflections occur only when h, k, l are all even or all odd (i.e., when $|\mathbf{F}_{hkl}|^2 = 16f^2$). For HCP crystals, reflections are not allowed for $H+2K=3$, L odd; thus (00.1) is not present in diffraction patterns (Fig. 16). Similar analyses gives the conditions for reflection for all crystal structures.

1.10 Representation of Diffraction Patterns

We have seen in the preceding sections that complete diffraction occurs when all three Laue conditions are satisfied. Thus in real three-dimensional lattices the diagrammatic representation of the interference resulting from such diffraction can be done by the superposition of the three spectra corresponding to each Laue condition. For each Laue condition, the period of identity of each atom row can be represented in space as a family of cones which, from Fig. 9, have the angles of reflection $2\alpha_2$, $2\beta_2$, $2\gamma_2$ respectively at the apices. The possible values of these angles are determined by the integral values of h, k, and l in the path differences. If we consider an ideal simple cubic crystal (as Fig. 12 without the body-centered atom) with an electron beam parallel to a cube edge and place a photographic plate normal to the incident beam, we will obtain a diffraction picture from rows normal to each other and chosen along the coordinate axes: two rows parallel to the plane of the photographic plate will give two families of mutually orthogonal hyperbolae, whereas the third row normal to the plate, i.e., parallel to the beam, will give a family of circles.[2] Figure 13 shows the superposition of the three spectra considered (exaggerated). Since the wavelength of electrons is so small, the hyperbolae actually approximate to nearly straight lines. It must be remembered that the intensities of the diffracted beams have a maximum only when all three Laue conditions are satisfied so that the maxima represented in Fig. 13 are formed at the intersections of the two hyperbolae and the circles. In any given case the probability of such intersection is not very great, since in most cases of electron diffraction the atom rows parallel to the beam are not resolved. The reason for this is discussed in section 1.13. In general, the electron diffraction patterns obtained from single crystals are made up of a two-dimensional system of spots resulting from the intersections of the hyperbolae shown in Fig. 13. These spot patterns are most

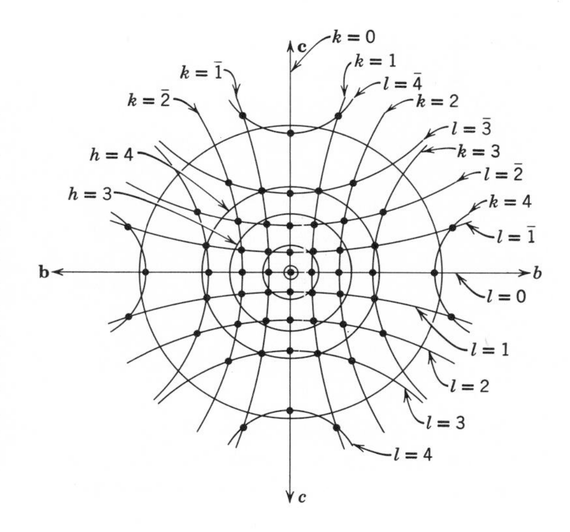

FIG. 13. Superposition of diffraction spectra from a simple cubic lattice (schematic). The hyperbolae occur for cube edges normal to the direction of the beam and the circles for the cube edge parallel to the beam. A three-dimensional pattern consists of spots and circles, and the two-dimensional pattern consists of only spots (e.g., from thin foils). The integers h k l (Laue numbers) correspond to maxima.

conveniently represented in terms of the reciprocal lattice and the reflecting sphere.

1.11 The Reciprocal Lattice

The interpretation and indexing of spot patterns from single crystals are nearly always carried out using the concept of the reciprocal lattice applied originally by Ewald and von Laue. The reciprocal lattice is one composed of a system of points, each of which represents a reflecting plane in the crystal and has the same indices as the corresponding reflecting plane. The reciprocal lattice is constructed from the real lattice by drawing a line through the origin normal to the corresponding reflecting plane of the crystal of length equal to the reciprocal of the crystal plane spacing. Thus if d_{hkl} is the distance of the (hkl) plane from origin in real space, the corresponding distance in reciprocal space is $1/d_{hkl}$. In this way the reciprocal lattice can be constructed as a system of points which will always form a space lattice, e.g., the face-centered cubic real lattice has a corresponding body-centered cubic reciprocal lattice as shown in Fig. 14. If a^*, b^*, c^* are the lengths of the reciprocal lattice cell edges and a, b, c those of the real lattice cell edges separated by angles α, β, γ respectively,

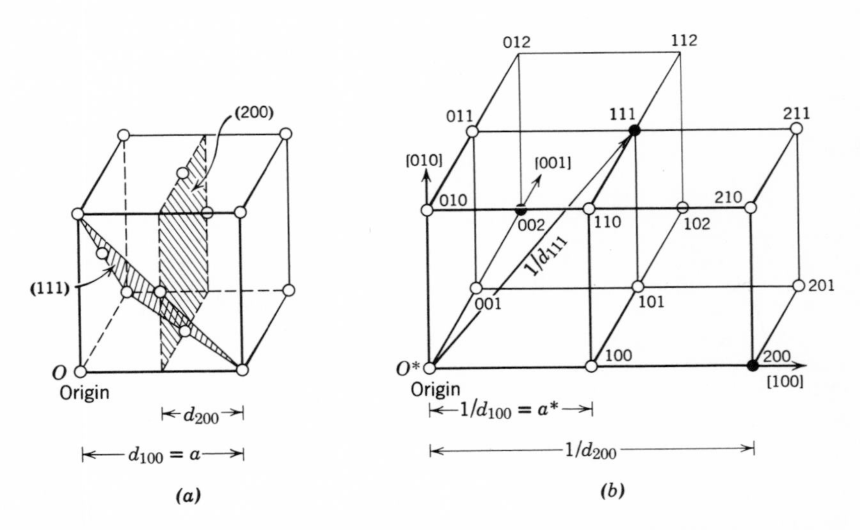

FIG. 14. The face-centered cubic lattice (*a*) showing two possible reflecting planes (111) and (200), and (*b*) part of its reciprocal lattice. Of the points marked only the {111} and {200} spots would appear in the diffraction pattern.

then

$$a^* = \frac{bc \sin \alpha}{V}, \quad b^* = \frac{ac \sin \beta}{V}, \quad c^* = \frac{ab \sin \gamma}{V} \tag{34}$$

where V is the volume of the unit cell (i.e., area of base $\times$ height). In vector notation these relationships can be written

$$\mathbf{a}^* = \frac{\mathbf{b} \times \mathbf{c}}{\mathbf{a} \cdot (\mathbf{b} \times \mathbf{c})} \text{ etc.,}$$

where the volume $\qquad V = \mathbf{a} \cdot (\mathbf{b} \times \mathbf{c}) \qquad (35)$

The vector product $(\mathbf{b} \times \mathbf{c})$ is a vector normal to $\mathbf{c}$ and $\mathbf{b}$ and of magnitude $bc \sin \alpha$. This means that $\mathbf{a}^*$ is perpendicular to both $\mathbf{b}$ and $\mathbf{c}$ and of magnitude

$$|\mathbf{a}^*| = \frac{1}{\text{spacing of planes in real space containing } b \text{ and } c}$$

and by definition the scalar or dot products $\mathbf{a}^* \cdot \mathbf{a} = \mathbf{b}^* \cdot \mathbf{b} = \mathbf{c}^* \cdot \mathbf{c} = 1$.

In real space a lattice point at position r_n is defined by

$$\mathbf{r}_n = u\mathbf{a} + v\mathbf{b} + w\mathbf{c} \tag{36a}$$

where $\mathbf{a}$, $\mathbf{b}$, $\mathbf{c}$ are unit translation vectors. Similarly in the reciprocal lattice a point at r^* has coordinates corresponding to reflecting planes hkl, so that we can define the reciprocal lattice point by

$$\mathbf{r}^*_{hkl} = h\mathbf{a}^* + k\mathbf{b}^* + l\mathbf{c}^* \tag{36b}$$

and the vector $\mathbf{r}^*_{hkl}$ is perpendicular to the hkl planes in the real lattice and of magnitude equal to the reciprocal of the spacing between hkl planes. Usually the reciprocal lattice vector is denoted by $\mathbf{g}$.

1.12 The Reflecting Sphere

A simple geometrical construction in the reciprocal lattice gives the conditions corresponding to reflection. If a sphere of radius equal to the reciprocal of the wavelength of the incident beam touches the origin of the reciprocal lattice and intersects any reciprocal lattice point, then, depending on the structure factor, the corresponding plane will be a reflecting plane. This can be seen from Fig. 15*a*, in which the reflecting plane AB has a corresponding reciprocal lattice point at r^*. Now r^* lies on the sphere (radius $1/\lambda$) if $\sin \theta = g/2/\lambda = \lambda/2d_{hkl}$, i.e., the Bragg law $2d_{hkl} \sin \theta = \lambda$ is then satisfied. In terms of vectors this can be written as $\mathbf{k}_1 - \mathbf{k}_0 = \mathbf{g}$, where $\mathbf{k}_0$ is the incident wave vector, $\mathbf{k}_1$ the diffracted

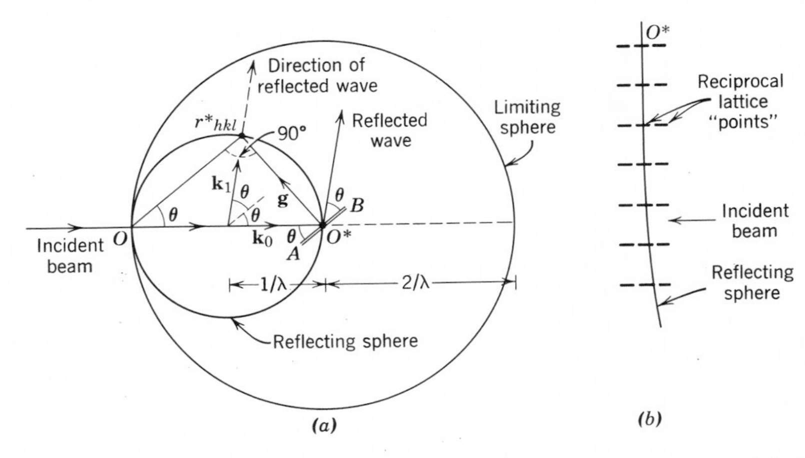

FIG. 15(*a*). Showing the sphere of reflection. If this intersects a reciprocal lattice point, reflection occurs. (*b*) For electrons, since λ is very small, there are many intersections.

wave vector, and $\mathbf{g}$ the reciprocal lattice vector (Fig. 15*a*). If $\mathbf{r}$ is a real lattice vector (Fig. 10) and $\mathbf{g}$ the corresponding reciprocal lattice vector, waves scattered by a point situated at $\mathbf{r}$ in the crystal are in phase when $\mathbf{g}\cdot\mathbf{r}$ or $(\mathbf{k}_1 - \mathbf{k}_0)\cdot\mathbf{r}$ equals an integral number times 2π. This number then gives the order of the reflection. This is another way of expressing the Laue equations. All planes that can be made to reflect will have their reciprocal lattice points within a distance $2/\lambda$ from the origin, i.e., within the limiting sphere shown in Fig. 15. It should be noticed that for electrons, λ may be as small as 0.04 Å, so that the diameter of the reflecting sphere may be 50 $Å^{-1}$, and thus approximates to a plane. This is why Fig. 13 is exaggerated. The sphere thus cuts many reciprocal lattice points, as shown in Fig. 15*b*, and for very thin specimens where the reciprocal lattice points are extended (see section 1.13), many orders of reflections will be obtained.

Since for each reciprocal lattice point *hkl* there will be a corresponding opposite point $\bar{h}\,\bar{k}\,\bar{l}$, and provided that the plane of the specimen is normal to the beam, an electron diffraction pattern will then show a system of spots symmetrical about horizontal and vertical lines through the center. This can be explained as follows: if a zone axis of the crystal (i.e., an axis common to a number of planes) makes a small angle with the beam, when viewed along the axis, the planes will be seen on edge and thus appear as lines. Figure 16*a* at (1) represents a face-centered cubic structure viewed along the [001] zone axis, i.e., a cube edge. Some of the directions of the principal planes are marked showing that the lines form the pattern of a cross-grating. If the beam makes a suitable angle with any of these planes, there will be a corresponding diffraction spot in the pattern. Other examples for face-centered cubic metals are shown in Fig. 16*a* and for hexagonal close-packed metals in Fig. 16*b*. These illustrate the change in form of the pattern with a change in orientation of the crystal.

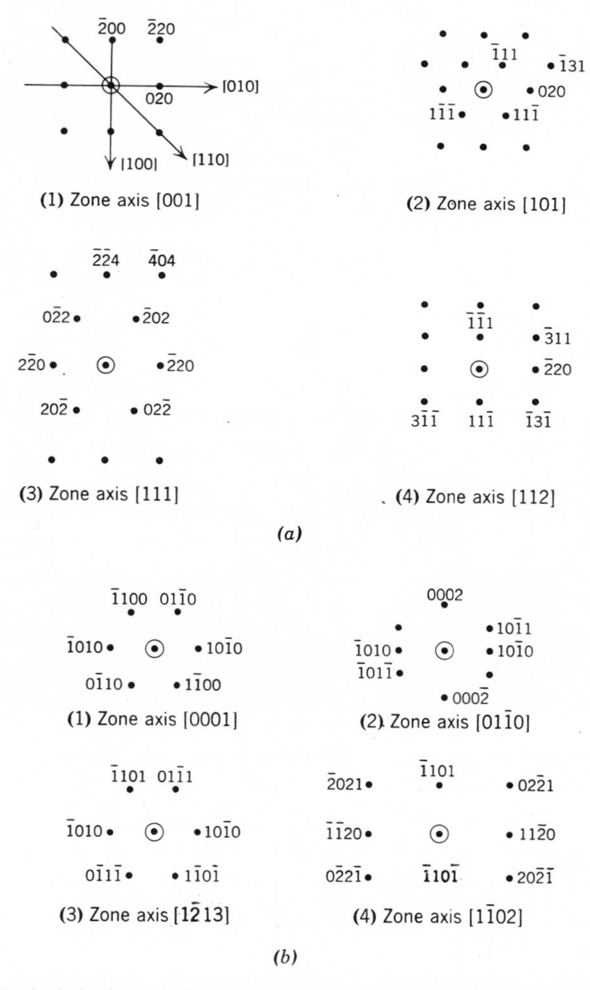

FIG. 16. Typical single-crystal spot patterns from (*a*) face-centered cubic crystals, (*b*) hexagonal close-packed crystals for different orientations. Hexagonal indices are written *HKiL* where $-i=H+K$. Note in (2) that the (0001) reflection is not allowed.

The zone axes are marked in each case and show the direction of the electron beam through the crystal. These patterns are typical of the single-crystal spot patterns commonly obtained in transmission electron microscopy. It will be seen that the diffraction pattern is a plane through the reciprocal lattice perpendicular to the incident electron beam.

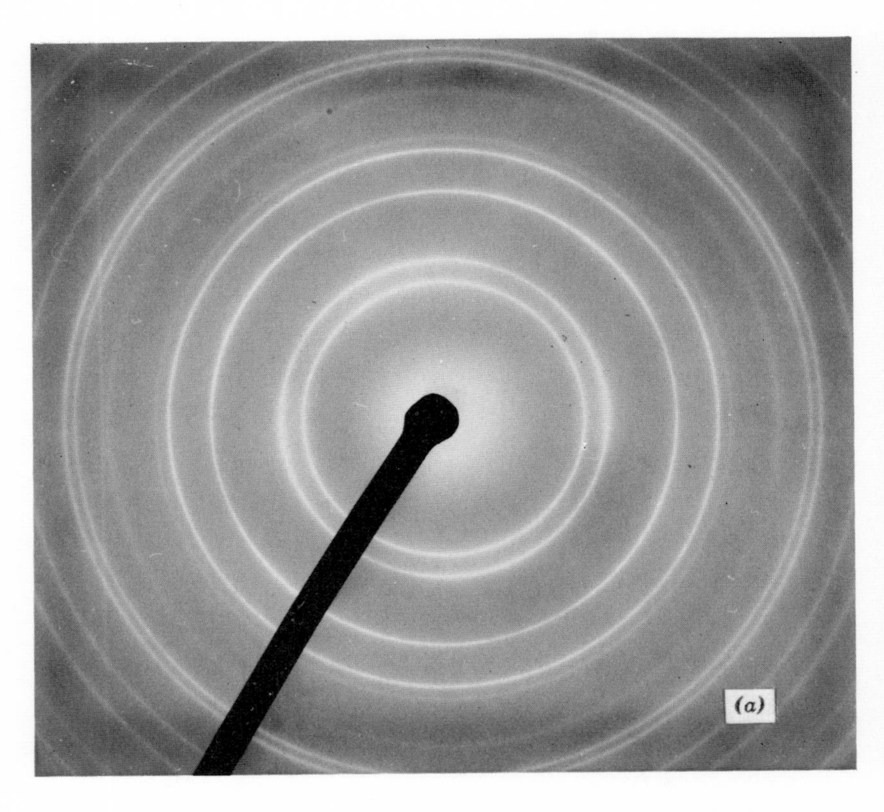

FIG. 17(*a*). Ring pattern from a thin film of polycrystalline gold obtained by evaporation. Arcing of the rings indicates the existence of a preferred orientation. (Note for FCC crystals, reflections allowed for $(h^2 + k^2 + l^2) = 3, 4, 8, 11, 12, 16$, etc.) The black line is the shadow of the beam stop.

If the electron beam is broad enough or the crystal size is small enough so that intersection of many randomly oriented crystals in a polycrystalline specimen occurs, the diffraction pattern is made up of a whole system of spots corresponding to the orientation of each individual crystal. If there are enough crystals present, these spots merge into a series of continuous rings whose diameters correspond to the spacings of the allowed reflecting planes in the crystal. Figure 17*a* is an example of a ring pattern from an evaporated gold layer composed of many small crystallites.† By reducing the area contributing to the electron diffraction pattern with the use of suitable apertures, fewer and fewer spots will appear in the pattern,

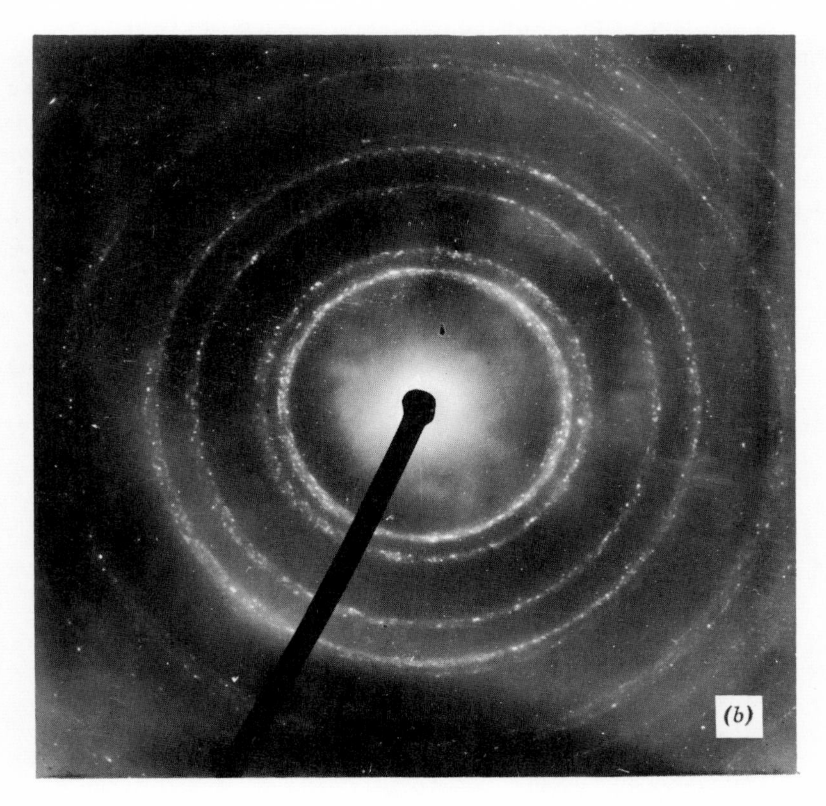

FIG. 17(*b*). As Fig. 17*a*, but area of specimen contributing to the diffraction pattern is now reduced. Note rings are composed of many spots.

e.g., as shown in Fig. 17*b*, and if the number of individual resolved spots is counted it is possible to estimate the number of crystals contributing to the pattern. When the electron microscope is operated for thin foil work, the illuminating beam is of the order of 10–20 μ in diameter, and areas

† All the electron diffraction patterns shown in this book have been obtained in the electron microscope.

down to 0.2 μ^2 can be observed (selected area technique, see section 2.1), so that we are mainly concerned with the single crystal spot pattern.

Diffraction patterns are fairly easily indexed when the zone laws governing crystal lattices are applied and when stereographic or other standard crystallographic projection techniques are used. (See Appendix A.) Before this can be done, it is sometimes helpful to know the camera constant of the diffraction apparatus, which can be obtained by considering Fig. 18. If a specimen AB is at a distance L from the photographic plate O^*, and r^* is a spot on the photographic plate corresponding to the beam Or^* which has undergone a Bragg reflection, it is clear that the diffraction pattern at O^*r^* is a magnified image of the plane PQ of the reciprocal lattice. Since PQ is the distance from the origin of the reciprocal lattice point Q, its length is equal to d^{-1}, where d is the spacing of the hkl plane giving rise to the reflection obtained at r^* on the plate. Thus from the similar triangles OQP, Or^*O^*, and since the Bragg angles are very small, it follows that

$$2\theta = \frac{1}{d}\Big/\frac{1}{\lambda} = \frac{r}{L} \tag{37}$$

hence

$$dr = \lambda L \tag{38}$$

where λL is known as the camera constant. In the electron microscope L can be varied by the use of suitable lenses (see Chapter 3).

It is easy to measure r directly from the photographic plate, and once r^* is indexed, λL can be established. This is not easy to do if the structure is not known. However, if a pattern from a known crystal is super-

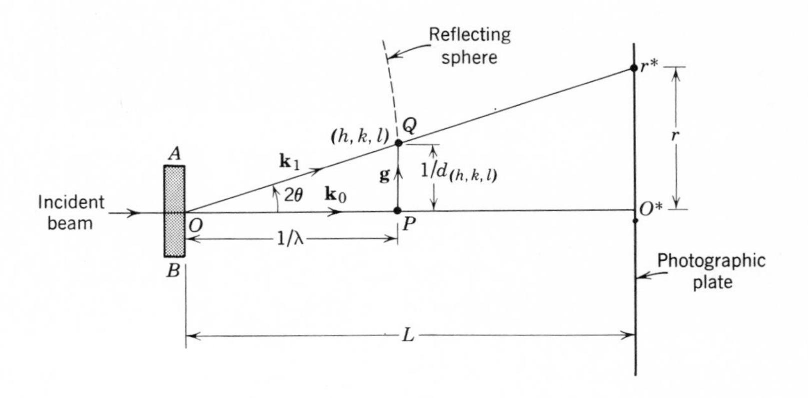

FIG. 18. The relationship between spacings of spots r in the diffraction pattern and the spacings of planes d in the crystal, where $dr = \lambda L$ (the camera constant).

imposed upon the unknown (e.g., by evaporating a small amount onto the surface of the crystal), λL can be worked out from the standard pattern and the unknown spots may then be indexed after working out to what crystal class the specimen belongs, i.e., whether it is cubic, hexagonal, etc. In most foil work, however, the crystal structure is already known, so that this procedure is unnecessary. Nevertheless, indexing of patterns is facilitated by evaporating a layer of the same metal onto the foil; an example of this is shown in Fig. 19. This is obtained from an aluminum thin foil (spot pattern), upon which is superimposed a thin layer of aluminum (ring pattern). Since the *d* values for the aluminum (*hkl*) reflections can be obtained from standard tables,† it is possible to calculate λL and so identify and index the spots as shown. However, because the position of the photographic plate and specimen may change for each exposure, λL may also change. The camera constant will also vary with the lens currents used so that λL must be worked out for each pattern that is obtained. A typical example of indexing a diffraction pattern is given in

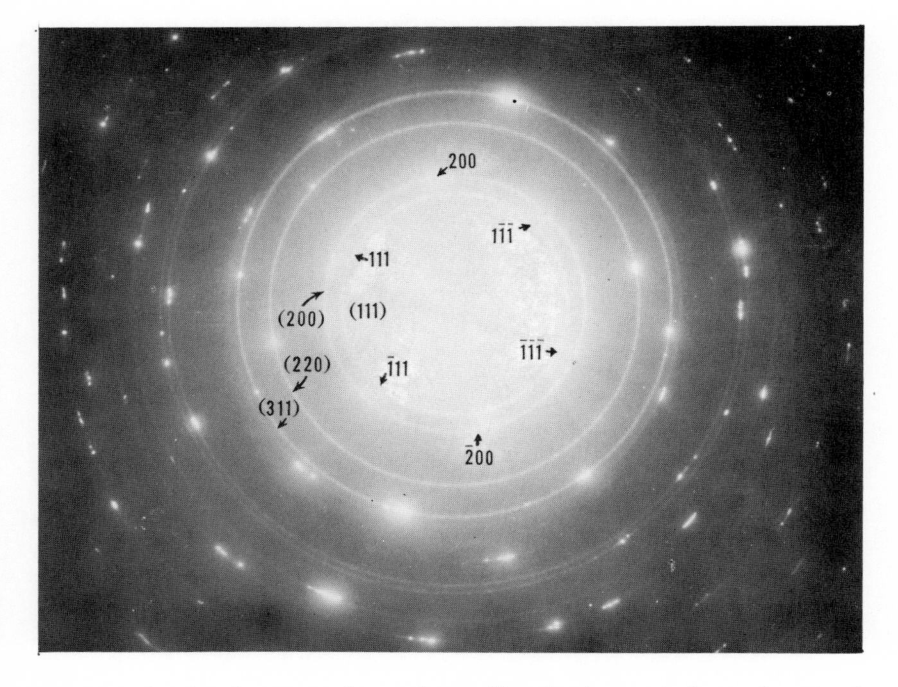

FIG. 19. Showing ring pattern from polycrystalline aluminum superimposed on a spot pattern from a single crystal of aluminum.

† See A.S.T.M. card index files.

Appendix A. As seen in section 1.8, many effects may be encountered in electron diffraction, so that great care must be exercised in analyzing the diffraction patterns. One of the most important of these effects arise when diffraction occurs from very thin crystals (of atomic dimensions), when broadening of spots and extra reflections may be produced.

1.13 Reciprocal Space for Small Crystals

1.13.1 *Introduction*

The electron diffraction pattern consists of a system of spots corresponding to those reciprocal lattice points which are intersected by the reflecting sphere. These spots have indices corresponding to the allowed reflecting planes in the crystal. However, when the intensities of the reflections are considered, we find that the reciprocal lattice point is not a point at all, but has a size and shape which depend mainly on the geometry of the crystal concerned.[13] In order to estimate the shape of the reciprocal lattice point the intensity of the diffracted electrons must be calculated. To do this we shall use the kinematical theory of diffraction, where only one diffracted beam is considered. It is convenient in the first instance to choose a one-dimensional model of a crystal, i.e., a lattice composed of a single row of unit cells such that only one Laue condition applies. The main assumption used in the kinematical theory is that a plane monochromatic wave is diffracted by a crystal in a direction which deviates slightly from the Bragg condition (section 1.6). A further stipulation must also be made, namely, that the crystal be thin enough so that there will be negligible rescattering or absorption.

1.13.2 *The Intensities of Diffracted Beams*

If a wave of intensity I_0 is incident parallel to a row of unit cells, each unit cell in the row will scatter a portion of I_0 so as to produce an angular distribution of scattered intensity I about the direction of the incident wave. In the kinematical approximation this scattered intensity is small compared to I_0 since the direction of scattering deviates from the exact Bragg condition. This means that successive unit cells do not scatter in phase. In any given direction the amplitude of the diffracted wave will be the result of the superposition of all the individual waves scattered by each unit cell at a given point, i.e., at the reciprocal lattice point corresponding to the reflection which is operating. The magnitude of the total amplitude depends upon the relationship of the phases of the individual waves.

In terms of the reciprocal lattice and reflecting sphere construction the condition for diffraction will then be given as shown by Fig. 20, where PP' is a thin column of crystal, length t, one unit cell thick, and containing n unit cells in the direction of the diffracted beam; 2θ is the exact Bragg angle corresponding to the reciprocal lattice point r^*, and $d\theta$ represents the deviation from the Bragg condition. It is assumed that each unit cell contains only one scattering point as in Fig. 9. It is clear that for the diffraction pattern to be obtained, the point r^* must be extended by an amount s in a direction parallel to the column being considered so as to intersect the reflecting sphere at R. This is shown greatly exaggerated in Fig. 20. The amplitude of the out-of-phase vibrations scattered by the column may now be calculated using expression (18). Now the phase factor ϕ, from Fig. 20, is given by $2\pi(\mathbf{g} + \mathbf{s})\cdot \mathbf{r}_n$; hence for unit incident amplitude, and for unit cells spaced unit distance apart the amplitude diffracted by the column is

$$A = \textstyle\sum_n f_n \exp\,[2\pi i(\mathbf{g} + \mathbf{s})\cdot \mathbf{r}_n] \tag{39}$$

where f_n is the scattering factor for electron waves of the contents of the unit cell situated at $\mathbf{r}_n$ in the column (expression 29). If all the atoms are the same, the f_n's are constant and can be taken outside the summation. Expression (39) can then be approximated by an integral, so that by taking the origin O at the center of the column in Fig. 20, and neglecting the factor f, the total diffracted amplitude is given by

$$A \simeq \int_{-t/2}^{+t/2} \exp\,[2\pi i(\mathbf{g} + \mathbf{s})\cdot \mathbf{r}_n]dr \tag{40}$$

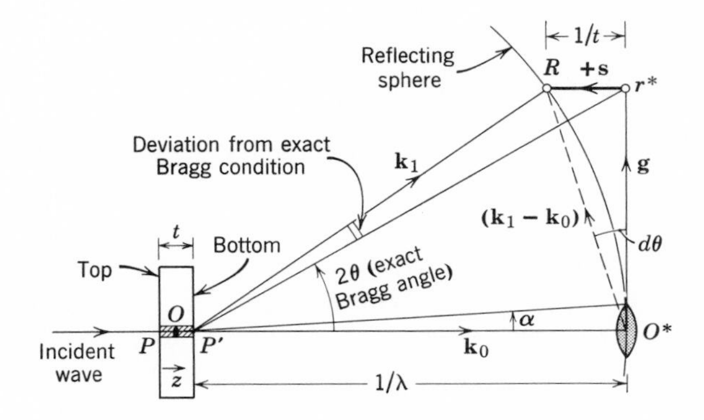

FIG. 20. The sphere of reflection construction for small crystals (exaggerated) illustrating the broadening $\mathbf{s}$ of a reciprocal lattice point and the corresponding beam divergence $d\theta$. (Also see Fig. 15a.)

Since $\mathbf{g}$ is a reciprocal lattice vector and $\mathbf{r}_n$ is a real lattice vector [given by expressions (36*a*, *b*)], $\mathbf{g} \cdot \mathbf{r}_n$ is an integer, so that $\exp(2\pi i \mathbf{g} \cdot \mathbf{r}_n)$ is unity. Hence we can write

$$A \simeq \int_{-t/2}^{+t/2} \exp(2\pi i \mathbf{s} \cdot \mathbf{r}_n)\, dr \tag{41}$$

$$\simeq \frac{\sin(\pi t s)}{\pi s} \tag{42}$$

The intensity is thus

$$I \simeq \frac{\sin^2(\pi t s)}{(\pi s)^2} \tag{43}$$

Hirsch et al.[14] have pointed out that expression (43) gives the amplitude scattered in one particular direction from a column one unit cell thick. For each reflection the diffracted rays from such a column are actually spread out over a whole range of directions. However, because of the interference in the crystal between rays diffracted from neighboring columns, the rays at the bottom surface will be concentrated in a small range of angles around that corresponding to the reciprocal lattice point for the particular diffracted beam. The total intensity diffracted from the column in this small range of angles may be obtained by calculating the average intensity per unit area of cross section from a column of finite lateral dimensions, and integrating the intensity scattered over the reflecting sphere. In this way the intensity is given by[14]

$$I = \frac{f^2}{k_0^2 V^2 \cos^2\theta} \frac{\sin^2(\pi t s)}{(\pi s)^2} \tag{44}$$

where $k_0 = \lambda^{-1}$ is the wave number of the incident electron wave (Fig. 20), V is the volume of the unit cell, and θ is the Bragg angle. Since θ is only one or two degrees, the column in Fig. 20 can be considered to lie parallel to the incident beam, and in subsequent discussions, since expressions (43) and (44) are identical, the proportionality factor given in (44) will be neglected.

The calculation of the diffracted intensity may also be represented in terms of the amplitude-phase diagram (section 1.9). If the unit cells are spaced unit distance apart in the column, and if they scatter in phase (i.e., exact Bragg condition) the amplitude-phase diagram will be a straight line, since the phase angle between successively diffracted waves is 2π. However, if we deviate the direction of scattering slightly away from the exact Bragg condition as in Fig. 20, successive cells no longer scatter in phase and for an element of the column lying between depths z and $z + dz$ in the direction of t, the amplitude is proportional to dz and the phase angle is $2\pi s\, z$. The amplitude-phase diagram will now be made up of a set of vectors representing the scattered amplitude from each element, but each vector rotated with respect to the preceding one by the phase angle $2\pi s\, z$. The resultant amplitude B for the first three elements of the column will then be given by Fig. 21*a*. The complete amplitude-phase diagram for the column choosing the origin at its center, is thus a circle of radius $R = (2\pi s)^{-1}$. The resultant amplitude $A = PP'$ is then obtained from Fig. 21*b*, where OP and OP' represent the amplitudes scattered by the top and bottom halves of the column shown in Fig. 20. The corresponding angle subtended by an arc of the circle at its center is then $2\pi ts$. The length of the secant joining the ends of the circle, i.e., the resultant amplitude A, is twice the radius multiplied by the sine of half the angle subtended at the center. Hence

$$A = \frac{2}{2\pi s} \sin\left(\frac{2\pi ts}{2}\right) \tag{45}$$

The intensity is thus

$$I = \frac{\sin^2 (\pi ts)}{(\pi s)^2}$$

which is the same result as that obtained in expression (43). Equation (43)

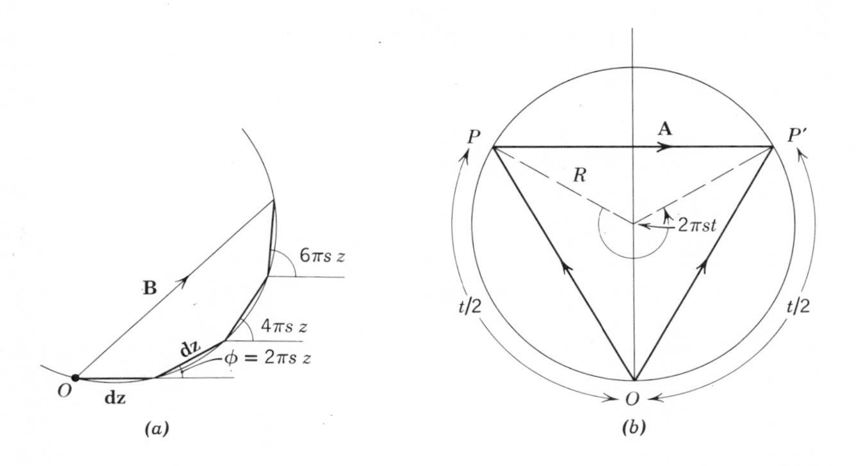

FIG. 21. Summation of amplitudes scattered by successive unit cells from the column shown in Fig. 20. (*a*) Amplitude-phase diagram for first three scattered amplitudes. (*b*) Amplitude-phase diagram for the perfect crystal thickness t. (After Hirsch et al.,[14] Courtesy the Royal Society.)

shows that the intensity of the diffracted wave varies sinusoidally with depth in the crystal, a result which is well known for the dynamical case.[14] If the expression for the intensity is plotted with s as abscissa, Fig. 22 is obtained. This gives the kinematical intensity distribution about the reciprocal lattice point. In terms of the amplitude-phase diagram, the principal maximum corresponds to n vectors **dz** lying in a straight line, i.e., $s = 0$. The first minimum corresponds to a complete revolution of the circle in Fig. 21*b*, i.e., when $s = \pm 1/t = \pm(t_0')^{-1}$. The wavelength of the intensity oscillations is thus $t_0' = s^{-1}$, so that at successive depths t_0' through the crystal the diffracted intensity is zero, i.e., extinction occurs. This result will be of particular interest when we come to consider the mechanism of contrast in electron microscope images (see Chapter 2).

The first subsidiary maximum in Fig. 22 is represented by one and one-half revolutions of the amplitude-phase diagram of Fig. 21*b*. With increasing phase difference between the vectors representing the scattering by successive cells the amplitude-phase diagram becomes smaller and smaller, (i.e., R decreases as s increases), until the smallest maximum $(1/t)^2$ of the principal maximum is reached; from this direction on, the amplitude-phase diagram will unwind with increasing values for the subsidiary maximum until the phase difference between successive atoms is again zero. This corresponds to a path difference of two wavelengths. At this position another principal maximum will occur. The width of the principal maximum is thus $2/t$ corresponding to a streak in reciprocal space along the direction of t (Fig. 20).

The value of s decreases for thicker crystals, i.e., for larger t, so that dynamical effects then become important (see section 2.3.3). *Hence the*

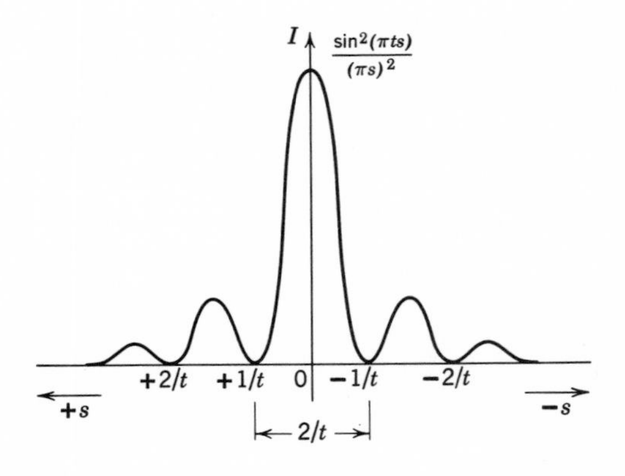

FIG. 22. The kinematical intensity distribution about a reciprocal lattice point. (After Hirsch et al.[14])

kinematical theory is only applicable to very thin crystals. The thickness of crystals corresponding to kinematical conditions are only of the order of a few hundred angstroms (see section 2.3.3), but for most purposes in electron microscopy the use of the kinematical theory seems justified.[14]

The resultant intensity distribution of the waves scattered by the individual elements of the crystal gives rise to a divergent diffracted beam having an angular width $d\theta$ corresponding to the magnitude of $s = 1/t$ (the diffraction error). This effect is exactly the same as if the waves had passed through a slit of width t. The magnitude of $d\theta$ thus determines the resolving power of the crystal, and hence the size of the spots (i.e., the reciprocal lattice "points"), in the diffraction pattern. The value of $d\theta$ may be obtained from Fig. 20; if the reciprocal lattice point r^* corresponds to a set of reflecting planes of spacing d, then from the triangle O^*r^*R, $d\theta = (1/t)/g$, where g is the length of the reciprocal lattice vector O^*r^*, equal in magnitude to $1/d$; hence $d\theta = d/t$. Thus the angular width, i.e., the size of the diffraction spot, decreases for smaller d, i.e., for larger angles of reflection. This result shows that higher orders of reflection and thicker crystals give sharper diffraction spots. For axial illumination the value of $d\theta$ depends only upon these factors. However, in the electron microscope (or a diffraction camera) owing to the focusing action of the condenser lens the illuminating beam converges onto the specimen. As a result, there will be an angular range of incident electrons giving a selection of possible angles for reflection. This range is small in magnitude compared with 2 or 3° corresponding to the appearance of many orders of reflection in the diffraction pattern. The resultant divergence then varies from s = minimum to s = maximum for each component beam reflected by a given plane.

This may be made clear from the following. For any given angle of incidence α_1, s is only a minimum when the Laue condition $a(\cos \alpha_1 - \cos \alpha_2)$ is equal to an integral number of wavelengths. If α_1 deviates from the exact Bragg angle, the optimum path difference changes, and so the value of α_2 changes from the direction of the maximum. If we differentiate the Laue equation with respect to α_1 and λ, it is clear that a change in α_1 of $d\alpha_1$ is inversely proportional to $\sin \alpha_1$. The maximum resolution of an atom row is thus obtained for incident angles $\alpha_1 = \pi/2$ and a minimum for α_1 = zero, i.e., for atom rows parallel to the beam. Thus in general, for a small range of angles $d\alpha_1$ in a converging beam, there will be a corresponding range of values of s given by $ds = (d/t)\, d\theta$. In electron microscopes having the facility of double condenser lenses (see section 3.5.3), the angular width of the beam may be reduced by defocusing the second condenser. The minimum value of the beam divergence is then about 10^{-3} radians, i.e., an order of magnitude smaller than the Bragg angles involved in forming the diffraction pattern from metals. A similar effect is obtained when a single incident wave passes through a buckled crystal, since the local misorientations will change the value of s for successively diffracted waves along the length of the column. This will be reconsidered in Chapter 2 because it gives rise to certain contrast effects in the image.

Let us now determine the angular width 2α of the transmitted beam for which the angle of reflection α_2 is zero. This beam corresponds to scattering from unit cells lying parallel to the beam. Since α_1 is also zero, the phase difference δ, between waves scattered from successive unit cells spaced unit distance apart, is given by substituting in the Laue equation (20), where

$$\delta = \frac{2\pi}{\lambda}(1 - \cos\alpha) \tag{46}$$

Replacing $\cos\alpha$ by $2\sin^2\alpha/2$ we can rewrite (46) as

$$\delta = \frac{2\pi}{\lambda} 2\sin^2\frac{\alpha}{2} \tag{47a}$$

$$\simeq \frac{\pi\alpha^2}{\lambda} \tag{47b}$$

Since expression (43) gives the intensity in terms of half the phase difference, by substituting for $2\pi s = \pi\alpha^2/\lambda$ we obtain †

$$I = \sin^2\left(\frac{\pi t\alpha^2}{2\lambda}\right)\Big/\left(\frac{\pi\alpha^2}{2\lambda}\right)^2 \tag{48}$$

The condition for the first minimum is clearly $I = \sin^2\pi$, i.e., when

$$\alpha = \sqrt{2\lambda/t} \tag{49}$$

Thus the total angular width 2α is $\sqrt{8\lambda/t}$. This shows that the diffraction maxima for the atom rows parallel to the beam become diffuse; the other maxima become gradually sharper as the diffraction angles increase. Thus in all transmission electron diffraction patterns the central spot is the broadest. From this we can now see why in thin crystals the diffraction maxima from rows parallel to the beam, i.e., the circles in Fig. 13, are not resolved.

1.13.3 *Effect of Crystal Geometry*

In the previous section we have only considered the behavior of a single column of scattering centers. In order to investigate the behavior of a

† We have chosen cells spaced unit distance apart so that the phase difference is $2\pi s$ radian; for cells spaced a distance a apart the phase difference is $2\pi sa$ radian.

complete crystal it is necessary to take into account the effect of its shape. In principle it is possible to estimate the scattering from a crystal of any shape using the method previously outlined.[13] Let us here consider a parallel-piped crystal made up of many cells, each of unit volume, along the axes *xyz*. Using a similar analysis to that which gave expression (43), the total scattered intensity along the three axes is given by

$$I \simeq \sin^2 \frac{(\pi t_1 s_1)}{(\pi s_1)^2} \sin^2 \frac{(\pi t_2 s_2)}{(\pi s_2)^2} \sin^2 \frac{(\pi t_3 s_3)}{(\pi s_3)^2} \tag{50}$$

where s_1, s_2, s_3 are the deviations from the reciprocal lattice point along the directions a^*, b^*, c^* in reciprocal space, and t_1, t_2, t_3 are the dimensions of the crystal along the three axes. The maximum value of the interference function given in expression (50) is for $s_1 = s_2 = s_3 = 0$, i.e., $(t_1 t_2 t_3)^2$.

Since $(t_1 t_2 t_3)$ is the volume of the crystal made up of unit cells of unit volume, the maximum value of the intensity is the square of the total number of unit cells in the crystal. For given distances s_1, s_2, s_3 from the reciprocal lattice point the intensity is greatest along the reciprocal axes, i.e., the intensity falls off least rapidly in directions normal to the crystal faces. Hence each plane face of a crystal is represented by a spike or streak in reciprocal space. The direction of the spike is perpendicular to the face. The smaller the values of t_1, t_2, t_3 the lower is the intensity and the greater is the length s of the streak, since s is inversely proportional to t. Thus for very small crystals the reciprocal lattice points are always extended in the direction of the finite extent of the crystal. For reciprocal lattice points of successive orders to be connected by a streak, it is clear that s must be equal to the interplanar spacing, i.e., the crystal must have dimensions of one unit cell.

If the crystal is platelike with t_2 and $t_3 > t_1$, e.g., a normal thin foil specimen where t_1 lies along the z axis parallel to the electron beam, the intensity will be concentrated in reciprocal space in the direction of the c^* axis along a line parallel to the electron beam; i.e., the streak is perpendicular to a photographic plate placed normal to the beam. This is the situation drawn in Fig. 20. These streaks are sometimes referred to as rel-rods. If the crystal is face-centered cubic and is only one unit cell thick in the direction of the beam, e.g., along the cube axis [001], the (111) reciprocal lattice point of Fig. 14*b* is elongated sufficiently to intersect the plane of the reflecting sphere at the (110) point. In this case a reflection of indices (110), forbidden by the structure factor, will appear on the diffraction pattern.

From the foregoing it can be seen that the formation of the two-dimensional spot diffraction pattern typical of those obtained from crystals (e.g., Fig. 16) is due to a considerable distortion of the reciprocal lattice

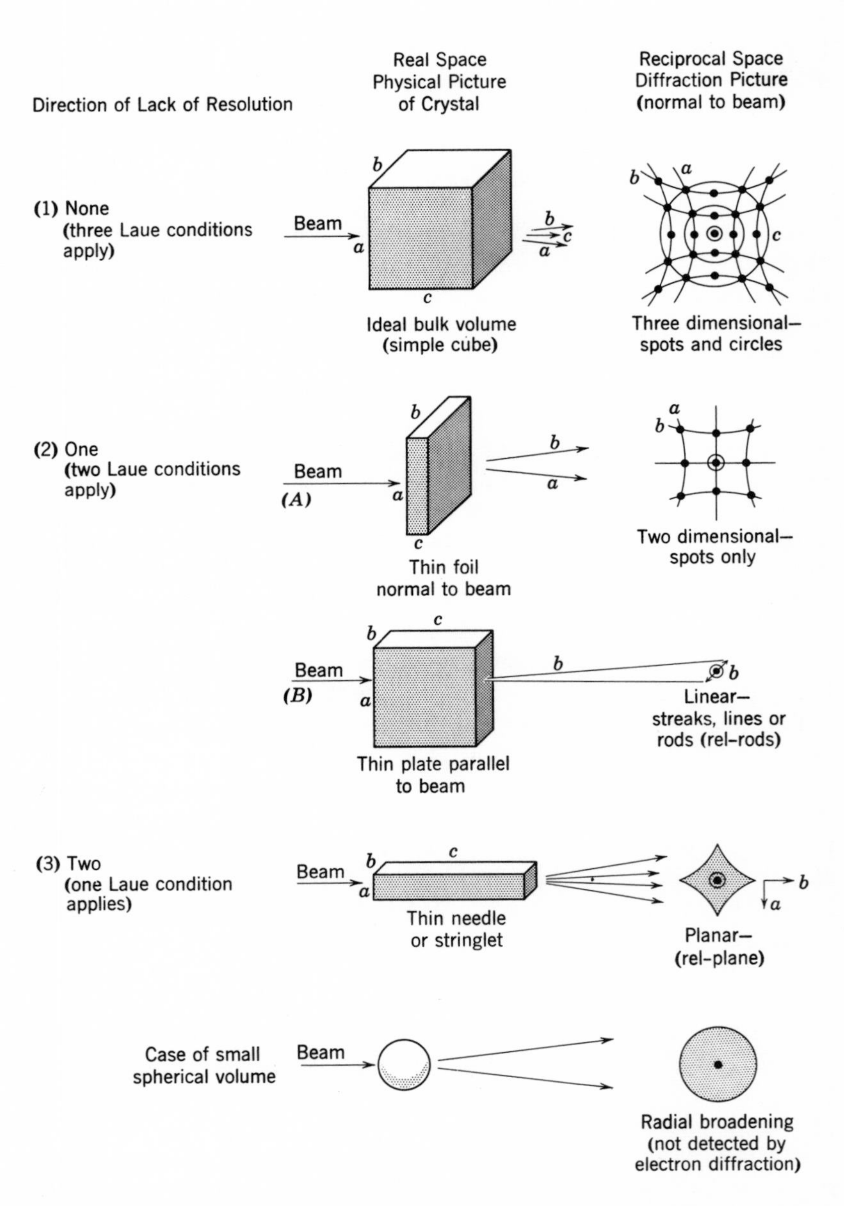

FIG. 23. Summary of diffraction effects from crystals. (After Geisler and Hill.[15])

points in the direction of the thickness of the specimen, i.e., parallel to the incident beam. This means that the corresponding Laue condition is relaxed. If the crystal is thin in a direction normal to the beam, streaks will be observed in the diffraction pattern parallel to this direction, provided that the thickness is of the order of the dimensions of the unit cell. Similarly for a needle-shaped crystal lying parallel to the beam the reciprocal lattice point will be extended in the plane of the reflecting sphere with streaks extending in the directions of t_2 and t_3, parallel to a^* and b^*. This may give rise to a plane of intensity at the reciprocal lattice point which is sometimes referred to as a rel-plane. Similar arguments can be applied to show that for spherical crystals the intensity is proportional to the diameter of the crystal.[13] Thus for small spherical specimens there will be a radial broadening of the reciprocal lattice point. These effects are summarized in Fig. 23.

One-dimensional diffraction is not observed in perfect pure metals. They may be observed in some cases where the crystal is deformed such that they contain stacking faults or thin twins. More often these effects are found in certain alloys where small regions of clustering or short range order occur.[15] A well-known example is that of Guinier-Preston zones in aluminum-4% copper alloys. (See also Chapter 5, Fig. 135.) These zones consist of platelets of copper, one or two atoms thick lying along {100}. When a crystal is oriented such that the beam is parallel to a cube axis, continuous streaks will be observed through the {200} reciprocal lattice points in ⟨200⟩. Thus for a foil in [001], streaks occur in reciprocal space parallel to [010] and [100] as shown in Fig. 24. Numerous other examples of these diffraction effects for X-rays have been given by Guinier.[16]

We have now considered in some detail how the diffraction pattern and the specimen are related, and we have shown that the patterns can provide much information regarding the size, shape, and orientation of a crystal. Before the picture can be completed, however, we still have to consider other examples of diffraction effects that may be obtained from metals. These will be summarized in the following sections.

1.14 Special Cases of Electron Diffraction

1.14.1 *Stacking Faults*

From the preceding discussions it is to be expected that distortions in the crystal will modify the electron diffraction pattern. If only the direct beam is disturbed, this means that the distortion is general, whereas if only

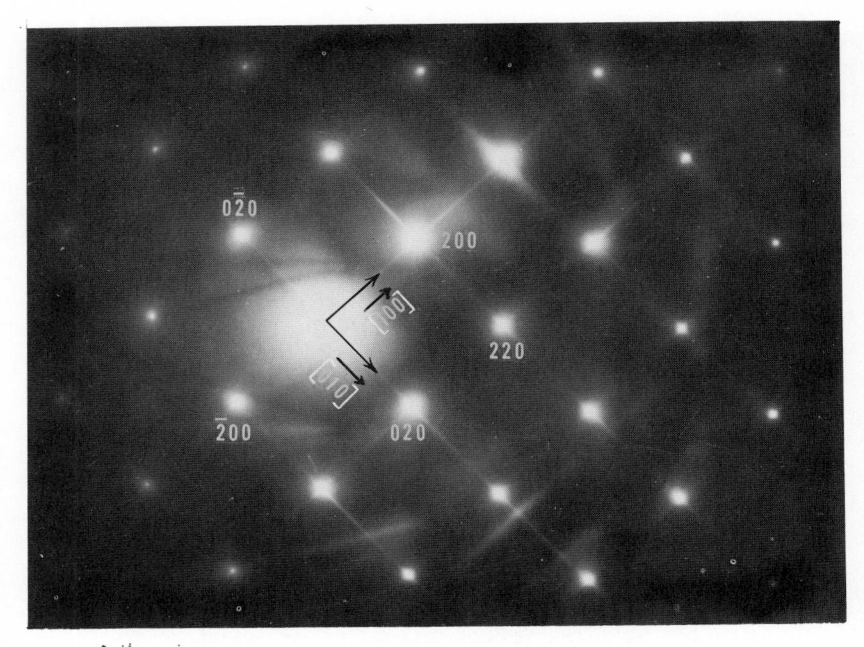

FIG. 24. Streaks in diffraction pattern from plates of G.P.1 zones one to two atoms thick in Al-4% Cu alloys. Zones form on {100}. Foil orientation [001], streaks parallel [100] and [010]. Diffraction effects correspond to condition 2*B*, Fig. 23. (Diffraction pattern obtained in the electron microscope from a selected area similar to that shown in Fig. 135, Chapter 5.)

certain diffracted beams are disturbed e.g., as a result of lattice displacements, it is possible to decide in which planes the distortion occurs. In this section we shall consider a crystal which is distorted because of mistakes in the stacking sequence of certain planes.

In face-centered cubic crystals a stacking fault occurs when the stacking sequence *ABCABC* etc., of (111) planes is interrupted (see also sections 5.2.1 and 5.2.14). This can be produced by the removal or addition of a whole or part of a (111) plane or by a shear of the crystal in a $\langle 112 \rangle$ direction. As a result of this a thin region of hexagonal close-packed structure is introduced in which the *c*-axis lies in [111]. The stacking sequence at the fault is then of the type *ABCABABC*, etc.[17] By using an analysis similar to that given in section 1.9, Paterson,[18] and Warren and Warekois[19] have shown that for stacking faults in (111) planes reflections are affected when $(h + k + l)$ is equal to $3N \pm 1$, where N is an integer; hence (111), $(2\bar{2}0)$, $(31\bar{1})$, etc., as well as (000) are unaltered. Whelan and

Hirsch[20] have also studied this problem in detail for the electron diffraction pattern and its corresponding image. If a stacking fault is formed by a shear of $\frac{1}{6}[1\bar{2}1]$ in (111), those reciprocal lattice points for which $(h + k + l) = 3N \pm 1$ are broadened and displaced. This displacement is due to a phase factor ϕ equal to $\pi/3\{h - 2k + l\}$ which is introduced into the scattered wave. The possible values of ϕ are 0 or $\pm 2\pi/3$, depending on the reflection operating (i.e., on crystal orientation). Clearly when $\phi = 0$ no disturbance of the reflection (h,k,l) will occur.† In the electron diffraction patterns streaks will be observed running through all the reciprocal lattice points for which $\pi/3\{h - 2k + l\}$ is not zero, i.e., for reflections where the phase angle is $\pm 2\pi/3$, and these streaks will lie in a direction parallel to the thin direction of the fault, i.e., in $\langle 111 \rangle$. Thus, depending on the orientation of the crystal with respect to the beam, streaks or extra spots will then be seen in the pattern.‡

For hexagonal close-packed crystals where stacking faults may occur in the basal planes, since the spacing of these planes is unaffected faulting, (00.2) reflections are unaffected. Since the hexagonal indices HK.L and the cubic indices h, k, l are related as follows (see Fig. 16):

$$H = -\tfrac{1}{2}(h - k), \quad K = -\tfrac{1}{2}(k - l), \quad L = (h + k + l) \tag{51}$$

it is clear that $(h - 2k + l) = -2(H - K)$, so that the phase difference is $-2\pi/3(H - K)$.†† In other words reflections from hexagonal lattices are unaffected when $(H - K) = 3N$ and is independent of L. A similar approach will be used in Chapter 5 in order to explain certain image contrast effects associated with stacking faults.

1.14.2 *Twinning*

Extra reflections are also found in diffraction patterns from twinned crystals. A perfect FCC twin corresponds to a succession of stacking faults on every (111) plane so that the structure is unaltered, whereas if every other (111) plane is faulted, a perfect HCP structure is formed. In cubic lattices, the twinned lattice is obtained by a rotation of 180° about any $\langle 111 \rangle$ twin axis. In the electron diffraction pattern twin spots will

† If the atoms in a crystal are displaced by a vector **R** the phase factor ϕ can be calculated for *all* types of distortions. In each case no disturbance of reflections occurs for $\phi = 0$. (See section 5.2.14.)

‡ Anomalous streaks may occur in patterns from alloys if the solute atoms segregate to the fault. In this case the periodicity may be interrupted for all reflections because of the change in composition. Hence streaks may be observed through all the reciprocal lattice points.

†† A fault in the basal plane of HCP lattices occurs by a shear of $\frac{1}{3}[\bar{1}1.0]$, which is the same as $\frac{1}{6}[1\bar{2}1]$ in the FCC lattice.

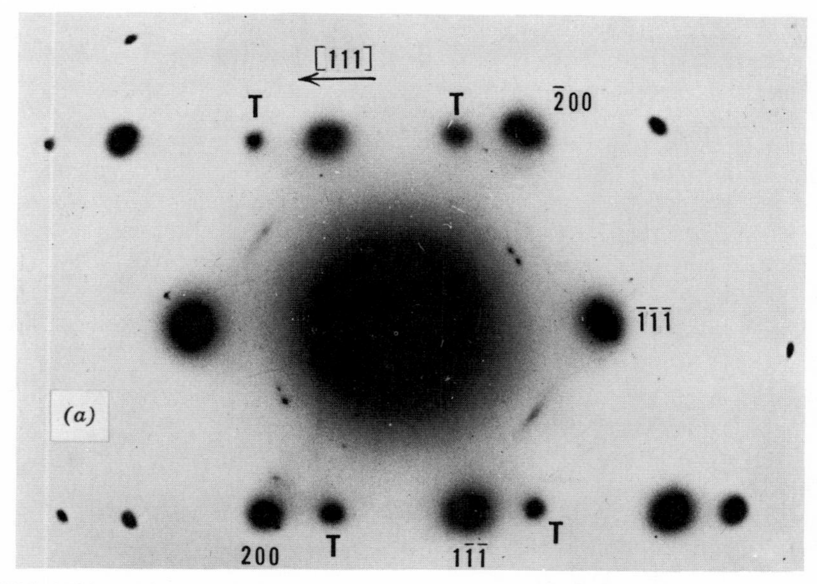

FIG. 25(*a*). Selected area diffraction pattern from a twinned FCC crystal (Cu-Be alloy), twin reflections at *T*. Streaks are due to small precipitates (see Fig. 145).

be observed at about one-third of the distance between the original spots of the untwinned lattice, and the pattern is rotated through 180°. Figure 25*a* shows an example of a selected area diffraction pattern taken from a specimen of Cu–2% Be alloy. This pattern corresponds to an area of

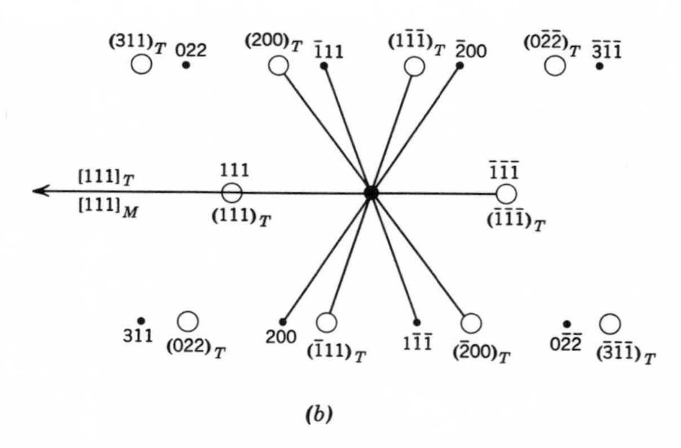

FIG. 25(*b*) Analysis of Fig. 25*a*. Open circles represent twin spots *T* in (*a*). A rotation of 180° about [111] makes points and circles coincide.

about 1 μ^2 of the specimen so that only one twin and the neighboring matrix are illuminated by the incident beam. By indexing the pattern using the procedure given in Appendix A it can be analyzed as shown in Fig. 25*b*. From this it is clear that the orientation relationships are

$$(111) \text{ twin parallel to } (111) \text{ matrix}$$

$$[0\bar{1}1] \text{ twin parallel to } [01\bar{1}] \text{ matrix}$$

and that a rotation of 180° about [111] will bring the twin and matrix reflections into coincidence.

1.14.3 *Double Diffraction*

In dynamical scattering a diffracted beam can be further diffracted when passing through the specimen, which results in reflections that are forbidden by the structure factor. Diffraction under these conditions is characterized by a considerable intensity of the primary reflections. For example, if a beam reflected by the $h_1k_1l_1$ plane is again reflected by the $h_2k_2l_2$ plane, extra reflections appear on the diffraction pattern having indices $h_1 \pm h_2, k_1 \pm k_2, l_1 \pm l_2$.

Superlattices, or ordered structures in alloys, also give rise to extra reflections in the diffraction pattern. In the superlattice, if atoms of one kind are preferentially arranged on certain planes, atoms of the other kind will be arranged on alternate planes. Thus, reflections will occur which, for the disordered alloy, would be forbidden by the structure factor. For

FIG. 26. Superlattice reflections from CuAu II. Note splitting of {200} spots into satellites (see also Fig. 150, Chapter 5). Bright field image formed from *A*, dark field from *B*. (After Glossop and Pashley,[21] Courtesy the Royal Society.)

example, in a face-centered cubic lattice if ordering of atoms of dissimilar scattering factors occurs on alternate {100} planes, extra reflections {100}, {300}, {500}, etc., would be observed on the pattern, provided the crystal is in a favorable orientation. Figure 26 shows a diffraction pattern from the CuAu II superlattice structure. (See also Fig. 150, Chapter 5.) The spots around the primary reflections are superlattice reflections—the splitting into four satellites being a direct consequence of the 20 Å periodicity of the domain boundaries in the superlattice.[21]

If two crystals with parallel lattices and similar lattice constants are superimposed, the beam diffracted by the first crystal is further diffracted by the second crystal, giving rise to double diffraction spots around the principal reflections. These diffraction effects are similar to those shown in Fig. 26, and an example is given in Fig. 126, Chapter 5. If these doubly diffracted beams are made to reunite so as to form an image of the specimen, a moiré pattern is obtained in which a magnified image of the lattice planes may be resolved.[22] This is further discussed in Chapter 2 and examples given in Chapter 5. (See also Guild's book.)

1.14.4 *Kikuchi Lines*

When the thickness of a sample is increased, the diffraction pattern changes from a system of only spots (two-dimensional pattern) to a set of spots arranged on circles (three-dimensional pattern), i.e., when all three Laue conditions are satisfied (Fig. 13). Very often in thicker films pairs of black and white lines may be observed, e.g., as shown in Fig. 27*a*. These lines, known as Kikuchi lines, are formed as a result of the inelastic scattering of electrons. A similar effect occurs with X-rays, which are known as Kossel lines.[13] The most convenient way of describing Kikuchi lines is by using geometrical arguments as first proposed by Kikuchi,[23] even though they are produced as a result of dynamical scattering.[2] Let us suppose that a beam of electrons is not immediately scattered coherently as it first enters a crystal, but rather that the electrons are inelastically scattered. However, some of these electrons impinge on planes inside the crystal at a Bragg angle and are consequently reflected. Suppose that in Fig. 27*b* the planes AB, $A'B'$ are reflecting planes; some of the inelastically scattered electrons which would have emerged in the direction P are reflected into the direction Q when they meet the plane AB at a Bragg angle θ. On the other hand, some which would have gone in the Q direction are similarly reflected by the plane $A'B'$ into direction P. Since the electrons reflected from AB have undergone a smaller deviation ab at the first scattering, they are more numerous than

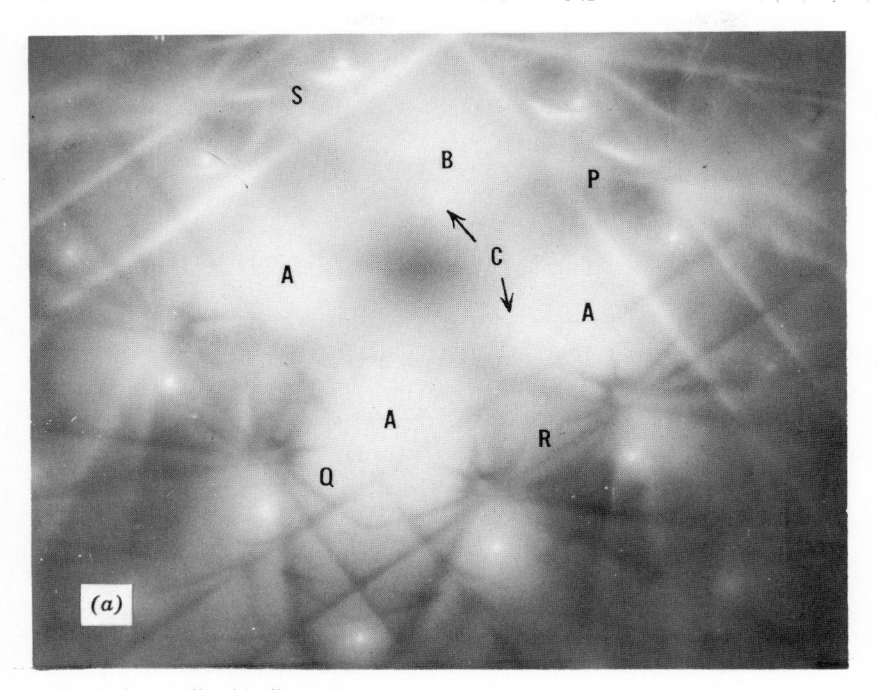

FIG. 27(*a*). Kikuchi lines in a diffraction pattern from a thick foil of MgO. Note the complementary pairs of black lines *Q*, *R*, and white lines *P*, *S*, (see Fig. 27*b*), and the high intensity of the diffracted beams *A* and the direct beam *B* compared to the higher order spots. Double diffraction has occurred at *C*.

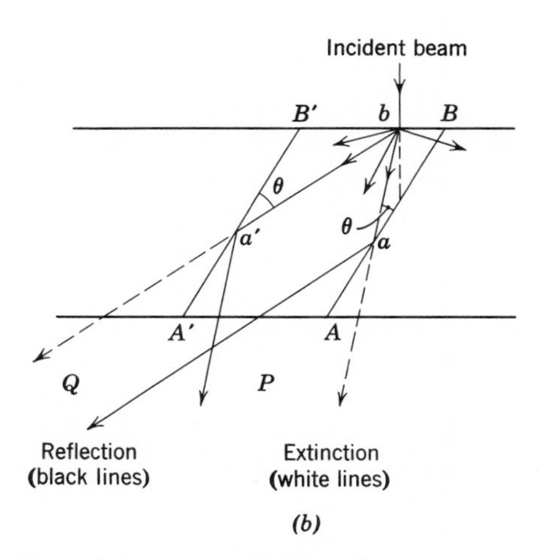

FIG. 27(*b*). Geometrical illustration of the origin of Kikuchi lines.[23]

those deviated along $a'b$. Consequently, there is a gain in electrons in the Q direction, and a loss in the P direction. The reflected directions form cones with the normal to AB as axis and the semivertex angles $(\pi/2 - \theta)$ on either side of it. These cones intersect the screen (or photographic plate) in hyperbolae which, as we have seen in the electron case where λ is very small, are actually very nearly straight lines. The angle between P and Q is 2θ, so that the distance S between the black and white lines on the plate is $S = 2\theta L = n\lambda L/d$ (λL is the camera constant) where n is the order of the reflection and d is the spacing of the planes. Consequently it is possible to determine the orientation of the crystal. For n greater than one there will obviously be several pairs of black and white lines (Fig. 27*a*). It will be seen that if the reflecting planes are parallel to the electron beam there should be no effect since the reflection to the two sides should be equal. However, in this case a band of enhancement is produced whose edges are approximately at positions where the lines would be if the primary beam were slightly deviated. In order to explain this fact and the intensity distribution in the diffraction pattern (e.g., notice the intense diffraction spots at A in Fig. 27*a*), it is necessary to use the dynamical theory of electron diffraction, a detailed account of which has been given by Pinsker.[2]

1.15 Refraction of Electrons

Bending of electron beams may occur when they traverse a medium in which there is a change in refractive index, e.g., due to a potential field[2] (analagous to the refraction of light). For example, from Fig. 28 the refractive index μ due to a potential field V_2 with respect to V_1 is:

$$\mu = \frac{\mu_1}{\mu_2} = \frac{\cos\theta_1}{\cos\theta_2} = \frac{V_1}{V_2} = \frac{\lambda_1}{\lambda_2} \tag{52}$$

As a result of the change in wavelength, the angle of reflection of a system of planes in a crystal will be changed, so that Bragg's law now becomes

$$n\lambda_2 = 2d\sin\theta_2 \tag{53a}$$

or

$$n\lambda_1 = 2d\sqrt{\mu^2 - \cos^2\theta_1} \tag{53b}$$

It follows from this that

$$\sin^2\theta_1 = \frac{n^2{\lambda_1}^2}{4d^2} - (\mu^2 - 1) \tag{54}$$

The beam reflected within the crystal will be able to leave it only when $\sin^2\theta_1$ is positive, i.e., when

$$\frac{n^2{\lambda_1}^2}{4d^2} < (\mu^2 - 1) \tag{55}$$

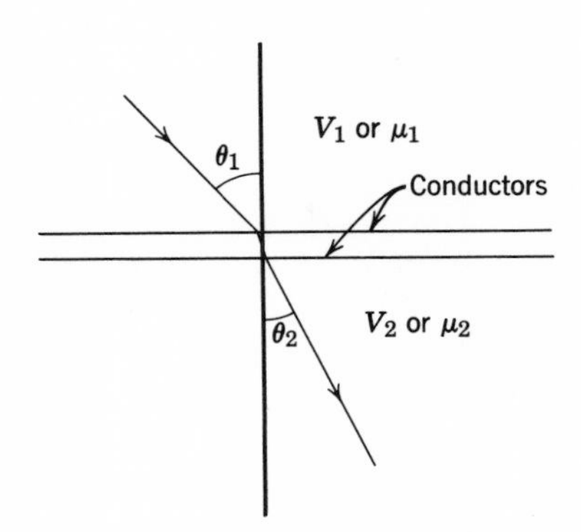

FIG. 28. The refraction of electrons on passing from one region where the potential is V_1 to another where the potential is V_2. The change in potential is sharpened by the two conductors. (After Hall.[6] By permission from *Introduction to Electron Microscopy*, by C. E. Hall. Copyright 1953, McGraw-Hill Book Co., Inc.)

there will be no reflection. This may happen for cleavage crystals where low-order cleavage faces are very smooth, but the effect is not important in transmission electron microscopy.[6] The abrupt change in potential shown in Fig. 28 is never realized in practice, and since the potential varies continuously in space the law of refraction is not as useful as its analagous form in classical optics (Snell's law).[3,4]

1.16 Summary

So far we have been concerned with the interactions which take place when an electron beam passes through a crystal and have considered the important factors governing the formation and intensities of the diffracted beams. The electron microscope image represents a magnified picture of the electron diffraction pattern, and consequently all that we have said concerning the resolution of the pattern applies equally well to resolving the structures which give rise to the pattern. In this, the orientation of the crystal with respect to the incident beam is extremely important. Our next step then is to consider the relationship between the diffraction pattern and the image, which will be the subject of the following chapter.

References

1. L. de Broglie, *Phil. Mag.* 1924, **47**, p. 446.
2. Z. G. Pinsker, *Electron Diffraction* (Butterworths Scientific Publications, London), 1953.
3. M. Born and E. Wolf, *Principles of Optics* (Pergamon Press, London), 1959.
4. F. A. Jenkins and H. E. White, *Fundamentals of Optics* (McGraw-Hill Book Co., New York), 1950.

5. L. Brillouin, *Wave Propagation in Periodic Structures* (McGraw-Hill Book Co., New York), 1946.
6. C. E. Hall, *Introduction to Electron Microscopy* (McGraw-Hill Book Co., New York), 1953, Chapter 9.
7. B. von Borries, *Proc. 3rd Int. Conf. Electron Microscopy* 1954 (Roy. Mic. Soc., London, 1956), p. 9.
8. F. Lenz, *Z. Naturf.* 1954, **9**(a), p. 185.
9. W. L. Bragg, *Proc. Camb. Phil. Soc.* 1913, **17**, p. 43.
10. N. F. Mott and H. S. W. Massey, *Theory of Atomic Collisions* (Clarendon Press, Oxford), 2nd Ed., 1949.
11. W. L. Bragg, *The Crystalline State* 1933, **1**, p. 254 (Bell & Sons Ltd., London).
12. J. A. Ibers, *Acta Cryst.* 1958, **11**, p. 178.
13. A. J. C. Wilson, *X-Ray Optics* (Methuen & Co. Ltd., London), 1949, p. 25.
14. P. B. Hirsch, A. Howie, and M. J. Whelan, *Phil. Trans. Roy. Soc.* A 1960, **252**, p. 499.
15. A. H. Geisler and J. K. Hill, *Acta Cryst.* 1948, **1**, p. 238.
16. A Guinier, *Solid State Physics* 1959, **9**, p. 293.
17. F. C. Frank, *Phil. Mag.* 1951, **42**, p. 809.
18. M. S. Paterson, *J. Appl. Phys.* 1952, **23**, p. 805.
19. B. E. Warren and E. P. Warekois, *Acta Met.* 1955, **3**, p. 473
20. M. J. Whelan and P. B. Hirsch, *Phil. Mag.* 1957, **2**, pp. 1121, 1303.
21. A. B. Glossop and D. W. Pashley, *Proc. Roy. Soc.* A 1959, **250**, p. 132.
22. J. W. Menter, *ibid.* 1956, **236**, p. 119.
23. S. Kikuchi, *Japan J. of Phys.* 1926, **5**, p. 83.

Bibliography

Elekroneninterferenzen, H. Raether, *Hanbuch der Physik* (Springer-Verlag Berlin), 1957, **32**, p. 443.

Theory and Practice of Electron Diffraction by G. P. Thompson and W. Cochrane (MacMillan and Sons, London), 1939.

Electron Diffraction by Z. G. Pinsker (Butterworths Scientific Publications, London), 1953.

Theory of Atomic Collisions by N. F. Mott and H. S. W. Massey (Clarendon Press, Oxford), 2nd Ed., 1949.

Principles of Optics by M. Born and E. Wolf (Pergamon Press, London), 1959.

X-ray Optics by A. J. C. Wilson (Methuen and Co. Ltd., London), 1949.

Optical Principles of the Diffraction of X-rays, by R. W. James (Bell & Sons Ltd., London), 1948.

Structure of Metals by C. S. Barrett (McGraw-Hill Book Co., New York), 2nd Ed., 1953.

Interpretation of X-ray Diffraction Photographs by N. F. M. Henry, H. Lipson, and W. A. Wooster (MacMillan and Sons, London), 1951.

Introduction to Solid State Physics, by C. Kittel (John Wiley and Sons, New York), 2nd Ed., 1957.

The Interference Systems of Crossed Diffraction Gratings (*Theory of Moiré Fringes*) by J. Guild (Clarendon Press, Oxford), 1956.

2

Image Formation and Contrast from Crystals

2.1 Introduction

In the electron microscope, as a result of the facility of selected area diffraction (see also section 3.7), it is a simple matter to convert from microscopy to diffraction, and vice-versa. Since crystalline specimens diffract electrons, the diffracted beams play a very important role in the formation of images, so that it is most important to be able to change quickly from the micrograph to the diffraction pattern. The principle of this operation is illustrated as follows. Figure 29 shows schematically the ray paths through a lens imaging a periodic structure. Diffracted beams are focused to a point in the back focal plane of the lens AA'. In the electron microscope it is possible to introduce a limiting aperture in the intermediate image plane CD to restrict to as small as 0.2 μ^2 the area contributing to the image. By reducing the strength of the intermediate (or first projector) lens it is then possible to form an enlarged image of the back focal plane of the objective lens on the final screen. Under these conditions one thus obtains an image of the diffraction pattern of a highly selected area of the specimen. If the strength of the intermediate lens is increased again, then an image is obtained of the same area contributing to the diffraction pattern. In this way it is possible to obtain all the information normally extracted from diffraction patterns, and to more readily interpret many of the contrast effects observed in images of crystals.

2.2 Imaging of Periodic Structures

The mechanism of image formation depends mainly on the spacing of the lattice planes in the crystal and thus on the Bragg diffraction angles. Direct imaging of lattice planes is only possible when the spacings in the

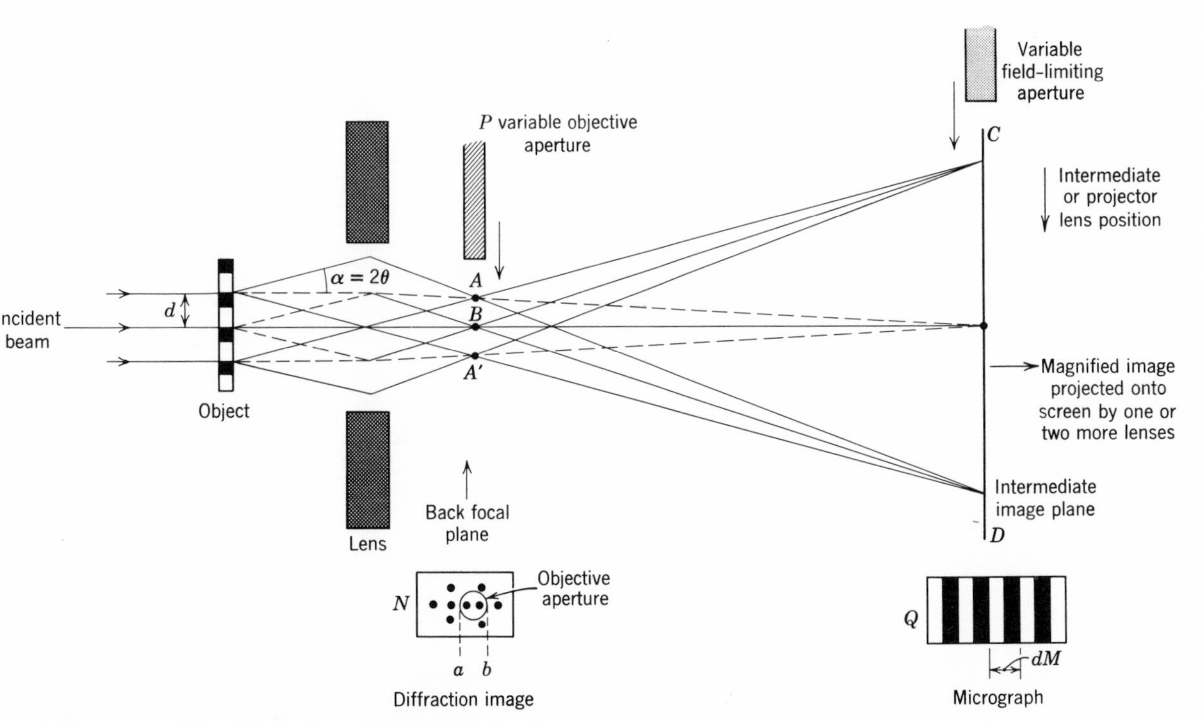

FIG. 29. Illustrating conditions for imaging periodic structures. If d is large enough and α small enough, the diffracted beam A or A' recombines with the direct beam B to give a magnified image of the planes d. The diffraction pattern is formed in the back focal plane AA'. N represents the diffraction pattern and Q the magnified image of this pattern for the beams A and B.

crystal are large enough so that the diffracted beams emerge from the crystal at a small enough angle to enter the aperture of the imaging, objective lens. The magnitude of this aperture is determined by the instrumental resolving power (see Chapter 3). The problem here is thus twofold. We must consider the resolving power of the crystal (discussed in Section 1.13) in relation to that of the instrument. The resolving power of the crystal is its ability to form sharp diffraction images, which depends upon such factors as orientation and the geometry of the crystal. Because of the divergence of the diffracted beam, it leaves the crystal with an angular width about the diffraction angle 2θ (corresponding to $d\theta$, Fig. 20). This is known as the diffraction error. In Chapter 1 we have seen that the angular width is a minimum for infinitely long atom rows normal to the incident beam and for axial illumination. However, since the value of $d\theta$ compared to 2θ is small, we can say that for the first order reflection, as shown in Fig. 29, the beam leaves the crystal at the angle $2\theta \simeq \lambda/d$ (as for a grating, Fig. 20).

In order that the diffracted beam may enter the aperture α of the objective lens, $2\theta < \alpha$. The value of α is determined by the lens errors and in particular by the spherical aberration error (see section 3.3). When this is equated to the diffraction error, an optimum value of $\alpha = 10^{-2}$ radian is obtained. This in turn sets a lower limit upon d of about 4 Å. However, other instrumental errors such as astigmatism may raise this value to 5 Å. For crystals of interplanar spacings larger than this and oriented as shown in Fig. 29, the first order diffracted beam can be made to recombine with the direct beam at the image plane of the lens. Then by the Abbé theory as in physical optics[1] a magnified image of spacing dM (M = magnification) will be formed at the image plane of the objective lens. This image is simply an interference pattern produced when the diffracted orders are recombined. For metal crystals where the lowest Bragg angles correspond to $d < 4$ Å, direct imaging of lattice planes is not yet possible; thus the mechanism of image formation is quite different. However, as we shall see in section 2.3 the diffracted beams are utilized in order to obtain contrast.

Experimentally, the arrangement for resolving crystal lattice planes is done as shown in Fig. 29. By inserting a physical objective aperture at the back focal plane of the objective lens, as shown at N, so that only one diffracted beam along with the direct beam is allowed to pass through, a one-dimensional image is obtained as shown at Q. A two-dimensional image, i.e., a cross-grating pattern of planes, is obtained when the crystal is oriented such that two different sets of reflecting planes lie parallel to the beam. In this case the corresponding two first order diffracted beams are made to recombine with the direct beam. As we saw in Chapter 1, a true three-dimensional pattern cannot be obtained owing to the lack of resolution of reflecting planes lying normal to the beam.

The contrast in the image depends entirely upon the relative intensities of the direct and diffracted beams (expression 50). High contrast is obtained only when the atomic scattering factor is large, when the beams are diffracted exactly at the Bragg condition, and when the beam divergence is a minimum, since this determines the breadth of the maxima.

Although the lattice planes in metal crystals have not yet been resolved, considerable success has been achieved in resolving planes in nonmetallic crystals, particularly the metal phthalocyanines and oxides.[2] Figure 30 is an example of the resolved $(20\bar{1})$ planes of spacing 12 Å in platinum phthalocyanine. These planes are densely populated with platinum atoms which are strong diffracting centers. By placing a suitable stop in the back focal plane of the objective lens, as illustrated in Fig. 29, imaging results from recombination of the direct beam and the $(20\bar{1})$ diffracted beam. More recently the (020) planes in molybdenum trioxide have been resolved where the spacing is only 6.9 Å[3] and if suitable crystals can be found, even better resolutions should be obtained. These materials are thus excellent specimens for testing the resolution limit of an electron microscope (see section 3.3).

One of the advantages of the direct resolution method is, of course, that any imperfections in the lattice planes (e.g., dislocations) will also be

FIG. 30. The $(20\bar{1})$ planes in platinum phthalocyanine resolved (11.9 Å spacing), using the technique illustrated in Fig. 29. (× 340,000.)

resolved, and many beautiful examples have been given by Menter.[2] However, direct imaging of lattice planes is possible only when the microscope is operated at very high resolutions and when the following conditions are satisfied:

1. Specimens prepared in the normal way must have a suitable habit such that there is a reasonable chance of finding the lattice planes to be resolved lying parallel to the beam. They are then favorably oriented to produce the essential diffraction spectra for image formation.
2. The structure factor for the reflections from the lattice planes to be resolved must be high enough to obtain adequate contrast in the image.
3. The crystal must be stable under the influence of the electron bombardment.
4. The crystals must be sufficiently thin in the direction of the electron beam in order to reduce to a minimum the effects of inelastic scattering that introduce chromatic aberrations in the final image and thus impair resolution (section 1.4).

From this it can be seen that resolving periodic structures is not a simple experiment, since the requirements for both instrument operation and specimen preparation are stringent. Fortunately, it is possible to indirectly resolve lattice planes in metal crystals by applying the well-known optical principle of forming moiré patterns to the imaging of crystal gratings with electrons.[4] Two basic types of moiré pattern may be formed as illustrated in Fig. 31. The parallel moiré pattern is formed by the parallel superposition of two unequal gratings of pitch d_1 and d_2, giving rise to a moiré pattern of lines parallel to the lines of the elementary gratings with a spacing $D = d_1d_2/(d_1 - d_2)$. The rotation moiré pattern is formed by the superposition of two equal gratings of pitch d with a small angular twist ϵ about the normal to the plane of the gratings. This leads to a moiré pattern, whose direction bisects the obtuse angle between the lines of the elementary gratings and has a spacing given by $D = d/\epsilon$. In the electron microscope moiré images are obtained by the action of double diffraction, i.e., a beam diffracted by the top crystal acts as a primary beam to be further diffracted by the second crystal.† Thus a diffraction pattern from the overlapping crystals consists of two primary patterns and doubly diffracted beams around the primaries.[4] An example is given in Fig. 126, Chapter 5. By placing an aperture in the back focal plane so as to allow only the zero order and its satellite reflections to pass through, a simple moiré pattern (bright field) will be imaged (see Fig. 125, Chapter 5). With a full aperture and aberration-free lens we would obtain an image where

† The theory of moiré images is discussed in detail by Dowell et al. (reference 79, Chapter 5). See also Guild's book (Chapter 1).

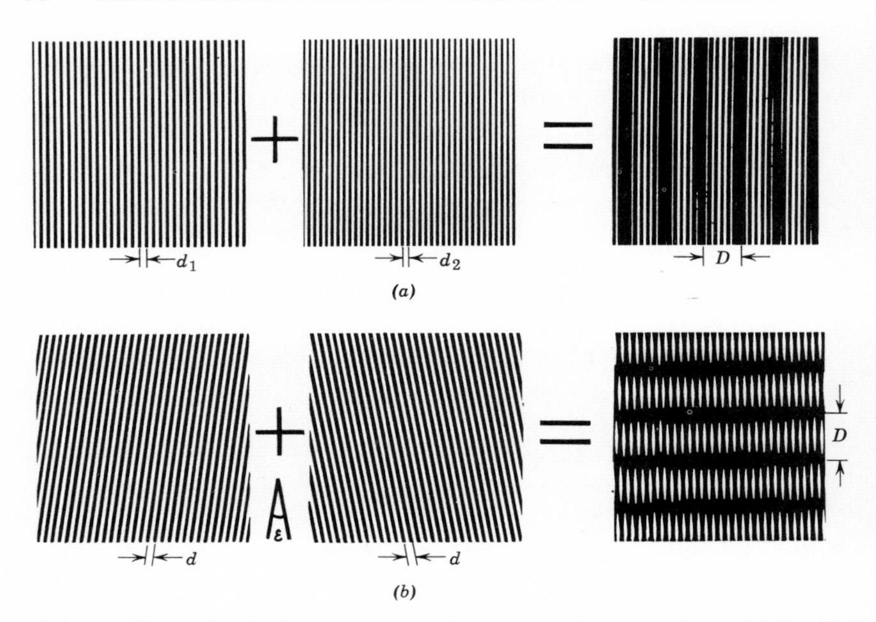

FIG. 31. Illustrating the formation of two basic types of moiré pattern. (*a*) Parallel moiré pattern where $D = d_1 d_2/(d_1 - d_2)$. (*b*) Rotational moiré pattern where $D = d/\epsilon$. (After Bassett et al.[4] Courtesy the Royal Society.)

all the fine detail of both gratings could be resolved. However, because the objective lens is imperfect it is necessary to allow only one beam and its secondaries to be imaged. This pattern is also reasonably easy to interpret. A similar demonstration may be made for rotated gratings. As we shall see in Chapter 5, moiré images from epitaxially deposited metal films provide valuable information regarding imperfections in metals, since now the lattice planes of the structure may be resolved in the magnified moiré image. The smallest resolved spacing so far reported is 5.8 Å from superimposed parallel single crystals of nickel and gold.[4] Other examples have been described by Menter.[5] The moiré technique may also be used to study the interfaces between coherent or partially coherent dispersed phases in metals (see section 5.3.2).

From the principles discussed here it can be seen that domains in ordered alloys may be directly resolved by similar techniques provided their spacing is > 5 Å. Thus if the direct beam and its satellites are made to recombine, then from Fig. 26 the bright field image is obtained with the aperture at *A* (see Fig. 150), and a dark field image with the aperture at *B*. However, better resolution is obtained in the dark field image if the electron beam is tilted so as to bring the reflections at B to the position A, leaving the objective aperture perfectly centered about the optic axis.

2.3 Images of Metal Crystals

2.3.1 *Introduction*

In the preceding section it has been shown that because of the small lattice spacings of metal crystals, and owing to the limitations set by the instrumental resolution, we are unable to form images of their lattice planes. However, since the diffracted beams fall outside the aperture of the lens, and if it is arranged that they do not contribute to the final image, dark contrast will result wherever local conditions in the thin specimen produce strong diffraction. This can be done by inserting suitably sized objective apertures in the back focal plane of the objective lens.† The contrast can be reversed by changing from bright to dark field illumination. Fig. 32 illustrates the arrangement for diffraction contrast for (*a*) bright

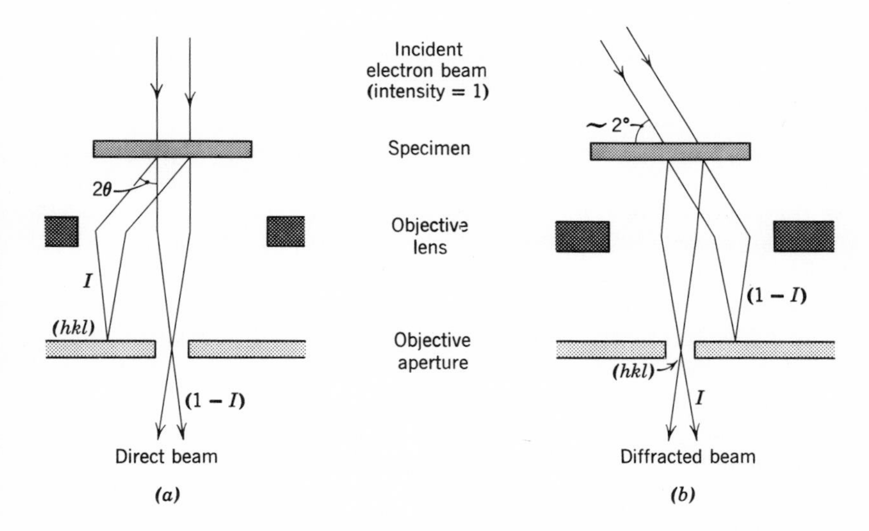

FIG. 32. Illustrating diffraction contrast for (*a*) bright field and (*b*) dark field illumination with tilted electron gun. Contrast in the bright field image arises through local variations of the intensities of the diffracted beams; the diffracted beam is stopped by the objective aperture.

† For example, with aluminum and using 80 kv electrons the first allowed reflection is (111); hence 2θ is 1.8×10^{-2} radian. If an objective aperture of 30 μ diameter is used (equivalent to 5.0×10^{-3} radian semiangular aperture for an objective focal length of 3 mm) it is seen that (111) and higher order reflections will not pass through the aperture.

field and (*b*) dark field illumination. This method of contrast formation has an advantage over the method of direct lattice resolution in that the atomic array is not resolved, so that resolution is not a limiting factor and specimen requirements are less stringent. Since contrast in the image depends upon the intensities of the electron beam leaving the specimen, the image simply represents the intensity distribution of electrons at the bottom surface of the crystal. From Fig. 32 it can be seen that the contrast in the bright field image is obtained by subtracting the intensity scattered into the Bragg reflections $\{hkl\}$ from that of the direct beam. Superimposed upon this is the contribution to the background intensity from the inelastically scattered electrons. For most purposes, however, this can be neglected. Thus in order to understand contrast effects in images of crystals we must consider in detail the intensities of electrons scattered by the crystal.

In alloys the contrast mechanism is similar to that of the pure metal except that the Bragg angles will be different, and, of course, there will be differences in the intensities of scattering due to the structure factor. Thus all the contrast effects predicted for the pure metal may be also used to explain contrast effects observed in alloys.† In the general case, therefore, in this chapter we need only discuss single-phase crystals. Since diffraction contrast is sensitive to the positions of the atoms in the crystal, any displacements from their normal positions will produce a phase difference between waves diffracted from successive scattering centers. Thus regions of strain in the crystal will produce contrast effects provided the displacements of atoms are normal to the incident wave. As a result, it is possible to observe defects such as dislocations,[6] grain boundaries,[7] stacking faults,[8] and strains due to precipitates,[9] etc. The case of strained crystals will be discussed in Chapter 5 when we shall consider some of the many applications of electron microscopy to physical problems.

2.3.2 *Kinematical Theory of Diffraction Contrast*

The kinematical theory of diffraction outlined in Chapter 1 provides a fairly simple method for obtaining the intensities of electrons scattered into Bragg reflections. This method is a graphical one whereby the path of the electron wave through the crystal is represented by an amplitude-phase diagram. For the perfect crystal we have shown that this is a circle of radius $2\pi s^{-1}$, where s is the deviation in reciprocal space from the exact Bragg condition (see Fig. 20, section 1.13). The assumptions involved

† Some contribution to contrast will also arise from differences in scattering from regions of different mass thickness (section 1.4.3). Thus where locally, atoms of higher atomic number are segregated, dark contrast is expected in the image. (See Fig. 134.)

in the kinematical theory are: (1) The incident beam is not exactly in the Bragg condition for reflection so that the diffracted intensity is small compared to the intensity of the direct beam. The incident intensity is taken as unity, so that if I is the diffracted intensity the intensity of the direct beam is $1 - I$ (Fig. 32*a*). (2) A two beam approximation is used, i.e., it is assumed that only one diffracted beam (together with the direct beam) is excited in the crystal. (3) The crystal is thin in the direction of the incident beam so that there is negligible rescattering or absorption.

In this theory we thus have to consider three variables: (1) orientation of the specimen with respect to the incident beam, (2) specimen thickness t, and (3) the deviation s from the Bragg condition.

Since the wavelength of the incident electrons is very small and the Bragg angles are also very small (~ 2 or $3°$), there is a high probability that for any orientation of the crystal the incident beam will be close to the Bragg angle of a set of reflecting planes. In general, therefore, some electrons are weakly diffracted, but others are strongly diffracted when local conditions are such that the Bragg condition is exactly satisfied. Thus in the crystal, grains, subgrains, precipitates, etc., can be clearly distinguished because of differences in their orientation with respect to the beam. Obviously, the contrast can be changed by changing the orientation, e.g., by tilting the illumination or the specimen (section 2.3.6).

Figure 33 is an example of a molybdenum thin foil showing many grains after an annealing treatment (bright field image). The point of interest here is the variation in contrast between the light areas A and the dark areas B. This contrast arises solely from differences in orientation from grain to grain; the specimen is uniformly thick over the area shown. Since the diffracted beams are stopped off by the objective aperture, dark grains correspond to areas which are in a strong diffracting position.

At C fringes can be observed running parallel to the intersection of the grain boundary with the top and bottom surfaces of the foil. Fringe contrast may be generally explained with reference to Fig. 34. Figure 34*a* shows a cross section of a foil containing an inclined grain boundary, a hole, and a wedge shape at its edge. Assuming kinematical conditions, then as shown in section 1.13, the intensity of the diffracted wave varies sinusoidally with depth t in the crystal as shown at (*a*). The diffracted wave is out of phase by $\pi/2$ with respect to the direct wave. For every depth periodicity or wavelength t_0', extinction of the diffracted wave occurs, i.e., corresponding to a complete revolution of the amplitude-phase diagram of Fig. 21*b*, where $t_0' \simeq s^{-1}$. We should notice that as s decreases (thicker crystals) the wavelength of the oscillations t_0' may increase indefinitely; in this case, however, dynamical effects will then become

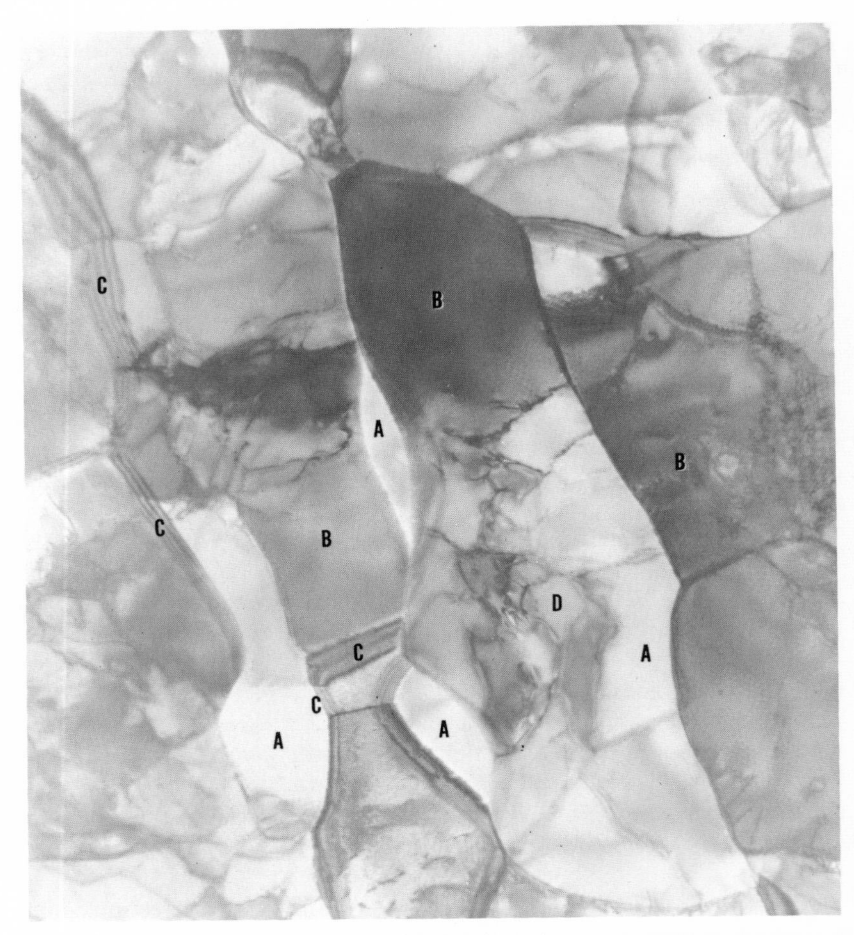

FIG. 33. Thin foil of molybdenum showing light grains *A*, dark grains *B*, fringes at grain boundaries *C*, subgrains *D*. Bright field image, dark contrast arises wherever there is strong diffraction; dark lines are dislocation images. (×25,000.)

important, and t_0' has an upper limit t_0 known as the extinction distance for the particular Bragg reflection being considered (section 2.3.3).

Because of the intensity oscillations of the two waves through the crystal, any defect plane inclined to the surface of the crystal will give rise to extinction fringes,[6,7] so that for grain boundaries, faults, holes and wedges, fringes will be observed at f as shown in Fig. 34*b*. In terms of the amplitude-phase diagram (Fig. 21*b*) this corresponds to the oscillation

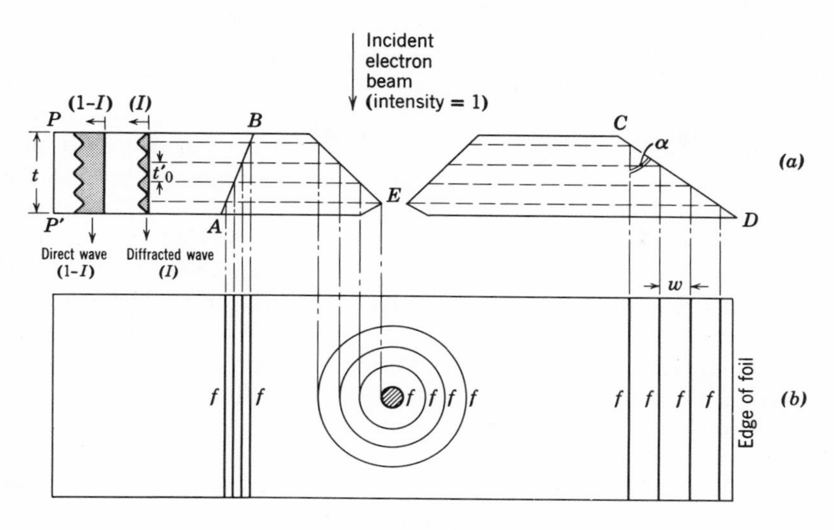

FIG. 34. Schematic illustration of the origin of fringe contrast in images of crystals. (*a*) Section through crystal showing kinematical intensity oscillations of direct and diffracted waves. The depth periodicity $t_0' = s^{-1}$, *AB* grain boundary or stacking fault *CD* wedge, *E* hole. (*b*) Section normal to beam representing bright field image showing dark fringes *f* (thickness extinction contours).

of the amplitude *PP'* as *t* varies. This accounts for the fringes at *C* in Fig. 33. Since the spacing of the fringes is proportional to s^{-1}, the fringe spacing decreases with increasing deviation from the Bragg angle.

If we look closely at Fig. 34*a*, it can be seen that as the beam first strikes the crystal, practically all of the incident intensity enters the image. After one-half wavelength, however, the diffracted intensity is a maximum so that the net intensity reaching the image plane is a minimum. This corresponds to the appearance of a dark fringe *f* as shown in (*b*). At this point in the crystal, however, the transmitted intensity $(1 - I)$ is never zero. The fringe contrast is thus a variation in intensity for a maximum to a minimum for every depth t_0' in the crystal. For the case shown in Fig. 34*a* the image corresponding to top and bottom surfaces always appear bright (bright field). The number of fringes observed depends only on the value of *s* and of t_0' compared to *t*; their spacing corresponds to a change in thickness equal to t_0'. Each fringe is a contour of equal thickness of the foil and is known as a thickness extinction contour.

If *n* is the number of fringes, the thickness of the foil at the *n*th fringe is $t_0'n$. Similarly t_0' is given by $w/\tan \alpha M$, where *w* is the spacing between fringes measured on the micrograph, *M* is the magnification, and α is the angle between the defect plane and the normal to the foil surface (Fig. 34). Fringes are commonly observed at holes and at the edges of specimens prepared by electropolishing (Chapter 4), because this technique often gives rise to wedge-shaped foils, and holes may be formed owing to preferential polishing, e.g., at impurities or precipitates. Of course, fringes are not observed when the defect planes are perfectly normal to or parallel to the crystal surface.

When the crystal is uniformly thick but may be buckled as a result of local bending, e.g., due to careless handling of the specimen, contours due to buckling are observed. Examples in a thin foil of molybdenum are shown in Fig. 35*a*. From this it can be seen that the position of a contour depends on the orientation of the foil since contours are not continuous across grain boundaries. These contours correspond to regions of the foil at constant inclination to the electron beam. Thus the diffracted beam follows the contours of buckling corresponding to the reciprocal lattice point shown in Fig. 20 sweeping through the reflecting sphere. Hence in the bright field image each diffracted wave gives rise to dark bands, called bend extinction contours. As the foil is tilted, these contours will sweep across the field of view.

FIG. 35(*a*). Thickness extinction contours and bend contours in a thin foil of molybdenum. The grain boundaries lie normal to the plane of the foil (no fringes). (× 15,000.)

In thicker regions of foils these bend contours are often observed to consist of wide bands containing subsidiary fringes. An example for aluminum is shown in Fig. 35*b*. These subsidiary fringes may be partly explained by considering the kinematical intensity distribution about the reciprocal lattice point shown in Fig. 22. If the reflecting sphere in Fig. 20 sweeps through this intensity distribution, then since s varies due to local misorientations produced by buckling, multiple fringes will occur corresponding to the various intensity maxima given by Fig. 22.

If the contrast in the image is due to Bragg diffraction, each dark region in the bright field image must correspond to a particular (hkl) reciprocal

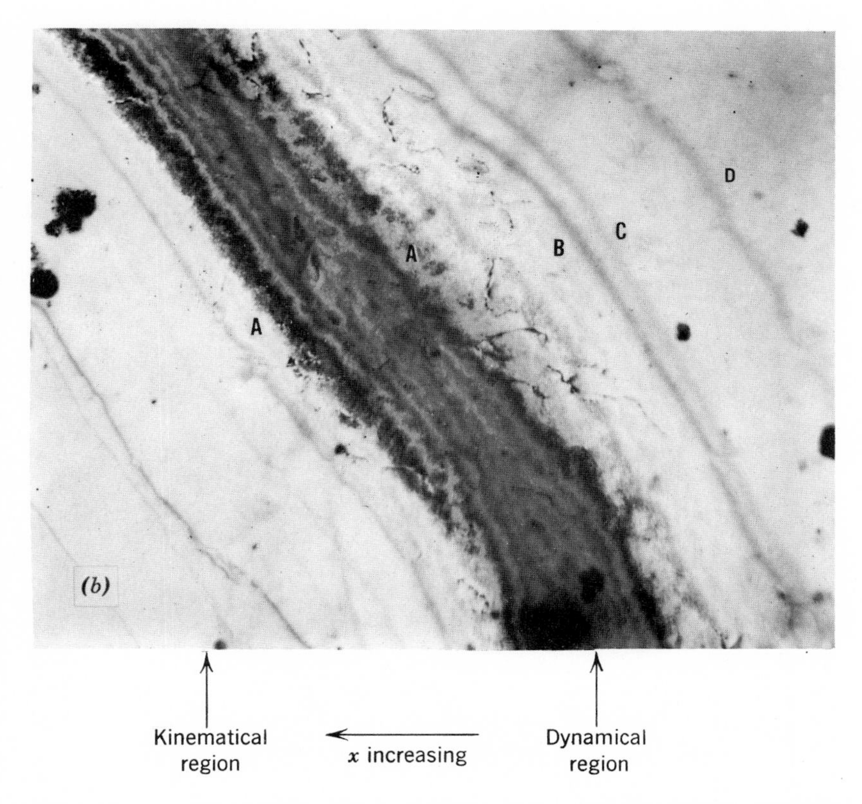

FIG. 35(*b*). Bend contours in a foil of aluminum. Note subsidiary contours between outermost fringes *A*, corresponding to two low order strong reflections hkl and $\bar{h}\bar{k}\bar{l}$. The single contours *B*, *C*, *D* correspond to higher orders of hkl, and can be indexed accordingly. Interruptions along the contours are probably due to dislocations. ($\times$23,000.)

lattice point, i.e., to a particular spot in the diffraction pattern. To check whether or not the contrast is obtained by a diffraction mechanism, it is necessary to change over from the bright field to the dark field image as illustrated in Fig. 32. If this change-over reverses the contrast from

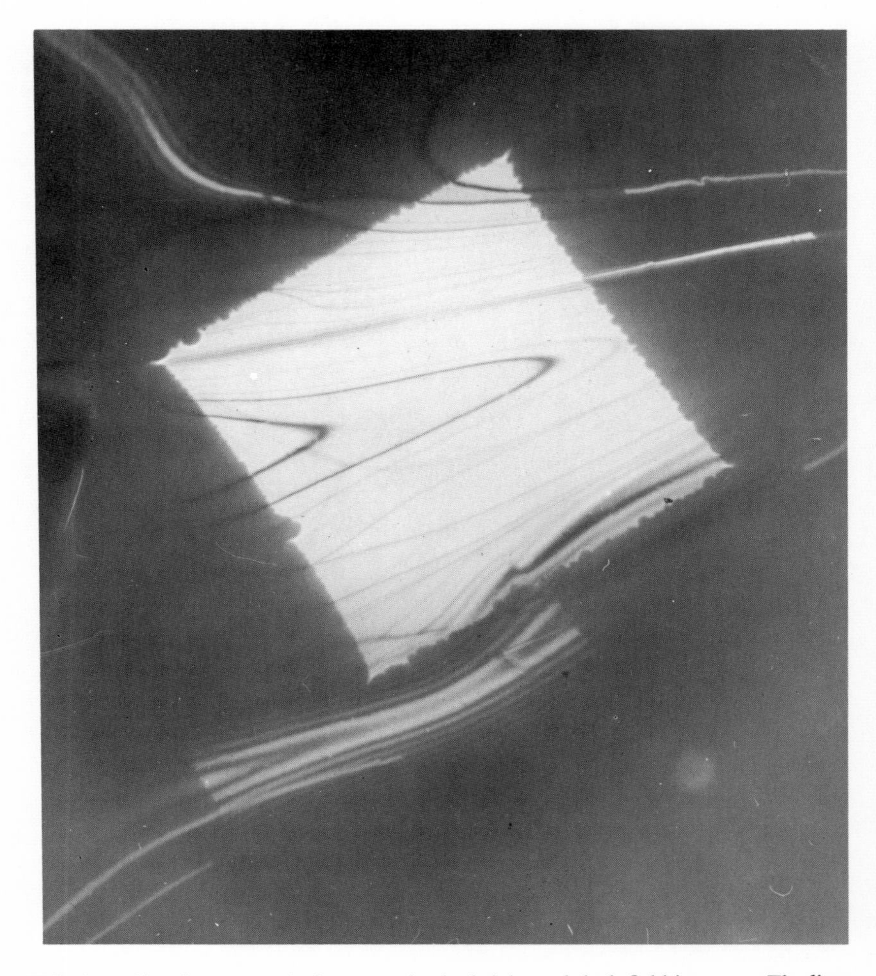

FIG. 36. Showing reversal of contrast in the bright and dark field images. The lines are bend extinction contours in a thin foil of aluminum. The rectangle is the image of the field-limiting (selected area) aperture. The area inside this is the bright field image and outside this the dark field image. By reducing the strength of the intermediate lens the lines become diffraction spots in the back focal plane of the objective. (×2,700).

black to white, it becomes obvious which Bragg reflection is responsible for the contrast. In this way the whole pattern may be used to analyze the contrast observed in the image.

Figure 36 illustrates the reversal of contrast in the bright and dark field images. The rectangle in the center of the micrograph is the image of the field-limiting aperture and the lines are bend extinction contours. Inside the rectangle, i.e., the bright field image, all the contours appear dark since no diffracted beams pass through. Outside this region those contours corresponding to the diffraction spots contributing to this area are seen to be in light contrast. Upon reducing the strength of the intermediate lens until the back focal plane of the objective is focused, each of these lines becomes a diffraction spot of different indices. Each extinction contour can thus be assigned its respective indices (*hkl*). *The image thus corresponds to a magnified picture of the diffraction pattern.*

When the diffraction pattern corresponding to Fig. 35*b* is carefully examined, it is found that the two outermost dark fringes correspond to magnified images of two low-order reflections of high intensity. In other words this contrast arises from dynamical scattering so that our explanation of the fringes at bend contours by kinematical theory is not strictly correct. Whenever the diffraction patterns contain intense diffraction maxima, we know that dynamical diffraction has occurred (e.g., see Fig. 27*a*). In terms of the kinematical theory, s is now very small, i.e., the crystal is very close to the exact Bragg condition. Let us now consider the effects which may be produced by dynamical scattering since this enables us to give a better explanation of contrast from relatively thick foils.

2.3.3 *Implications of the Dynamical Theory*

The limitations of the kinematical theory should now be well realized, but nevertheless the kinematical approach provides a practical and convenient method for understanding many of the contrast effects observed in transmission electron microscope images of crystals. We shall not here go through the complex mathematical treatments involved in dynamical theory but will limit our discussion to its principles and implications. The basis of the theory is to account for the exclusive interactions which take place between an incident electron beam and the atoms in the lattice, and includes from the beginning the possibility of the interaction of the diffracted beams with each other and with the transmitted beam. The close analogy with the theory of the dynamical scattering of X-rays[10–15] has played an important role in the development of the dynamical theory for electrons. In this case, however, the theory is more essential than for the

X-ray case because of the interactions of electrons with the positive nuclei. This is well demonstrated by the experimental fact that the diffracted electron beams often have the same intensity as the direct beam (e.g., in Kikuchi patterns, Fig. 27*a*).

In the theoretical derivation of the diffracted intensities certain boundary conditions must be defined by taking into account the distribution of potential within the lattice and at the crystal surfaces. This problem was first treated by Bethe[12] and later by MacGillavry.[13] More recently the theory has been used by Heidenreich,[7] and Whelan and Hirsch,[8] to account for the fringe patterns observed in transmission electron micrographs, but these treatments neglect effects due to absorption. In dynamical theories the starting point is the solution of the time-free, nonrelativistic Schrödinger equation (section 1.6), in which the lattice potential E is represented as a three-dimensional Fourier series. In its usual approximate form the theory assumes that only two strong beams (one direct and one diffracted) are excited in the crystal. The basis of the theory is outlined in Chapter 8 of Pinsker's book[16] and more specific details may be found in references 7 to 22 given at the end of this chapter. Numerical solutions of the dynamical equations for several simultaneous reflections have been developed by Howie and Whelan[17] using the method of systematic reflections due to Hoerni.[18] In this, a high speed digital computer is required to evaluate the terms developed from Schrödinger's equation. These are expressed as a system of linear equations. This treatment also takes into account effects due to absorption of electrons by the specimen, which are considered in section 2.3.5.

Whelan and Hirsch[8] have shown that the intensity of a diffracted beam through a perfect crystal in the dynamical case (neglecting absorption) is given by

$$I = \frac{\sin^2 \pi(t/t_0)(1 + x^2)^{1/2}}{(1 + x^2)} \tag{56}$$

where x is a parameter indicating a small deviation from the Bragg angle and t_0 is the extinction distance for the reflection being considered. In terms of the kinematical deviation s, $x = t_0 s$ and in the kinematical region of the crystal $x > 1$. The value $x = 1$ for (111) in stainless steel corresponds to a 0·5° deviation from the Bragg angle.[8]

The amplitude-phase diagram representing dynamical scattering from a perfect crystal is also a circle similar to that shown in Fig. 21*b*. In this case however, corresponding to small values of s, the radius of the circle is $t_0/2\pi$. The intensity profile of the diffracted wave about the reciprocal lattice point, calculated from equation (56), is shown in Fig. 37, which is to be compared with Fig. 22. Thus on the dynamical picture a continuous

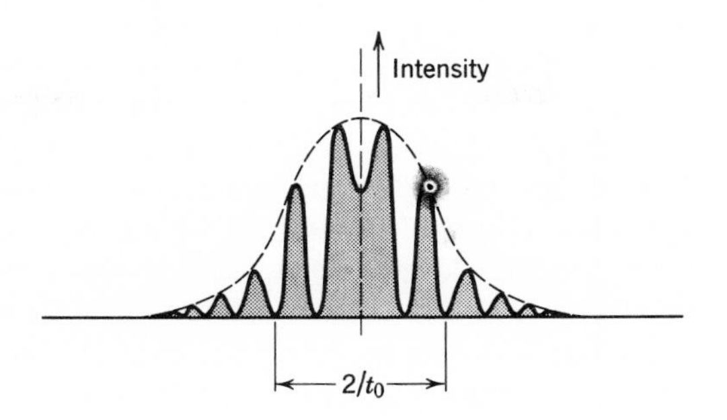

FIG. 37. Schematic illustration of the intensity profile of diffraction spot on the dynamical theory. Secondary maxima give rise to subsidiary fringes in the image (see between *AA* of Fig. 35*b*). (After Whelan and Hirsch.[8] Courtesy *Phil. Mag.*)

variation in intensity from maximum to zero is not obtained, but rather a series of secondary maxima. These secondary maxima give a better explanation of the number of subsidiary extinction contours shown in Fig. 35*b*, although the complete interpretation is still to be discussed (section 2.3.5).

The intensity variations of the direct and diffracted beams at the reflecting position are shown schematically in Fig. 38. In this case it is clear that crystals of thickness $(m + \frac{1}{2})t_0$ will have zero transmitted intensity, whereas crystals of thickness mt_0 have maximum transmitted intensity (m is an integer) when the electron beam is incident at the Bragg angle. Thus for wedge-shaped crystals, holes, grain-boundaries (Fig. 33), or faults along an inclined plane (Fig. 104), fringes of equal thickness will be observed in the bright field image such that along the defect plane *ad* (Fig. 38), at depths *a*,*c*, etc., the fringes show black contrast (extinction contours), and at *b*,*d*, etc., the fringes show light contrast. Unlike the kinematical case shown in Fig. 34 the intensity at the image corresponding to each extinction contour is zero since the direct and diffracted waves are equivalent in the sense that when one has zero intensity, the other has an intensity of unity. However, although the details of the dynamical and kinematical theories are different, the main qualitative features are similar. Both Figs. 34 and 38 predict intensity oscillations due to

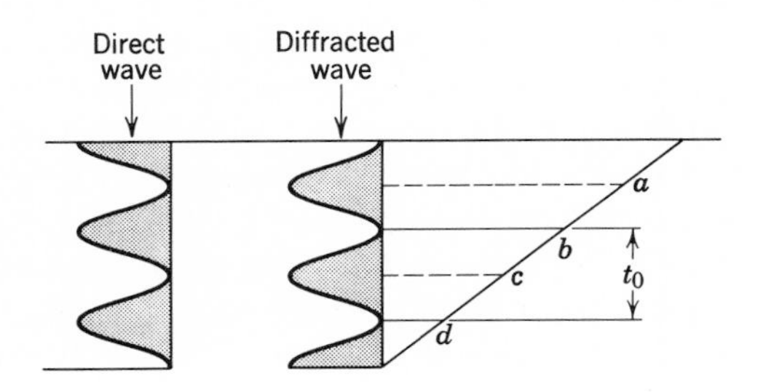

FIG. 38. Intensity variations of the direct and diffracted waves on the dynamical theory, neglecting absorption (schematic). t_0 is the extinction distance. (After Whelan and Hirsch.[8] Courtesy *Phil. Mag.*)

variations in thickness, and as outlined in section 2.3.2 can be used to explain fringe contrast in the image. It should be pointed out, however, that in the dynamical case illustrated in Fig. 38, at the points *a,c* and *b,d* respectively, the intensities of the direct and diffracted beams are zero. Hence the spacing of the fringes for the case shown of a fault along *ad* (stacking fault, grain boundary, etc.), is one-half that of the corresponding wedge fringes if one-half the faulted crystal is removed.

Whelan and Hirsch,[8] in their dynamical theory of fringe contrast at stacking faults (see also section 5.2.14), showed that as the orientation of the foil gradually changes, with x decreasing, the fringe separation $t_0' = t_0/(1 + x^2)^{1/2}$, where $t_0' < t_0$, increases, particularly for $x > 1$ on going from the kinematical to the dynamical regions. Furthermore, the subsidiary maxima will gradually become more prominent leading to splitting of the minima into doublets (corresponding to Fig. 37). This is characteristic of the dynamical region. At $x = 0$ (exact Bragg case), the maxima are all equal in intensity and the maxima and minima are equally spaced. Thus, for the dynamical equivalent of Fig. 34*b*, the spacing between dark and light fringes is $\frac{1}{2}t_0 \tan \alpha$, where α is the angle between the defect plane and the normal to the foil surface (Fig. 34*a*). The number of fringes depends only on x and t (foil thickness) and is independent of the angle α. Examples of stacking fault fringes may be seen in Fig. 104, Chapter 5.

The general intensity of fringe patterns and other contrast effects, and the contrast relative to the background intensity also change considerably with the value of x. The theoretical calculations[8] show that in the dynamical region ($x < 1$) the difference in intensity between the maxima and minima of the fringes can be of the order of unity, i.e., of the intensity of the incident beam. Hence, *the contrast in the region of a foil near a reflecting position is very great*, i.e., *at an extinction contour*. The contrast effects due to dislocations and thin precipitates shown in Figs. 35*b*, 111, and 136 are clearly strongest at the extinction contours, and these illustrations also show the decrease in contrast with increasing distance from the contours, i.e., as x increases (Fig. 35*b*). Since the intensity and contrast vary as x^{-2}, the contrast eventually may become unobservable for large values of x. In this case, it is necessary to tilt the specimen into a strong reflecting position in order to obtain good contrast (section 2.3.6).

In order to decide what magnitude of crystal thickness gives rise to dynamical effects, the extinction distance t_0 can be calculated corresponding to the various Bragg reflections from the X-ray atomic scattering factors, by using the Born approximation to obtain the atom factor for electrons (equation 29, section 1.8). The values of t_0 are then obtained using the following expression given by Hirsch et al.:[6]

$$t_0 = \frac{\pi k_0 V \cos\theta}{f} = \frac{\pi V \cos\theta}{\lambda f} \tag{57a}$$

where $\lambda(= 1/k_0)$ is the wavelength of the incident beam, V is the volume of the unit cell, and f is the scattering factor for electron waves of the contents of the unit cell. The extinction distance can also be obtained from the following expression:[7, 8]

$$t_0 = \frac{\lambda E}{Vg} \tag{57b}$$

where E is the electron energy in volts, Vg is the Fourier coefficient (in volts) of order **g** of the complex lattice potential, and **g** is the reciprocal lattice vector corresponding to the Bragg reflection. Expression (57*a*) is in fact equivalent to (57*b*).

As we saw in section 1.8, since f falls off rapidly with increasing scattering angle, the extinction distance increases for higher order reflections.

For a unit cell containing only one lattice point, f in expression (57*a*) is the atomic scattering factor for electrons. For unit cells containing more than one atom, f should be replaced by F, i.e., the structure factor. For face-centered cubic and body-centered cubic crystals $F_{hkl} = 4f$ and $2f$ respectively for allowed reflections $\{hkl\}$. In hexagonal crystals, however, not all reflections have the same structure factor F. When all possible values of $HK.L$ for allowed reflections are considered, F_{HKL} can be summarized as follows:

$H + 2K$	L	F	Example ($HK.L$)
$3n$	odd	0	none allowed
$3n$	even	$2f$	(00.2)
$3n \pm 1$	odd	$\sqrt{3}f$	(01.1)
$3n \pm 1$	even	f	(01.0)

The volume V of the unit cell is always taken as n times the atomic volume, where n is the number of atoms in the unit cell. Hence V is given by

$$n \times \frac{\text{atomic weight}}{\text{Avogadro's number} \times \text{density}} \ (\text{cm}^3)$$

For FCC, BCC, and HCP crystals $n = 4$, 2, and 2 respectively.

For unit cells containing mixed atoms F must be evaluated in terms of the different values of f for each kind of atom, e.g., in ordered alloys, intermetallic compounds, ionic crystals, etc.

To assist the reader to calculate t_0, the following two numerical examples are given for aluminum (FCC (111) reflection), and magnesium (HCP (00.2) reflection).

Firstly, the factors needed to obtain $f(\theta)$ are as follows:

	a_0 (Å)	Z	($\sin\theta/\lambda \times 10^{-8}$)	($\lambda/\sin\theta \times 10^8)^2$	f_x (by interpolation)
Al	4.049	13	0.214	21.84	8.80
Mg	3.209	12	0.192	27.165	8.752

Hence from expression (29*b*) we obtain

(1) Al $f(\theta)$ electrons $= (13 - 8.8) \times 21.84 \times 2.38 \times 10^{-10}$
$= 2.18 \times 10^{-8}$ cm (per atom);
hence $F = 4(2.18 \times 10^{-8})$ cm

(2) Mg $f(\theta)$ electrons $= (12 - 8.752) \times 27.165 \times 2.38 \times 10^{-10}$
$= 2.0999 \times 10^{-8}$ cm (per atom);
hence $F = 2(2.0999 \times 10^{-8})$ cm

The values of f_x (atomic scattering factor for X-rays) are obtained from tables (section 1.8). The volume of the FCC unit cell is four times the atomic volume, whereas the volume of the HCP unit cell is twice the atomic volume. For Al the volume V of the unit cell is

$$4\left(\frac{26.97}{6.02 \times 10^{23} \times 2.699}\right) (\text{cm}^3)$$

and for Mg V is
$$2\left(\frac{24.32}{6.02 \times 10^{23} \times 1.74}\right) (\text{cm}^3)$$

Thus from equation (57*a*), for 100 kv electrons ($\lambda = 3.7 \times 10^{-10}$ cm), and taking $\cos\theta = 1$, t_0 is obtained as follows:

(1) Al (111)

$$t_0 = \frac{3.142 \times 4(26.97) \times 1}{6.02 \times 10^{23} \times 2.699 \times 3.7 \times 10^{-10} \times 4(2.18 \times 10^{-8})} (\text{cm})$$
$$= 646 \text{ Å}$$

(2) Mg (00.2)

$$t_0 = \frac{3.142 \times 2(24.32)}{6.02 \times 10^{23} \times 1.74 \times 3.7 \times 10^{-10} \times 2(2.0999 \times 10^{-8})} (\text{cm})$$
$$= 938 \text{ Å}$$

It is most important to evaluate t_0 for different reflections for each material being examined in the electron microscope in order to establish the thickness limitation over which kinematical theory may be used and thus to enable a correct interpretation to be made of contrast effects. For example, when t_0 is greater than the foil thickness no fringes will be observed at wedges, grain boundaries, stacking faults, etc. However, values of t_0 greater than normal specimen thicknesses only occur for light metals and large Bragg angles (see Table II). Using the method shown above t_0 can be calculated for any crystal structure, and some values are listed in Table II for $\lambda = 0.037$ Å (100 kv electrons). The results for the FCC metals are taken from Hirsch et al.[6]

Table II

Metals Face-Centered Cubic†	Atomic Number	Reflection Extinction Distance (t_0) Å		
		(111)	(200)	(220)
Al	13	646	774	1240
Ni	28	258	302	468
Cu	29	268	308	472
Ag	47	250	285	403
Pt	78	165	188	262
Au	79	181	204	281
Body-Centered Cubic		(110)	(200)	(211)
αFe	26	296	444	582
Mo	42	267	373	467
Hexagonal-Close-Packed		(00.2)	(01.1)	(01.0)
Mg	12	938	1155	1774

† Courtesy of Drs. P. B. Hirsch, A. Howie, M. J. Whelan, and The Royal Society.

From Table II it can be seen that dynamical effects occur even in very thin crystals, which accounts for the anomalous intensities often found in electron diffraction patterns.

The values of t_0 given in Table II are only approximate for the following reasons:

(1) The expressions for t_0 are calculated using the two-beam approximation (i.e., for only one diffracted and one transmitted beam),[6–8] and when multiple diffraction is allowed for, t_0 is found to be about 10% smaller than the values given in Table II.[17]

(2) The expressions for t_0 are obtained by using the classical formulae for the potential energy of the electron, and neglect relativistic effects. This also leads to an error of about 10% at 100 kv (see section 1.3).

(3) The use of the Born approximation for metals of high atomic number is not strictly valid. However, Ibers[19] using Thomas-Fermi-Dirac potentials, has recently made accurate calculations of $f(\theta)$ for elements of atomic number up to 80, where $f(\theta)$ is expressed as a function of $(\sin \theta/\lambda)$ (Å^{-1}). (See Appendix E.) The values of $f(\theta)$ for a given reflection may be obtained by interpolation of Ibers' data, and can be used directly in equation (57*b*) to obtain t_0. This also simplifies the calculation since it is no longer necessary to use the Born approximation. Again account must be taken of the structure factor when dealing with crystals containing more than one atom per unit cell.

It is well to remember that in order to obtain values for the extinction distance experimentally the deviation x from the Bragg condition must be known since as we saw in section 2.3.3 the spacing of the fringes increases as x decreases. Thus when comparing measured and calculated values of t_0, low values (measured t_0') will indicate deviations from the exact Bragg case,[8] i.e., when $x \neq 0$. The value of x may be obtained from the expression $t_0' = t_0/(1 + x^2)^{1/2}$ provided t_0 has been calculated.[8]

For most practical purposes estimates of the extinction distance by the methods given above are quite satisfactory.

2.3.4 *Effect of Beam Divergence*

One other point that must be considered is the effect of the beam divergence. In all our calculations we have assumed an incident plane wave of electrons. However, as discussed previously in section 1.13, owing to the focusing action of condenser lenses in the electron microscope, the beam converges onto the specimen at an illuminating aperture, depending on the strength of the lens and upon the size of any physical apertures which may be used (section 3.5.3). As a result of this the beam divergence may be appreciable. In the best present-day instruments the divergence of the incident beam is $\sim 10^{-3}$ radian. The beam thus consists of a narrow cone of beams incident at slightly different angles. Each component beam produces a diffraction pattern corresponding to a particular value of the divergence from the Bragg condition, i.e., s (kinematical)[6] or x (dynamical).[8] The complete diffraction pattern therefore consists of a superposition of patterns of slightly different s or x. As a result some of the fine detail of the pattern calculated for a parallel beam may be lost (e.g., resolution of the intensity oscillations predicted by Figs. 22 and 37). In Chapter 1 we saw that the divergence decreases for increasing angles of reflection; at the same time, in the dynamical case, the extinction distance t_0 increases. The range dx of values of x corresponding to the divergence of the incident beam therefore increases with increasing order of reflection, since by analogy with Fig. 20, $dx = (t_0/d)d\theta$, where d is the spacing of the operating reflecting plane. Thus for silver and the (111) reflection $t_0 = 250$ Å, $d = 2.86$ Å, and taking $d\theta \sim 10^{-3}$ radian, dx is 0.87, which is quite appreciable. This accounts for the fact that subsidiary extinction contours are not observable for higher order reflections, e.g., at B,C,D in Fig. 35b and for the decrease in contrast at fringes from inclined boundaries, faults, etc., with increasing order of reflection. It also follows that in the kinematical region ($x > 1$) subsidiary contours will not be observed, whereas they will be observed in the strong central (dynamical) region of the reflection ($x \sim 0$). Whelan and Hirsch[8] have considered this effect in some detail in their dynamical theory of contrast from stacking faults. However, as we shall see in the following section a more complete account of fringe patterns is obtained by taking into account effects due to absorption of electrons by the crystal.

2.3.5 *Absorption*

In the foregoing discussion no mention has so far been made of effects arising from the absorption of electrons by the specimen. Since this is

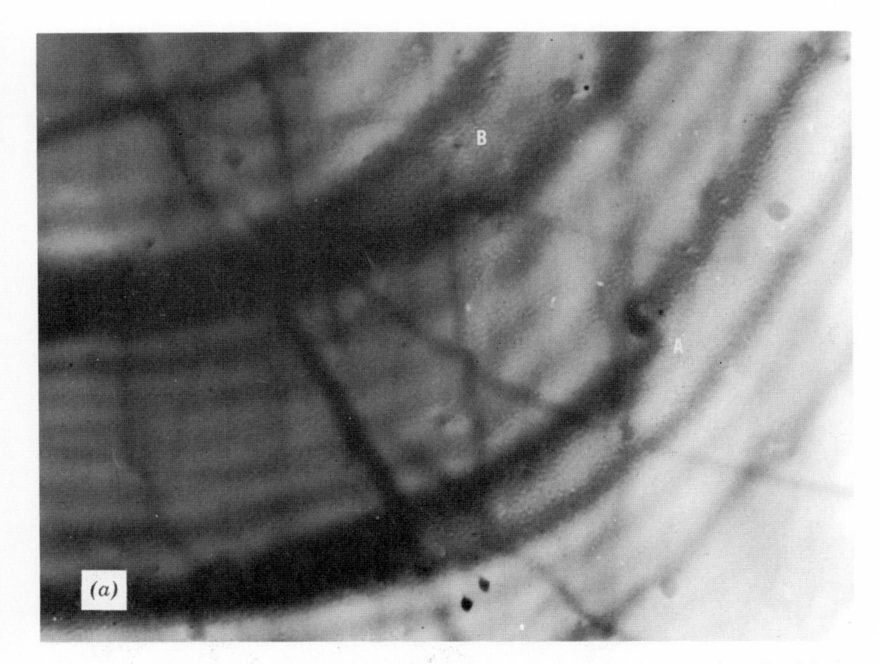

FIG. 39(a). Bright field image of a bend extinction contour in aluminum. The fringes AB correspond to (111) and ($\bar{1}\bar{1}\bar{1}$) reflections. (×67,500.)

known to give rise to contrast, particularly in thick areas of foils, an attempt is now being made to provide a satisfactory theory of the contribution to contrast from the absorption of electrons by accounting for the inelastic electron scattering from the crystal. For example, Hashimoto et al.[20] have recently shown that a bend extinction contour, such as that illustrated in Figs. 35*b* and 39*a*, is produced by the two lowest order Bragg reflections (*hkl*) and ($\bar{h}\,\bar{k}\,\bar{l}$). In Fig. 39a, the outermost fringes *A* and *B* occur for (strong) (111) and ($\bar{1}\bar{1}\bar{1}$) reflections. Figure 39*b* is the corresponding dark field image of the (111) reflection. The asymmetry of the intensity transmitted on either side of the Bragg reflection (corresponding to the outermost fringes *AB* in Fig. 39*a*) increases with increasing foil thickness while the visibility of subsidiary maxima decreases. In thicker regions a bend extinction contour appears as a broad dark band with no subsidiary maxima. This observation is explained by taking into account an absorption parameter τ_0 which is obtained from dynamical theory by including a term in the complex lattice potential to account for inelastic scattering.[17, 20] In the solution of the dynamical equations τ_0 is derived from the same formula as that given in expression (57*b*) by substituting

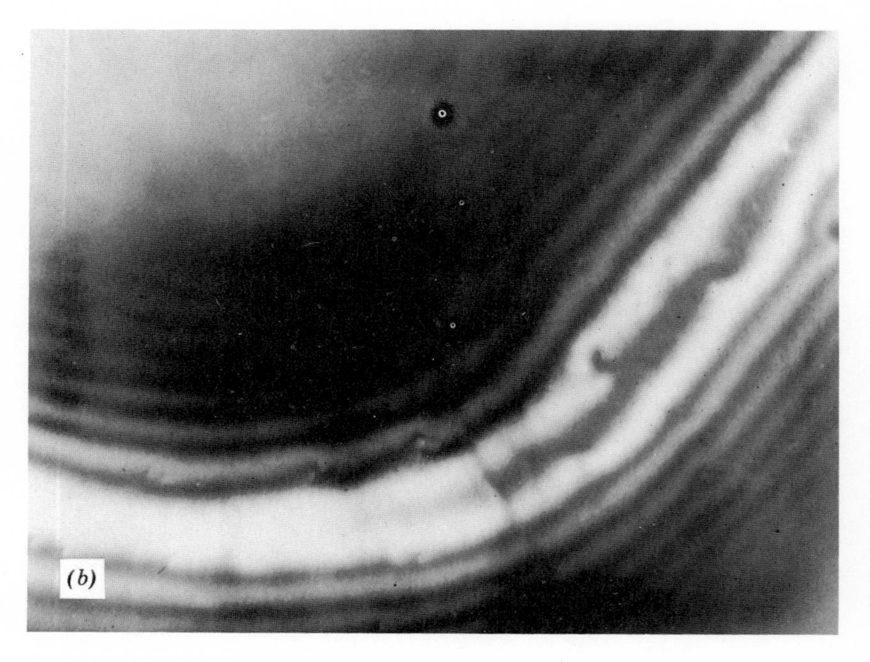

FIG. 39(*b*). Dark field image of Fig. 39*a* obtained from (111) reflection. (×67,500.)

for Vg, the Fourier coefficient of order **g** of the imaginary part of the complex lattice potential. The spacing between the subsidiary maxima in the fringes is then given by t/t_0 where t is the foil thickness and t_0 is the extinction distance, whereas the degree of asymmetry in the bright field image depends on t/τ_0. As t/τ_0 increases, the asymmetry of the bright field image increases, whereas the amplitude of the subsidiary oscillations decreases. Hashimoto et al.[20] showed that by this theory the "reversal" of the extinction line from the bright to the dark field image and the symmetry of the dark field image (Fig. 39*b*) can also be explained as well as other contrast effects observed at stacking faults and dislocations. These workers stress the importance of considering absorption effects in order to explain the *fine details* in transmission electron micrographs of thick metal foils. At present, however, there is no theoretical treatment of inelastic scattering in metals which leads to a reliable estimate of the absorption parameter. The observations of Hashimoto et al., on aluminum and stainless steel foils can be explained on the assumption that τ_0 is about ten to fifteen times t_0. The detailed mechanism of the absorption process is also unknown. There is little doubt, however, that in the near future a complete dynamical theory explaining absorption phenomena will be made available.

2.3.6 *Effect of Tilting the Specimen*

It should be quite clear that the major contribution to contrast in electron images of crystalline specimens is due to diffraction. Consequently, contrast is extremely sensitive to the orientation of the specimen in relation to the direction of the incident beam. Tilting the specimen may therefore produce contrast in one orientation and not in another. To illustrate this point let us consider the micrographs shown in Figs. 40*a* and *b*. These are taken from a thin foil of Cu–8% Al alloy prepared by electropolishing (see Chapter 4). In Fig. 40*a*, before tilting, we see that the twins AAE are black, whereas the untwinned areas *B* and *C* are in light contrast (bright field image). Upon tilting ~2°, the twins *A* are now in light contrast (Fig. 40*b*), and the contrast at *G* has been completely reversed. More details are also revealed after tilting, e.g., fringes at *D*, an extinction band *F*, and dislocations in the areas *C* and *E*. In each case dark contrast results whenever the diffracted beams fall outside the objective aperture.

High contrast is always obtained whenever there is strong Bragg diffraction, e.g., near extinction contours. These are the areas which must be looked at carefully if the microscopist is hoping to find features of interest. Thus during tilting, as the extinction contours sweep across the field of

FIG. 40(*a*). Bright field image of thin foil of Cu–8% Al alloy showing twins. (×36,000.) (Courtesy P. R. Swann.)

view, contrast effects may be shown up in areas where previously they were invisible.† Tilting the specimen will help to bring it into a strong

† Contours should never be confused with dislocation images. The former are broader and extend across the length of the grain. Long dislocation lines are only observed when they lie parallel to the surface. On tilting, the contours remain in contrast and can be observed to sweep across the field of view, whereas the dislocation image will not move appreciably; rather it will go in or out of contrast.

FIG. 40(*b*). Same area as Fig. 40*a* after 2° tilt of specimen. Note reversal of contrast at *A*, *E*, and *G* and appearance of fringes *D* and dislocations *C*. *F* is a broad thickness extinction contour. (× 36,000.) (Courtesy P. R. Swann.)

diffracting position; likewise although areas may appear opaque upon first viewing, tilting through a few degrees may result in transparency.

It cannot be emphasized too strongly that it is essential to use a tilting device for all work with crystals—not only to obtain good contrast but also to avoid false interpretation of results (see section 3.9).

Having considered many of the factors associated with the interaction of the beam with the specimen, let us now turn our attention to the electron microscope itself.

References

1. E. Abbé, *Arch. Mikr. Anat.* 1837, **9**, p. 413.
2. J. W. Menter, *Proc. Roy. Soc.* A 1956, **236,** p. 119.
3. D. W. Pashley, J. W. Menter, and G. A. Bassett, *Nature* 1957, **179**, p. 752.
4. G. A. Bassett, J. W. Menter, and D. W. Pashley, *Proc. Roy. Soc.* A 1958, **246,** p. 345.
5. J. W. Menter, *Adv. in Physics* 1958, **7**, p. 299.
6. P. B. Hirsch, A. Howie, and M. J. Whelan, *Phil. Trans. Roy. Soc.* A 1960, **252,** p. 499.
7. R. D. Heidenreich, *J. Appl. Physics* 1949, **20**, p. 993.
8. M. J. Whelan and P. B. Hirsch, *Phil. Mag.* 1957, **2**, pp. 1121, 1303.
9. R. B. Nicholson and J. Nutting, *ibid.* 1958, **3**, p. 531.
10. C. G. Darwin, *Phil. Mag.* 1914, **17,** pp. 315, 675.
11. P. P. Ewald, *Ann. Phys. Lpz.* 1916, **1**, p. 117.
12. H. A. Bethe, *Ann. d. Physik* 1928, **87**, p. 55.
13. C. H. MacGillavry, *Physica* 1940, **7**, p. 329.
14. W. H. Zachariasan, *Theory of X-ray Diffraction in Crystals* (John Wiley and Sons, New York), 1945.
15. R. W. James, *Optical Principles of the Diffraction of X-rays* (Bell & Sons Ltd., London), 1948.
16. Z. G. Pinsker, *Electron Diffraction* (Butterworths Scientific Publications, London), 1953.
17. A. Howie and M. J. Whelan, *European Regional Congress Electron Microscopy*, Delft, Holland 1960 (Nederlandse Vereniging voor Electronen-microscopie, Delft, 1961), **1**, 181.
18. J. Hoerni, *Phys. Rev.* 1956, **102**, p. 1534.
19. J. A. Ibers, *Acta Cryst.* 1958, **11**, p. 178.
20. H. Hashimoto, A. Howie, and M. J. Whelan, *Phil. Mag.* 1960, **5**, p. 967.
21. H. Hashimoto and R. Uyeda, *Acta Cryst.* 1947, **10**, p. 143
22. N. Kato, *J. Phys. Soc.* (Japan), 1952, **7**, pp. 397, 406; *ibid.* 1953, **8**, p. 350.

3

The Electron Microscope

3.1 Electron Optics and the Electron Microscope

It is not our intention to go into details on the subject of electron optics since this is well covered in the literature.† We shall, however, consider the paths taken by electrons in electrostatic and magnetic fields in order to understand the focusing action and imaging characteristics of lens systems normally encountered in modern microscopes. For most purposes it is convenient to regard the motion of electrons in a similar way to the propagation of light waves, i.e., as rays or normals to the wave fronts. The velocity of electrons is proportional to the square root of the applied potential and is, like the index of refraction for a given light ray, a function of position only. This shows the exact correspondence between ray (or geometric) optics and Newtonian particle mechanics. The chief difference between electron beams and light rays is that only the former are influenced by electric and magnetic fields. One other factor is important, too. Since electrons are strongly scattered when passing through matter (e.g., 0.1 mm thick aluminum sheet or 20 cm air will stop a 50 kv beam), it is necessary for any electron optical system to be under a high vacuum.

For the refraction of electron beams by a potential field, it can be seen from Fig. 28 that there is no simultaneous reflected electron ray (as for light rays) because the change in potential producing refraction is gradual.

3.1.1 *Deflections in Electrostatic Fields*

In a uniform electrostatic field a beam of electrons moves in a parabolic path, so that its motion is entirely analogous to that of a material particle

† A practical and theoretical treatment is given by Cosslett[1] including theory of lenses and their aberrations. References 3, 4, 5, 11, should also be consulted for information specific to electron microscopy.

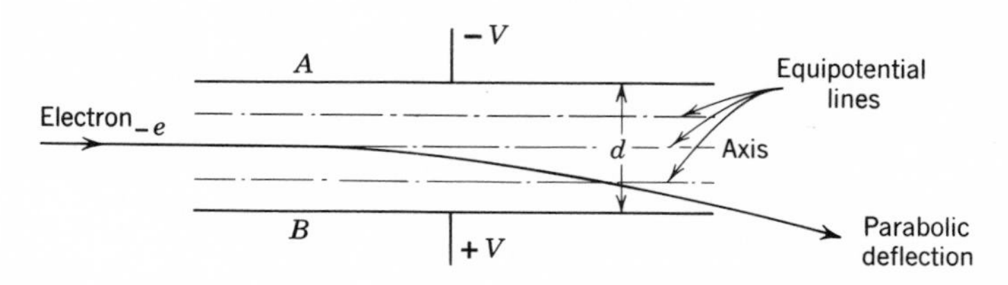

FIG. 41. Motion of an electron in a transverse electrostatic field. (After Cosslett.[1])

in a uniform gravitational field. This is illustrated in Fig. 41, in which a uniform potential V is kept between two plates AB. If these are separated by a distance d, then the electric field E in the gap is $E = V/d$. The force on the moving electron is Ee acting transversely to its initial path, so that a constant acceleration a occurs equal to Ee/m, where e is the charge and m is the mass of the electron. The transverse displacement toward the positive plate B after time t is thus $\frac{1}{2}at^2$, and since there is no change in the forward velocity, the electron path is parabolic. Usually, electric fields between parallel plates are not uniform, mainly as a result of distortion at the edges. However, electric lenses can be formed by any set of electrodes shaped so as to be symmetrical about a common axis. The total deflection of rays incident from a common center of divergence, e.g., an object point on the axis at different heights on the lens, is proportional to the height of incidence (due to the transverse force component of the field). The same is true for emergence; thus rays converge at a new point on the axis †—the image point—or they diverge in such a way that they appear to come from a point on the lens. In the ideal case this focusing action corresponds to that of a thin glass lens in light optics, as illustrated in Fig. 42. Here the lens L focuses an incident beam from the object O

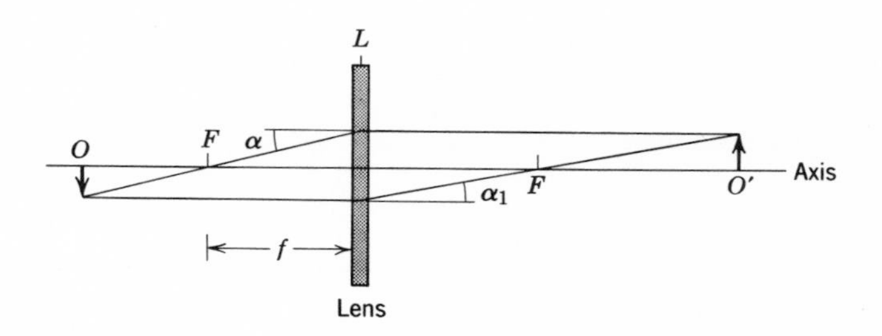

FIG. 42. Focusing action of a thin lens.

† Only for objects in field-free space.

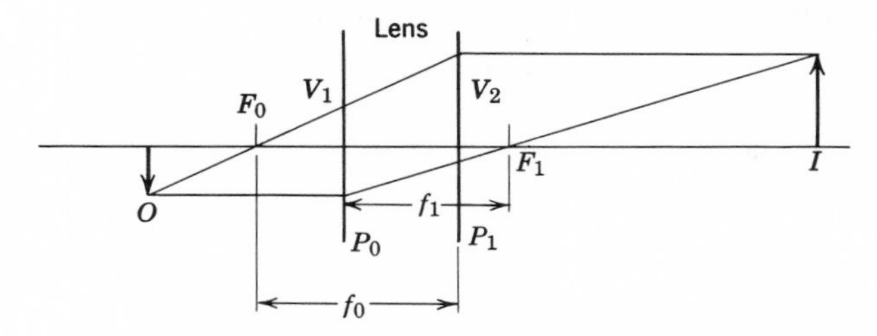

FIG. 43. Focusing action of a thick lens.

forming an image at O'. As in the light optical case, the linear magnification is LO'/OL and the focal length f of the lens is

$$\frac{1}{f} = \frac{1}{OL} + \frac{1}{LO'} \tag{58}$$

The angular magnification is $\tan \alpha_1/\tan \alpha$. In general, however, the situation in electron lenses is more closely related to geometrical light optics in the thick lens case, where it is necessary to know the principal planes P_0 and P_1 at the object and image side of the lens (Fig. 43). In this case the magnification is $\dfrac{P_0 I}{OP_1}\sqrt{\dfrac{V_1}{V_2}}$, where V_1 and V_2 are the potentials in object and image space respectively. Since magnetic lenses are almost invariably used in modern electron microscopes, no more need be said about electrostatic fields. However, the focusing action of thin and thick magnetic lenses may also be represented as shown in Figs. 42 and 43.

3.1.2 *Deflections in Magnetic Fields*

Magnetic fields with axial symmetry may be used as electron lenses, but in this case there is no close analogy between the motion of electrons in magnetic fields and the paths of light rays in any known medium. The thin magnetic lens is essentially a solenoid carrying current. However, before dealing with this let us consider the influence of a uniform transverse magnetic field on a moving electron. An electron moving into a uniform field H with velocity u (proportional to $\sqrt{V}$ where V is the accelerating voltage (see Table I, section 1.3)) experiences a force Heu normal to both the field and the original direction of motion of the electron. The electron is then deflected in a plane normal to the field, intersecting the lines of force at right angles. Thus the force on the electron is always perpendicular to its motion at any instant, so that the electron will

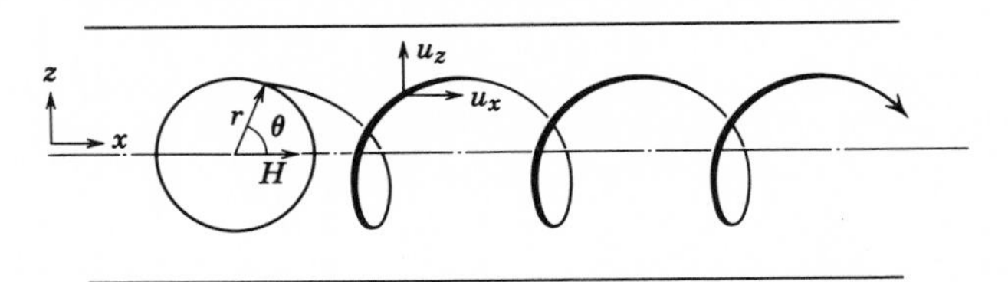

FIG. 44. Helical path of an electron in a magnetic field, e.g., a solenoidal lens. (After Cosslett.[1])

move in a circular path of radius r determined by the equivalence of the centripetal and deflecting forces, i.e.,

$$Heu = mu^2/r \tag{59}$$

Hence a 1 volt electron describes a circle of 3.37 cm radius in a field of 1 gauss, so that by replacing u with $\sqrt{V}$, we can write

$$r\,(\text{cm}) = \frac{3.37\sqrt{V}\,(\text{volts})}{H\,(\text{gauss})} \tag{60}$$

The frequency of rotation depends only upon the field and is 2.8 Mc/sec for 1 gauss. It is important to notice from this that a magnetic field parallel to the direction of motion of an electron exerts no deflecting force upon it.

If a point source of electrons is situated in a uniform longitudinal field so that the emitted electrons make an angle θ with the field, the velocity u_x along the field is uniform, but the velocity normal to the field, u_z, is under the influence of a force Heu_z. The resultant path of the electron will be helical as shown in Fig. 44. When u_x is zero (i.e., $\theta = 90°$), the path is a circle of radius r (uniform transverse field). Uniform longitudinal fields occur in a solenoidal lens, as illustrated in Fig. 45, where (a) represents a

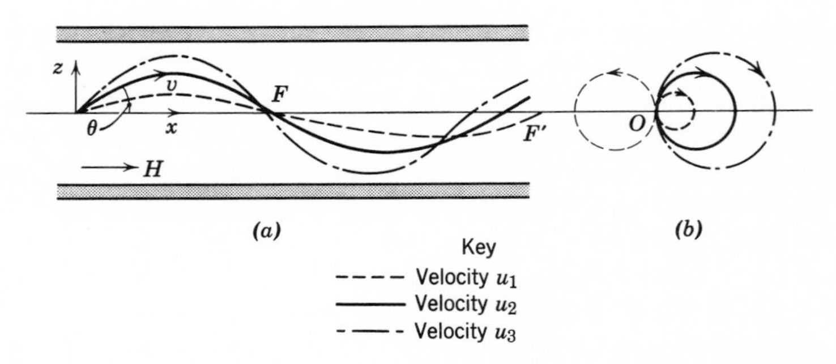

FIG. 45. Electron paths in a solenoidal magnetic lens. Section (a) along the field, and (b) normal to the field. (After Cosslett.[1])

section parallel to the field, and (*b*) a section perpendicular to the field. The tangential velocity components of the individual electrons cause a helical deflection in which the pitch of the helix is determined by the magnitude of these components, e.g., u_1, u_2, u_3 in Fig. 45.

If the axial velocities of the electrons are the same, e.g., in a monochromatic beam, the pitches of their helices are identical and all electrons passing through a given point will, after one revolution, come together at a second point at a distance equal to the pitch from the first point in the direction of the magnetic field. In this way focusing occurs, e.g., at F, F', etc. (Fig. 45*a*), and the images may be made visible by placing a suitable fluorescent screen at these points. The focal length of such a lens is directly proportional to the accelerating voltage on the electrons. Electrons of differing velocities will be focused at different points, thus giving rise to chromatic aberration.[1-4]

It can be seen from Fig. 45 that the motion of the electron in a magnetic field results in a rotation of the image with respect to the object, and is always clockwise when viewed along the field in the positive z direction (left-hand rule). This rotation vanishes only in a thick or compound lens where the magnetic fields act in opposite directions (e.g., by winding the coils in the two halves of the lens in opposite senses). Since the degree of rotation is proportional to the energizing current in the lens, it is necessary to calibrate this when carrying out selected area diffraction work (see section 3.6).

3.1.3 *Short Magnetic Lenses*

Figure 46 shows the focusing action of a short lens. This is different from the solenoidal case because here both object and image points lie outside the field. This type of lens corresponds to the short focal length lenses used for objective, intermediate, and projector lenses in electron microscopes. Shielding the lens with soft iron pole pieces concentrates the magnetic field over a very small region so that the radial component of the field is great. It is this component and not the longitudinal component that chiefly causes deviation of the electrons. The field in this type of lens is constant only over a small region close to the axis. In the second half of the lens the electrons are radially accelerated towards the axis with a rotation of ϕ which is proportional to H/u. The focal length f of the lens is proportional to u/H^2, where u is the velocity of electrons and H the field strength or $f \sim IN/\sqrt{V}$, i.e., f is proportional to the number of ampere-turns IN in the coil and thus can be changed simply by varying the lens current (see section 3.5.4).

Among the properties of magnetic lenses, one should always be

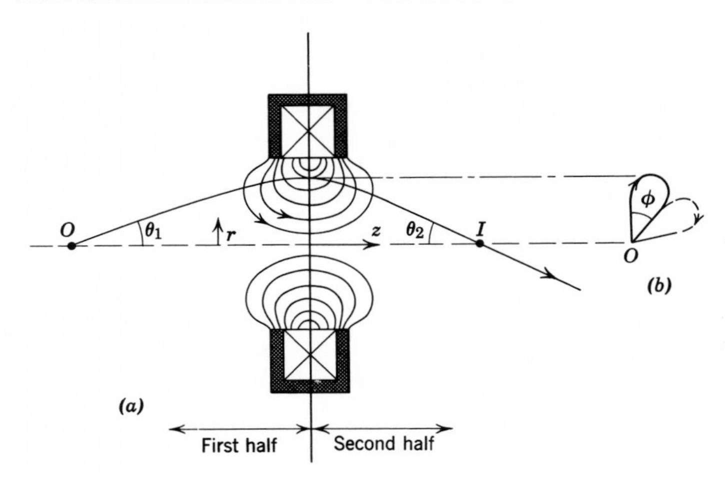

FIG. 46. Electron paths and focusing action in a short magnetic lens. (*a*) Section in meridian plane rotating with electron about axis. (*b*) Projection normal to axis *OI*. Notice rotation ϕ. (After Cosslett.[1])

remembered, viz., all magnetic lenses are convergent whether the object is in the magnetic field or not.

3.2 Image Defects in Electron Lenses[1-4]

Undistorted images may be obtained only when the following conditions are satisfied:

1. The extent of object and image is small.
2. The divergence of the image-forming rays is small.
3. The velocity of electrons is uniform.
4. The electron concentration at all points of the path is small, so that mutual repulsion of electrons is negligible.

Besides these conditions, we must also bear in mind that there should be (1) no defective construction in lens pole pieces, (2) a high vacuum always maintained, (3) perfect optical and magnetic alignment, and (4) no contamination in the path of the electron beam.

If we assume axial illumination and that the velocity of the electron along the axis in the lens field is the same as the actual velocity of the electrons (determined by the accelerating voltage), then as we saw in section 3.1.1 the angular magnification of an ideal lens is given by $\tan \alpha_1/\tan \alpha$ (Fig. 42), where α and α_1 are the angles subtended at the lens on the object and image side respectively. In classical optics, it can be shown that a lens is perfect only when $\tan \alpha \sim \sin \alpha \sim \alpha$ (condition 2, above). This approximation is equivalent to neglecting powers of α equal to three and greater with respect to α, since actually $\tan \alpha = \alpha + (\alpha^3/3) + (2\alpha^5/15) + \ldots$, etc. Deviations from an ideal lens thus correspond to a breakdown in the approximation $\tan \alpha = \alpha$, in which case imperfect images will be formed. Only the third order aberrations (proportional to α^3) are important in electron microscopy since they outweigh higher order terms. The third order aberrations for electron lenses are (1) spherical aberration and distortion, (2) curvature of field and astigmatism, and (3) coma. In addition to these, other factors such as chromatic aberration and space-charge distortion can result in image defects. We shall now consider the more important of these and how they can be corrected.

3.2.1 *Spherical Aberration*

Although the "shapes" of magnetic lenses cannot be represented in the same terms as those of glass lenses, spherical aberration is the only geometrical aberration which causes unsharpness of the image on the optic axis. This aberration corresponds to "bad shaping" of the lens surfaces such that a circle of confusion occurs about every image point (see Fig. 47). The radius r' of this circle is $\sim C_s(\alpha)^3$ where C_s is the coefficient of spherical aberration and α is half the vertex angle of rays leaving the object. The outermost rays are refracted more strongly than those close to the axis, but this can be reduced by using very small apertures in front of the objective lens, as shown at *AB* in Fig. 47. In projector lenses spherical aberration causes pincushion or barrel distortion of the image at low

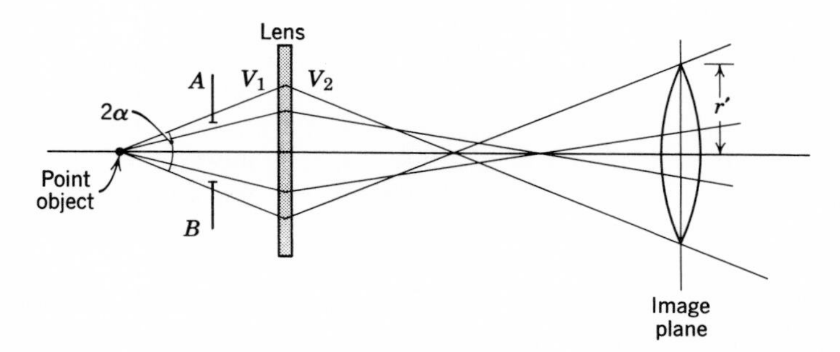

FIG. 47. Illustrating spherical aberration in lenses. *AB* is an aperture to stop off outermost beams which are refracted more strongly than those close to the axis; r' is radius of disc of confusion about an image point.

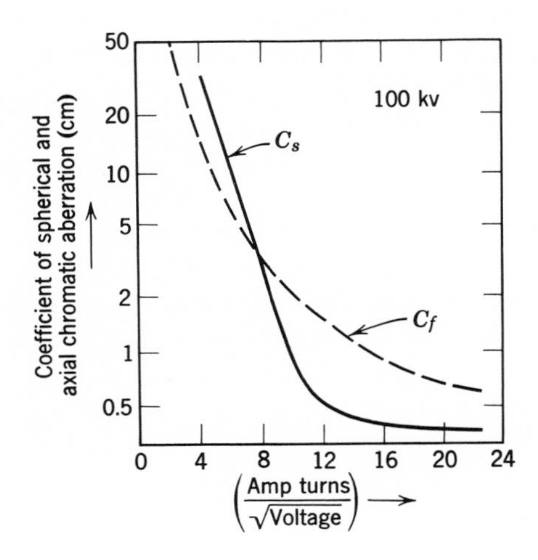

FIG. 48. Showing variation of spherical and axial chromatic aberration with excitation lens current. (Courtesy Hitachi Ltd.)

magnifications. This defect, however, may be compensated for by using an intermediate lens in conjunction with the projector (section 3.5.5).

The higher the exciting current in the lens, the lower is the spherical aberration (Fig. 48). The minimum value of C_s at present attainable in electron microscopes is about 0.03 cm. In order to minimize fluctuations in the field strengths of the lens, it is necessary to use current stabilizers. The limit of resolution due to spherical aberration is about 2.8 Å[5] (see section 3.3).

3.2.2 *Chromatic Aberration*

Electrons of different wave lengths are refracted by different amounts, and since $\lambda = \sqrt{150/V}$ (neglecting relativistic effects), any fluctuations in the accelerating voltage V lead to chromatic aberration. Since most microscopes operate up to 100 kv, the high tension supply (i.e., the accelerating voltage) must be regulated to fluctuations of less than 1 in 10^5 volts by voltage stabilizers. Chromatic defects are illustrated in Fig. 49 where the radius of the circle of confusion is proportional to $C_f\alpha\Delta E/E$, where C_f is the coefficient of axial chromatic aberration, $\Delta E/E$ the energy spread in the imaging electrons (because of voltage ripple or drift, energy losses from inelastic scattering by the object, or fluctuations in the lens current), and α is the aperture subtended at the lens.[4] The variation of C_f with energizing current in the lens is also shown in Fig. 48. However, when specimens of light atoms are employed, the ratio of the inelastic to elastic scattering is high (see Fig. 3), so that most of the inelastically scattered electrons remain within the objective aperture and the resultant

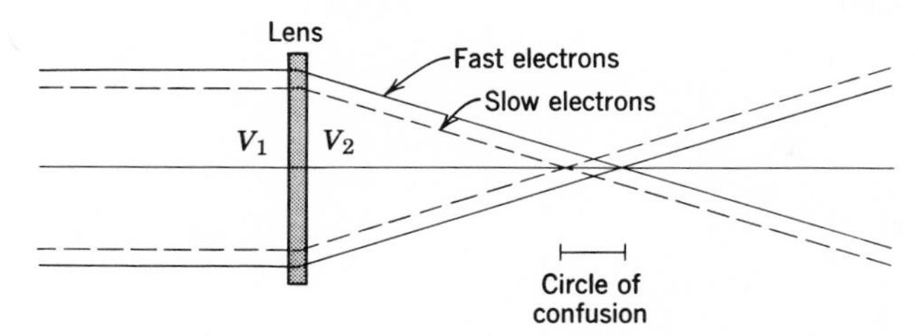

FIG. 49. Chromatic aberration resulting from focusing of electrons of different velocities (wavelengths). The same effect is produced by fluctuations in lens current.

chromatic aberrations may become an essential limit of resolution (see section 3.3).

3.2.3 *Curvature of Field: Astigmatism*

Curvature of the field occurs when rays adjoining the principal rays from the several off-axis object points converge on the latter ahead of or behind the image plane as illustrated in Fig. 50. More commonly these off-axis rays do not converge at a point, but in two mutually perpendicular line segments giving rise to astigmatism, e.g., as in Fig. 51. This defect is the most difficult of all to control because it arises from asymmetry in the pole pieces. Even with precision machining, the minimum astigmatic distance in lenses is not much smaller than 0.2μ, which sets the ultimate limit on resolution at about 4 Å[5]. It is obvious that any contamination in the pole pieces (e.g., decomposition of hydrocarbons in the impure vacuum, cleaning residues, "lost" specimens, etc.) will give rise to astigmatic defects, so that it is essential to keep all parts of the microscope

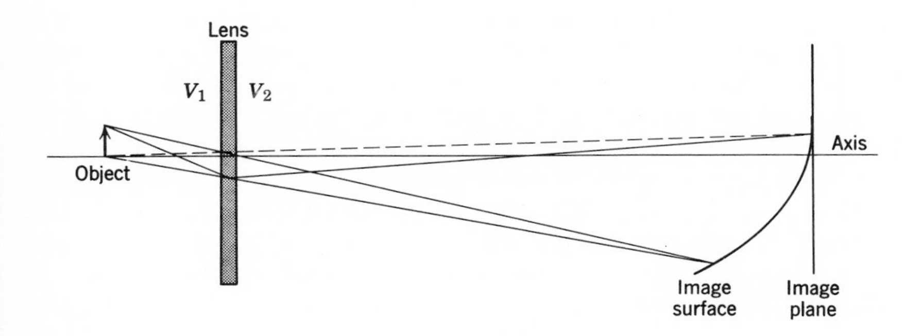

FIG. 50. Illustrating curvature of field.

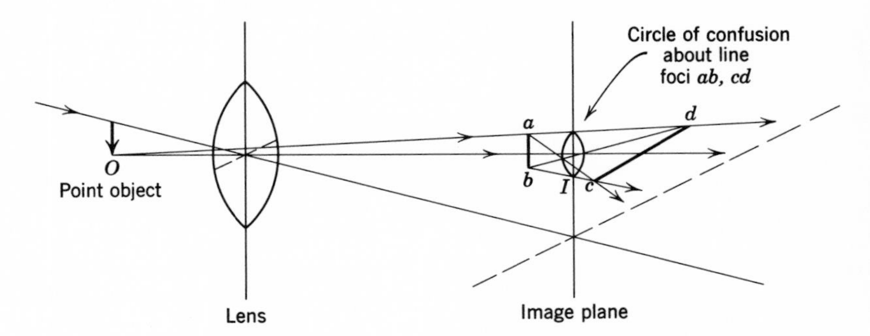

FIG. 51. Origin and effect of astigmatism. (After Hall.[4] By permission from *Introduction to Electron Microscopy* by C. E. Hall. Copyright 1953, McGraw-Hill Book Co.)

column in contact with the electron beam immaculately clean. Stigmator correctors are usually provided on most high-resolution instruments. These correctors simply superimpose, across the gap in the pole pieces, an additional weak, elliptical field, which can be varied in strength and direction to compensate for the astigmatism, and is particularly valuable when contamination of the objective aperture occurs. One type consists of a nickel-coated, brass, hollow cylinder which can be raised or lowered and rotated inside the objective pole pieces. A suitable test object is the Fresnel fringe that occurs in an out of focus image.

In physical optics,[6] diffraction effects are classified as Fresnel diffraction when either the illumination source or the screen, or both, are at a finite distance from the diffracting aperture or obstacle so that the illumination cannot be considered parallel. The origin of Fresnel diffraction fringes in an out-of-focus electron image[4,7] can be found by considering Fig. 52. If the edge of the specimen lies in a plane normal to the optic axis and a coherent beam of electrons passes through the free space at the edge of the specimen, then beyond the specimen the electron beam may be regarded as being made up of two waves. One is a plane wave (free beam), and the other is a cylindrical wave with the specimen edge as source. For an aberration-free lens the original plane wave will converge in the lower focal plane of the objective and diverge as a spherical wave beyond it. Similarly, the objective will convert the cylindrical wave into a cylindrical wave converging on a line in the plane conjugate to the plane of the edge at a distance b from the image-side principal plane of the objective. If the objective is focused not on the edge but on a plane situated at a distance a below it, then the two electron waves interfere at the screen plane at a distance L from the image-side principal plane of the objective

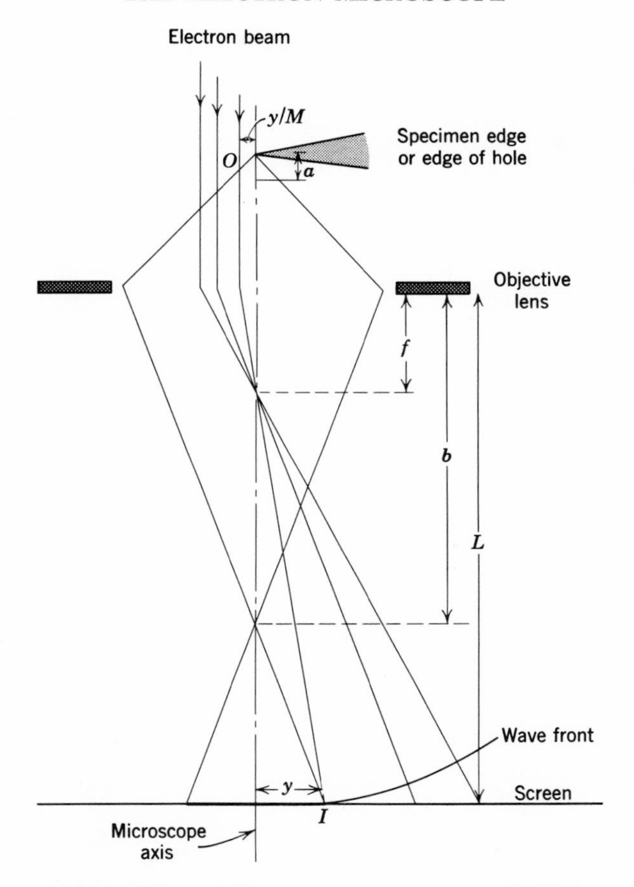

FIG. 52. The origin of Fresnel fringes in an out-of-focus image of specimen edge (overfocused case).

(Fig. 52). As a result of this interference, the intensity on the screen consists of a series of maxima and minima decreasing in amplitude with distance normal to the optic axis. The maximum occurs at a distance equal to the width of the fringe approximately as[8]

$$y = \sqrt{a\lambda n} \tag{61}$$

where $n = 1,2,3$ the order of the fringes. Since the source is of finite size ($=y/M$ where M is the magnification), each point on the source produces a separate Fresnel pattern. If the lens is free from aberrations and is exactly focused so that the object plane coincides with the edge, $a = 0$

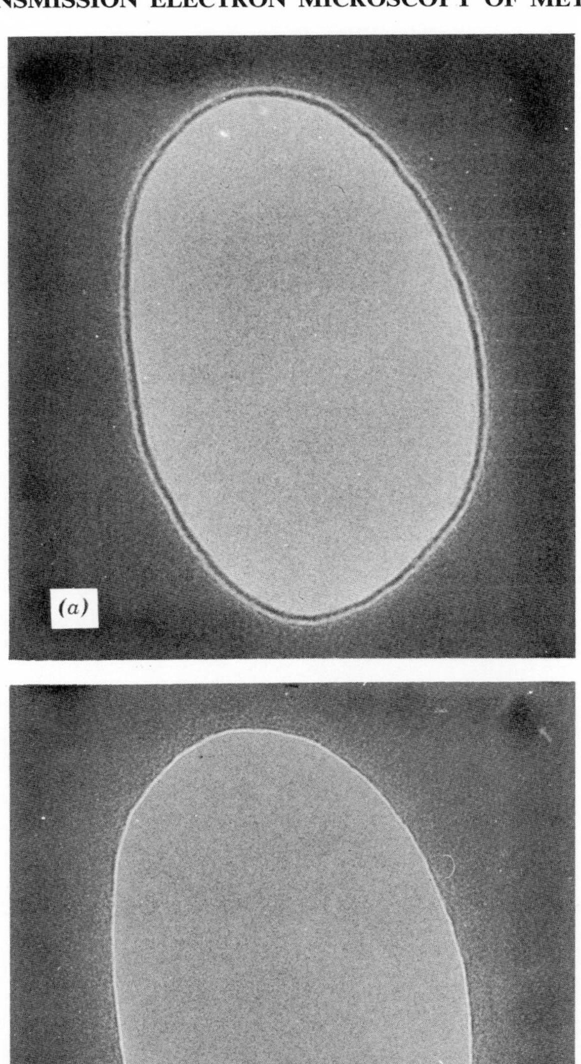
(a)

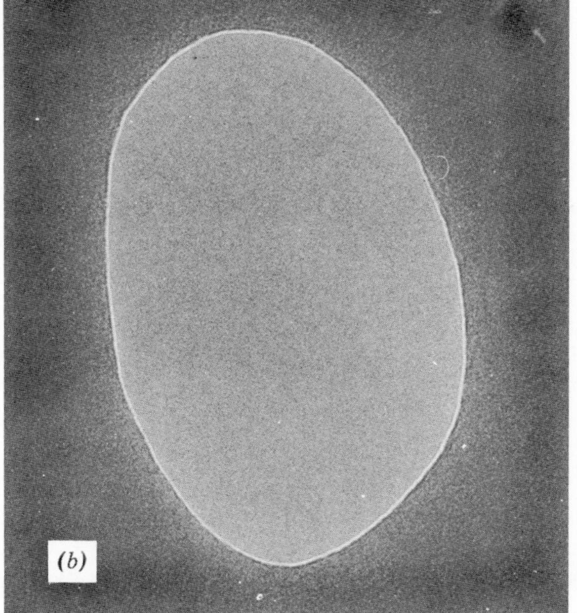
(b)

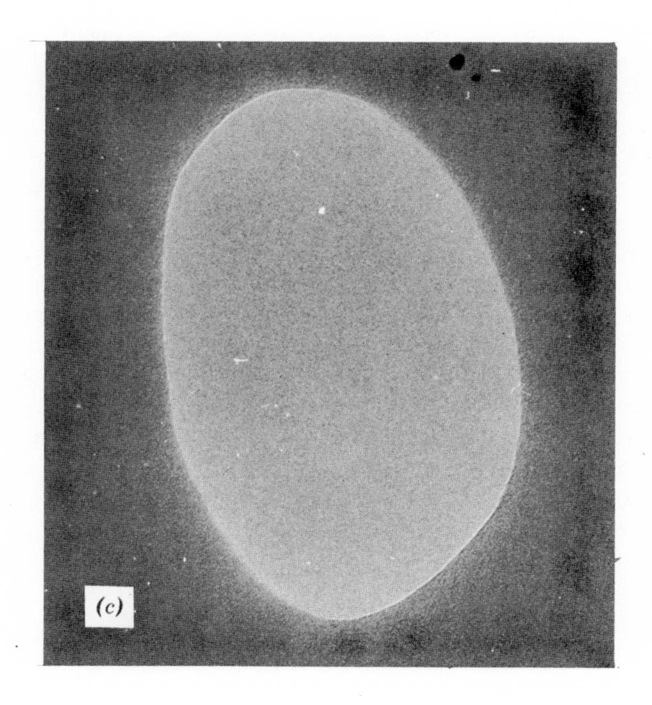

FIG. 53. Fresnel fringe from a hole in collodion film $\times 570{,}000$. (*a*) Overfocused. (*b*) Underfocused. (*c*) Just focused.

and so no fringes are observed. Deviations from this are due to lens aberrations.

Fringes can be observed in the microscope if the off-focus distance ($=a$) is adjusted by changing the lens current to give a fringe width on the fluorescent screen of about 50 μ. For the first order fringe and at a magnification of 50,000 the actual fringe width is thus 10 Å. In order for this to be clearly visible, a magnification of at least 80,000 times is therefore recommended. Defects in the fringe can arise from the following effects:

1. Asymmetric blurring—usually due to stage or column vibration.
2. Radial or symmetrical blurring—usually due to fluctuations in the high voltage or lens currents.
3. Asymmetry in the fringe width—usually due to astigmatism in the objective.

The Fresnel fringe can thus be utilized as a suitable test object for image defects; it has high contrast, and, being extended in one dimension, is much more easily seen under the low brightness conditions (at high magnifications) of the electron microscope image than is a point object. A convenient object is a specimen film perforated with small holes, e.g., as shown for a collodion film in Fig. 53 *a–c*. Notice that the fringe is more contrasty in the overfocused condition (objective lens current too strong), and that in case *c* the fringe has not fully disappeared even though the film is just-focused. This is due to residual astigmatism in the objective. The astigmatism is corrected by varying the direction and strength of the applied stigmator field until the residual astigmatism is a minimum.[8] For the case shown in Fig. 53*c* this is about 10 Å (i.e., the width of the out-of-focus fringe).

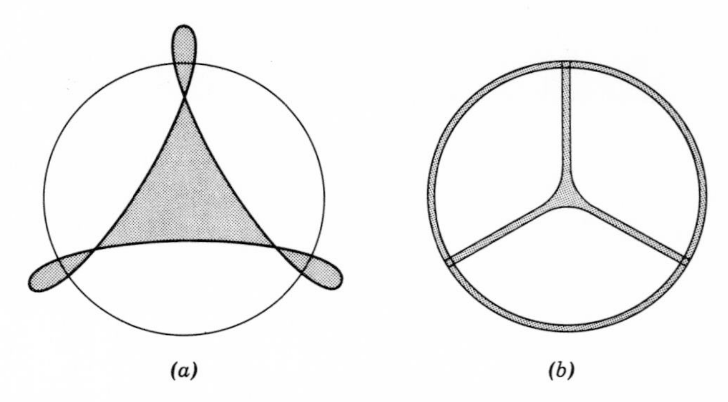

FIG. 54. Caustic patterns in images of a defocused electron beam (schematic). (*a*) Astigmatic. (*b*) Corrected.

Astigmatism in condenser lenses may be observed by the appearance of the caustic pattern of a beam defocused by the second condenser. This can be corrected by changing the condenser stigmator field strength until the caustic pattern becomes symmetrical; e.g., Fig. 54*a* represents an astigmatic condenser lens and Fig. 54*b* corrected for astigmatism. Not all electron microscopes are provided with stigmator correctors for condenser lenses, but since these lenses are of longer focal lengths than objective lenses, the problem is not serious. Nevertheless, absolute cleanliness in the condenser system is essential to minimize this defect, especially if high-resolution experiments are to be carried out.

3.2.4 *Space-Charge Effects*

This is peculiar to electron optics and occurs for slow electrons when the electron concentration at any point may be so strong as to cause mutual repulsion of electrons. The concentration is proportional to the reciprocal of the velocity of electrons, and thus in low velocity, high current beams, this effect becomes important. Space-charge effects result in spreading of the beam and this is analogous to the action of a negative (concave) lens. For the moderate beam intensities normally occurring in conventional electron microscopes, however, space-charge effects are not of serious proportions.

3.3 Resolving Power

Because of the wave nature of the electron beam and the finite aperture of the lens, the image of a point object would still be of finite dimensions even if the errors described in the previous section were reduced to zero. This is due to the divergence of the beam leaving the object and gives rise to a diffraction error previously discussed in Chapter 2. The image of a point consists of a disc (the Airy disc), in which the intensity drops from a maximum to zero at a distance, say, r' from the center (similar to Fig. 22). Let us suppose that this radius subtends an angle γ at the lens as shown in Fig. 55. In light optics, if R is the radius of a perfect lens, then similar to the case of a grating (section 1.13)

$$\gamma = \frac{1.22\lambda}{2R} \tag{62}$$

From Fig. 55

$$r' = L\gamma \tag{63a}$$

and

$$\beta = R/L \tag{63b}$$

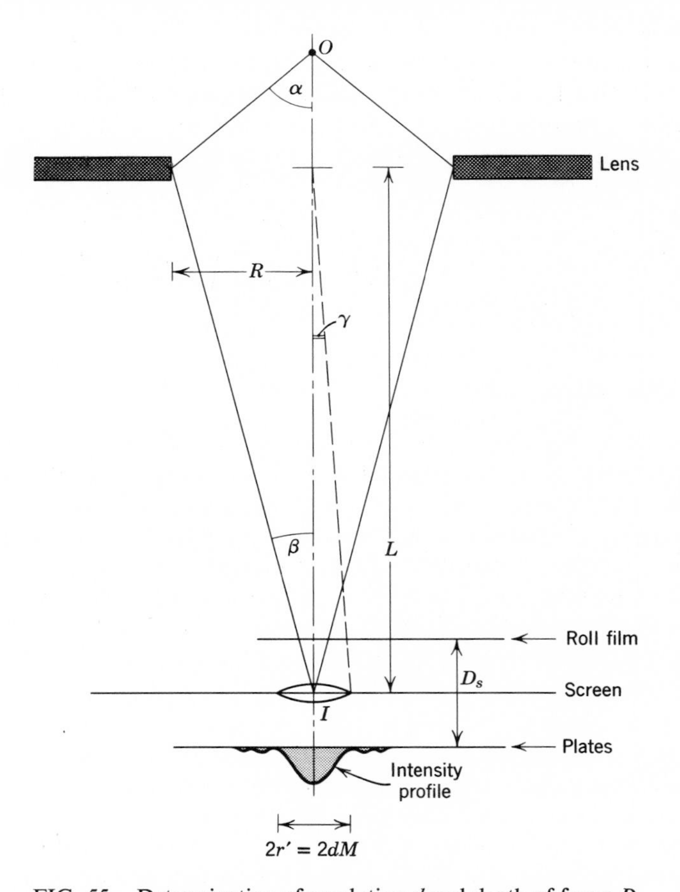

FIG. 55. Determination of resolution d and depth of focus D_s.

Hence, we obtain
$$r' = \frac{0.61\lambda}{\beta} \tag{64}$$

From Fig. 55 we see that $\beta = \alpha/M$ where M is the magnification and α is the lens aperture, thus

$$r' = \frac{0.61\lambda M}{\alpha} \tag{65}$$

In these derivations we have assumed that the sine and tangent of an angle is equal to the angle, and this is permissible in electron optics since the angles are very small. (However, see section 3.2.) For light optics it is more correct to write

$$r' = \frac{0.61\lambda M}{\mu \sin \alpha}, \quad \text{i.e., in object space } d = \frac{0.61\lambda}{\mu \sin \alpha} \tag{66}$$

where $d = r'/M$ is the apparent radius of an image point due to diffraction (neglecting the intensity outside the radius d), and μ is the refractive index. The term $\mu \sin \alpha$ $(=N)$ is known as the numerical aperture.

If two points are separated over a distance smaller than d, it is clear that the image will appear as a single spot, and thus d is the limit of resolution. Obviously, the smaller the value of d, the greater is the resolving power. In light optics d can be decreased by making λ as small as possible (ultraviolet light) and by making N as large as possible, e.g., with oil immersion $N = 1.4$. Thus for white light, $\lambda = 4200$ Å and for $N = 1.4$, $d = 1800$ Å. Using ultraviolet light, $\lambda = 2537$ Å and for $N = 1.4$, $d = 1100$ Å.

From this, it can be seen that the optimum resolution for light microscopy is about 1000 Å. For electron beams, however, λ may be as small as 0.037 Å (see Table I), so that the resolution could be better than 10^5 times that in light optics, and it should be possible to resolve details of less than atomic dimensions. However, this theoretical resolution is not obtained in the electron microscope because so far electron lenses cannot be made with large aperture systems which are correctable in the way that glass objectives are for light. Very small apertures are therefore used (of the order of 10^{-2} to 10^{-3} radian), and the electron optical system is so designed as to minimize the combined effects of the four most important sources of error in the image: diffraction error, spherical aberration, astigmatism, and chromatic aberration, which were described in the preceding paragraphs.

In the problem of resolving a periodic structure of spacing d we have already seen in Chapter 2 that in order for this to be done, the first order diffraction angle $\alpha = d/\lambda$ (Fig. 29) must be small enough to enter the aperture of the objective lens. For recombination α must be small enough so that the effect of spherical aberration is negligible. Because of spherical aberration, the apparent image of d is proportional to $C_s\alpha^3$. The optimum value of α is such that the spherical aberration and diffraction errors are about equal, i.e., $C_s \alpha^3 = 0.61\lambda/\alpha$ (equation 66), thus $\alpha\,(\text{opt}) = (0.61\lambda/C_s)^{1/4}$. However, for the first order diffracted beam to fall within the aperture of the lens $\alpha\,(\text{opt}) < (d/C_s)^{1/3}$. Hence by substituting for $\alpha = \lambda/d$ we obtain that for resolution, the minimum value of d is given by $C_s^{1/4}\lambda^{3/4}$. The same argument applies to the imaging of two points separated by a distance d where the first order maximum occurs at the angle λ/d. The breadth of the maxima is determined by the beam divergence (section 1.13).

Haine[11] has given more accurate expressions for α (opt) and d (min), as follows:

$$\alpha\,(\text{opt}) = 1.4\left(\frac{\lambda}{C_s}\right)^{1/4} \tag{67}$$

and

$$d\,(\text{min}) = 0.43 C_s^{1/4} \lambda^{3/4} \tag{68}$$

For 80 kv electrons ($\lambda = 0.04$ Å) and for $C_s \sim 0.05$ cm, d (min), i.e., the theoretical instrumental resolution, is about 3 Å, and α (opt) $\sim 10^{-2}$ radian. As discussed in Chapter 2, with metal crystals as specimens, the Bragg angles are greater than α (opt); consequently, lattice planes in metals cannot be directly resolved. However, we can see that by choosing suitable crystals of larger interplanar spacings (smaller Bragg angles), it is possible to use these as specimens in order to make an accurate determination of the resolving power of the electron microscope. Suitable crystals are the metal phthalocyanines, e.g., in platinum phthalocyanine the {201} planes have spacings of 11.94 Å.[14] If a thin single crystal of this material is oriented such that $(20\bar{1})$ lies parallel to the beam and provided the microscope is capable of resolutions better than 12 Å, it is possible to image these planes as described in section 2.2. An example is shown in Fig. 30.

Platinum phthalocyanine may be prepared by heating phthalonite with platinous chloride. Single crystals are obtained by sublimation in a stream of CO_2,† after which they are ground with a glass rod in the bottom of a test tube and suspended in ethyl alcohol. For examination in the electron microscope it is convenient to allow one drop of this suspension to evaporate on a microscope grid covered with a thin carbon support film (see section 4.1.2). The images of the crystals are seen to best advantage when not overlaid with the structure of the support film, so that it is necessary to look for areas where the crystal is suspended across holes in the film. The procedure outlined in section 2.2 should then be adopted for imaging the lattice planes using photographic plates of the highest possible contrast.‡

If resolutions below 10 Å need to be verified, it is convenient to use single crystals of molybdenum trioxide, in which the (020) planes have a spacing of 6.93 Å. (See reference 3, Chapter 2.) These can be prepared by thermal decomposition of ammonium molybdate followed by resublimation. Specimens may then be mounted for examination as described above.

For most practical purposes it is not necessary to attain resolutions of the order of 10 Å or so. Nevertheless, if the instrument is always main-

† See also P. A. Barrett, C. E. Dent, and R. P. Linstead, *J. Chem. Soc.* 1936, p. 1719.

‡ For example, Ilford or Kodak special contrasty lantern slides (fine-grained).

tained at its optimum resolution limit (say 10 Å), one can expect to obtain the best possible results even if imaging objects of dimensions considerably in excess of this.

The attainable practical resolution is somewhat worsened from the theoretical instrumental resolution as a result of chromatic aberration and astigmatism. Menter[5] has recently considered this problem with the following results. If the astigmatic focal length difference is Δf, the diameter of the image (Fig. 51) from a point object referred to the object plane is $\delta_A = \alpha \, \Delta f$. For this to be negligible compared with the instrumental resolving power say $\delta_A = 0.2 \, d\,(\text{min})$, i.e., $\Delta f \sim 6 \times 10^{-7}$ cm. The relative astigmatic current difference $\Delta I/I$ in the lens windings should then satisfy the condition

$$\Delta f = \frac{\Delta I}{I} \leq 1.1 \times 10^{-6} \tag{69}$$

Even with stigmator correctors (section 3.5.3), the best adjustment that can readily be made in practice corresponds to $\Delta I/I \sim 2 \times 10^{-5}$, so that astigmatism makes a very real contribution to loss of resolution.

If the accelerating voltage V changes by ΔV and the current I in the winding of the objective lens by ΔI, then the focal length of the lens changes to cause a chromatic disc of confusion (Fig. 49) of diameter

$$\delta_F = 2C_f\alpha\left\{\left(\frac{\Delta V}{V}\right)^2 + \left(\frac{2\Delta I}{I}\right)^2\right\}^{1/2} \tag{70}$$

where C_f is the chromatic aberration constant. Taking C_f at 2.2 mm, then for these to have negligible effect on the resolution $\Delta V/V \leq 1.5 \times 10^{-6}$ and $\Delta I/I \leq 0.75 \times 10^{-6}$. Such stabilities are not readily achieved for the necessary focusing and exposure times, and since chromatic errors also arise from energy losses in imaging electrons due to inelastic scattering, there seems to be no practical advantage to be gained from using elaborate stabilizing equipment. Chromatic aberrations due to inelastically scattered electrons may be minimized by using very thin specimens, but cannot be entirely eliminated without the use of some electrostatic filtering device. Even then severe difficulties may arise as a result of loss of brightness in the image. Thus we can say that because of the combined effects of the four aberrations in objective lenses, a reasonable value for the practical instrumental resolution is 5 to 6 Å. As we shall see in Chapter 5, monatomic planar aggregates of copper atoms only ~4 Å thick (G.P. zones) have already been detected in aluminum-copper alloys due to certain diffraction conditions.[10] Thus we may now look to even greater improvements in lens design and specimen preparation techniques, so that eventually even the individual atom may be resolved.[11]

From these considerations it can be seen that the use of suitable apertures, stable voltage, and lens current supplies and clean microscope parts are an essential requirement for the maintenance of high resolution. Some hints on cleaning are given in section 3.10. Because stray fields due to nearby high voltage units may affect both the beam and the lens currents, the column of the microscope must be shielded and the instrument placed well away from other units. Mechanical stability is also essential, so that sources of shock and vibration are to be avoided. Due consideration must be given to this when installing a microscope in a laboratory (see Appendix D). One other effect arises from specimens which do not conduct well and so become charged up in the beam. This gives rise to a vibrating image and may be controlled by using a "charge neutralizer" in the specimen chamber.† It is also important to have as good an evacuating system as possible, thereby minimizing the chemical reactions which occur in the residual vapors and which cause deposition of semi-conducting substances on objects in the path of the beam.

When accounting for the effect of the electron beam on the specimen, two things must be considered: the alteration of the specimen itself, and the deposition of substances from the impure vacuum. As a result of heating, excitation, or ionization processes within the specimen, chemical changes or recrystallization may occur. Such alterations may take place with inorganic materials, e.g., in the form of dehydration or the production of lower degrees of oxidation. Organic substances and alkali halides are practically always destroyed by the beam. In metal specimens, however, because of their good thermal properties, heating effects due to the beam are not very important, although prolonged exposure may result in damage by ion bombardment. The second pronounced effect on the specimen originates from hydrocarbon molecules contained in the imperfect vacuum inside the microscope, which are absorbed at the microscope parts and also on the specimen. There they are destroyed by the electron beam, the volatile constituents are evaporated, and carbon is left behind. The result is that during irradiation a slowly increasing layer of amorphous carbon is formed on the specimen surface, resulting in thickening and a loss of contrast and resolution. Specimen contamination from this may be reduced by using cold fingers or by externally heating the specimen (if this does not produce an unwanted change in structure).

† A charge neutralizer is simply an ion gun—the build-up of charge on the specimen can thus be neutralized by a beam of low energy (5 kv) ions which are sprayed from the gun onto the surface of the specimen. The gun is usually placed opposite the specimen holder.

3.4 Depth of Field and Depth of Focus

The electron microscope has another advantage over the light microscope in that the depth of field is much greater. This can be seen by reference to Fig. 56. Rays from the position A illuminating the object over a disc of diameter $2r$ images the center of this, C, as an Airy disc at C' on the final screen. The point A, however, is imaged on the screen as an out-of-focus disc since its image position occurs at A'. The diameter $2r'$ of this disc represents the image of the cross section at C of the rays from A, and this can be said to be in satisfactory focus provided

$$2r' \leq 2dM \quad \text{or} \quad r \leq d \tag{71}$$

where M is the magnification and d is the radius of the disc of confusion for

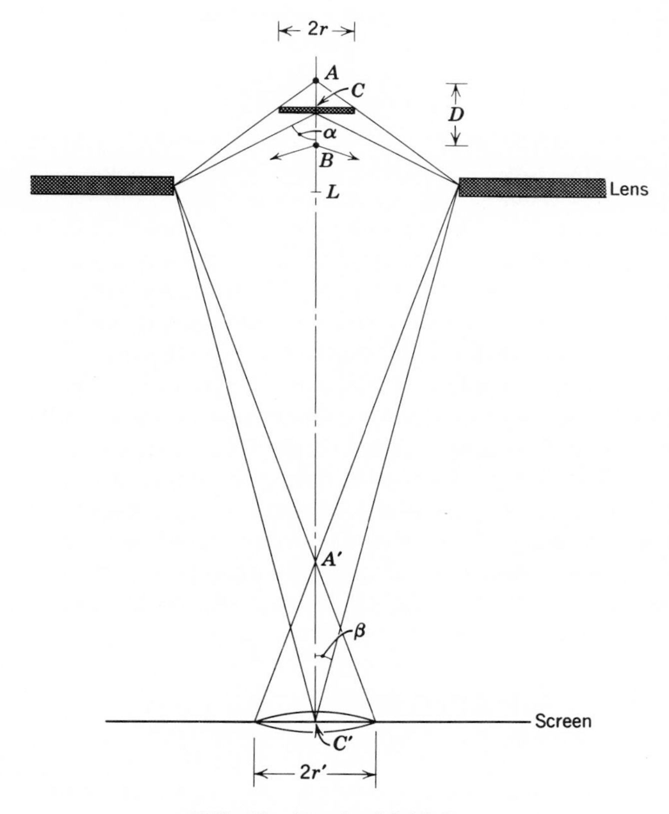

FIG. 56. Depth of field D.

C' referred to C (corresponding to the resolution limit). Similarly, on the other side of C the point B will also be in satisfactory focus on the screen. The distance $AB = D$ is the depth of field about the object point C. Since D is small compared with the object distance (CL), the rays leaving A,B,C subtend the same angle α, so that $r = D\alpha/2$. Hence we can write

$$D = 2d/\alpha \tag{72}$$

If we take $d = 10$ Å and $\alpha = 10^{-3}$ radian, D works out to be 2 μ. This is to be compared with the light microscope where the angle α may be as great as 70°, so that D may approach the least resolvable distance.

We may refer again to Fig. 55 to estimate the depth of focus. When the point O is in geometrical focus at the screen, it will be a diffuse Airy disc of diameter $2dM$. Thus the screen may be moved through a distance (we shall say D_s) along the axis without appreciably affecting the focus. This distance is called the depth of focus. From the geometry of Fig. 55, we get $D_s = 2dM/\beta$, but $\beta = \alpha/M$; hence

$$D_s = 2dM^2/\alpha \tag{73}$$

It is to be noted that D_s and D (depth of field) are simply related through M^2, the longitudinal magnification. On account of the factor M^2 the depth of focus can be quite large, e.g., with the use of just a single projector lens of magnifying power M_2 the depth of focus in the final image is $2dM^2\,(M_2{}^2/\alpha)$. When $M = M_2 = 100$ and for $d = 10$ Å and $\alpha = 10^{-3}$ radian, the depth of focus is 2×10^3 cm. Since this is only true, however, when D_s is small compared to L, this estimate must not be taken too literally. Nevertheless, for all practical purposes, we see that the depth of focus is infinite. This is why it is not necessary to adjust the focus for the purpose of recording the image on a photographic plate or film even though this may be some inches below (or above) the viewing screen (Fig. 55). Because of the enormous depth of focus with electron illumination, it is sometimes convenient to have an arrangement whereby a roll-film camera can be used at a position above the fluorescent screen (Fig. 55) if many exposures are required without disturbing the normal arrangement of having plates located below the viewing screen.

3.5 Magnetic Electron Microscopes

Commercial electron microscopes are manufactured by Siemens and Halske, Radio Corporation of America, Metropolitan Vickers, Phillips, Hitachi, Japan Electron Optics, Trüb-Täuber, Zeiss-Jena, etc., and a

Russian high-resolution instrument was demonstrated at the Brussels World Fair in 1958. The use of high-resolution microscopes is thus world-wide.† Since each company may manufacture more than one type of instrument, describing each one individually would take up far too much

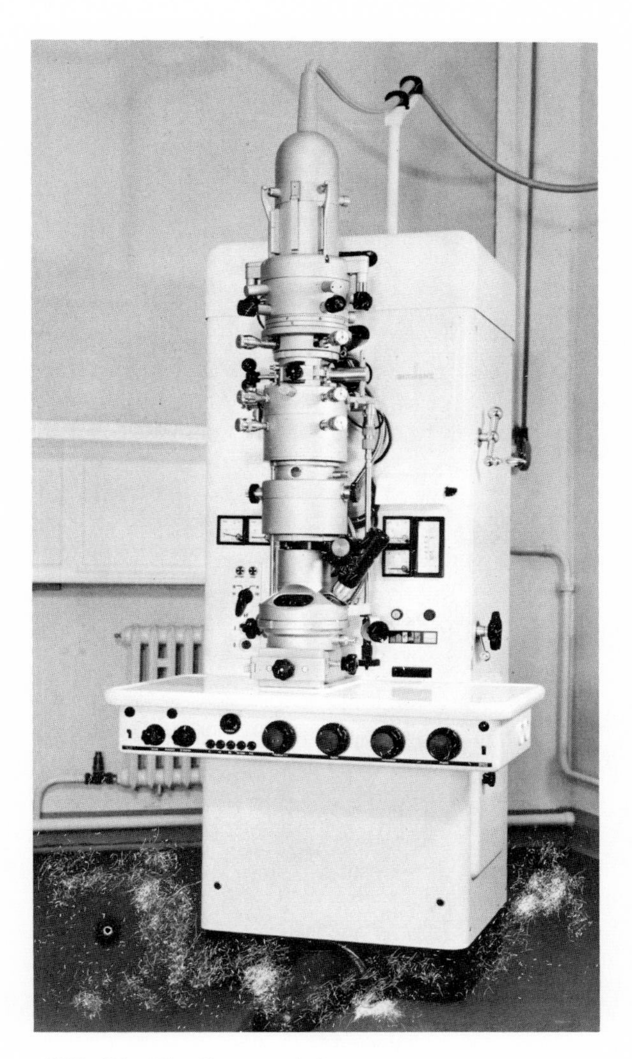

FIG. 57. The Siemens Elmiskop 1 electron microscope.

† For detailed accounts of design and construction see reference 11.

space. However, the main features are similar in each case. Two examples of five-lens (double condenser), high-resolution units are shown in Figs. 57 and 58, and Fig. 59 shows a cross section through the Siemens Elmiskop 1 which we shall use for descriptive purposes. The ray paths in an electron microscope are analogous to those in the light microscope, as shown in Fig. 60. In the five-lens microscope an intermediate lens is

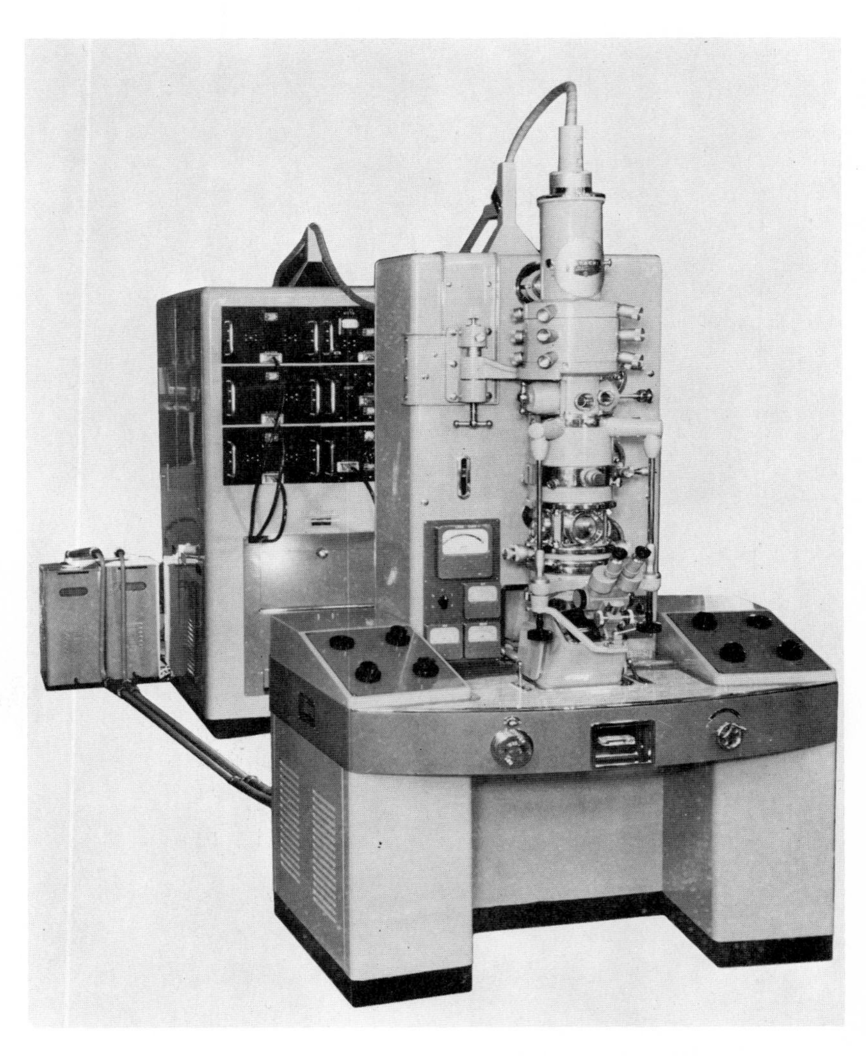

FIG. 58. The Hitachi H.U. 11 electron microscope and power supply cabinet.

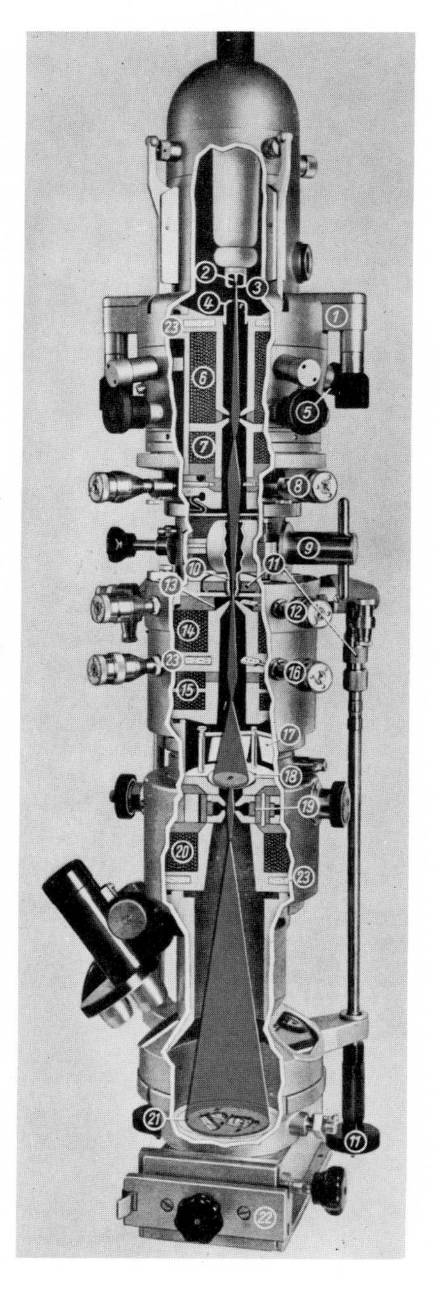

FIG. 59. Cross section of the column of a Siemens Elmiskop 1 electron microscope.
(1) Beam adjusting device
(2) Cathode
(3) Control grid-cap
(4) Anode
(5) Control knobs of electron gun (tilting device)
(6) First condenser coil
(7) Second condenser coil
(8) Variable condenser aperture control
(9) Specimen airlock handle
(10) Specimen cartridge
(11) Object stage and controls for moving specimen
(12) Objective aperture controls
(13) Location of stigmator
(14) Objective lens coil
(15) Intermediate lens coil
(16) Field limiting selected area or intermediate aperture controls
(17) Mirror for observing intermediate image
(18) Intermediate image screen
(19) Projector pole-piece drum and controls
(20) Projector lens coil
(21) Final fluorescent screen
(22) Plate camera
(23) Water cooling

placed between the objective and projector lenses (Fig. 60) to provide a three-stage magnification system. In this, it is usual to fix the projector lens current and to vary the magnification by adjusting only the intermediate lens current. Smaller microscopes (for work not requiring more than 20 Å resolution) are usually made without condenser or intermediate lenses, or with an intermediate and only one condenser lens. An example of this latter type is the Phillips E.M.100B shown in Fig. 61 where the column is arranged in a horizontal position.

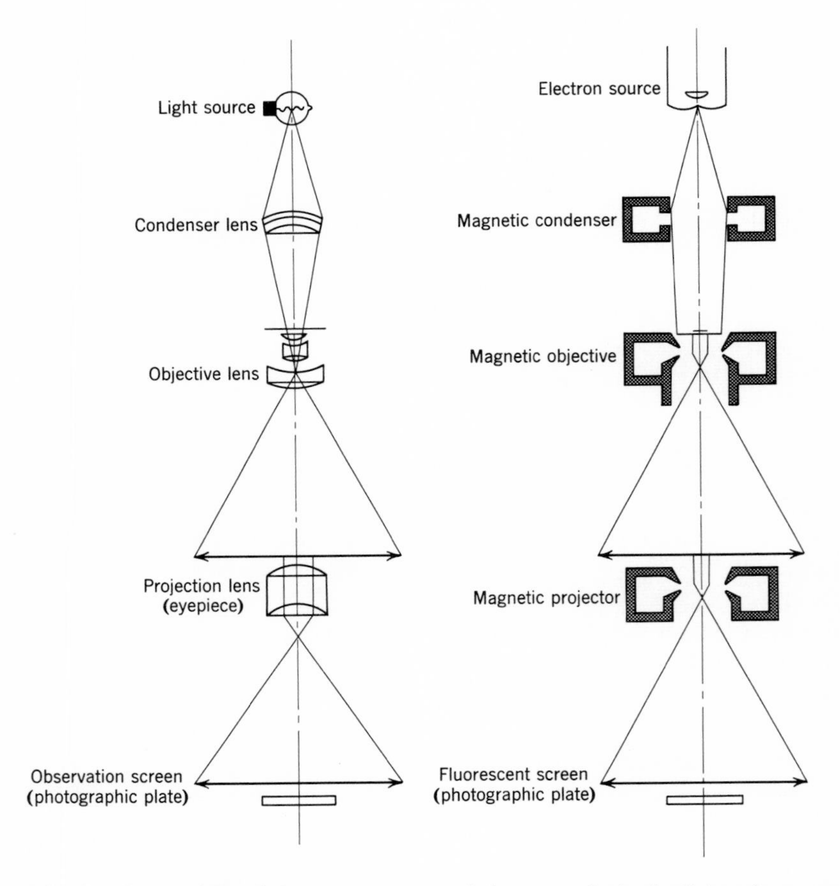

FIG. 60. Comparison of the arrangement and the ray paths in the light microscope and the magnetic electron microscope.

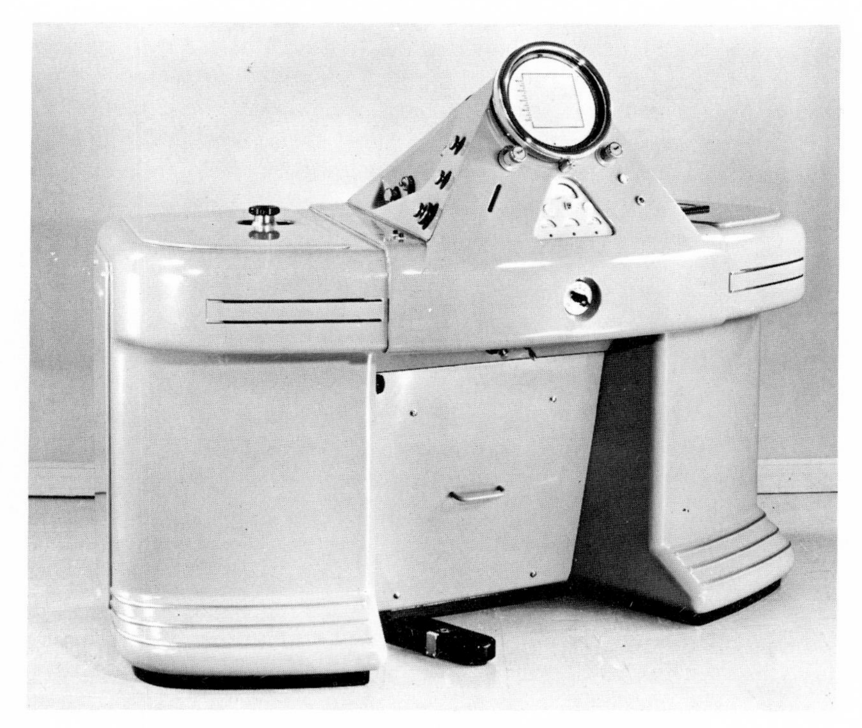

FIG. 61. The Phillips E.M.100B electron microscope.

3.5.1 *General Description*

Figure 59 shows a cross section through a typical high-resolution instrument. The electron source is usually a tungsten wire cathode which when heated emits electrons. These are then accelerated up to -100 kv through the anode. The condenser lenses focus the electron beam to a given diameter (which may be varied in a two condenser lens system) to illuminate the object. The position of the object may be changed by moving the objective stage through controls placed outside the column. Specimens may be quickly changed without loss of vacuum if specimen airlocks are provided (as is usually the case). The objective lens produces a focused, enlarged image which is then further enlarged by the intermediate and projector lenses to give a final image on the fluorescent screen.

Cameras are provided below the screen to allow photographic recording of the image on plates or roll-film. The column is connected to a vacuum system of diffusion and rotary pumps to ensure that a pressure of less than

10^{-5} mm Hg is always maintained during operation. Figure 59 shows the positions of the controls which allow centering of the cathode and alignment of the beam through all the lenses so as to coincide with the axis of the microscope. In many cases the whole of the condenser system can be tilted for reflection microscopy or for dark field illumination. Variable apertures may be inserted below the condenser and objective lenses, and apertures to select any required area of the specimen for diffraction may be positioned between the objective and intermediate lenses. Because of the high currents used in the lens coils, it is usually necessary to water cool the lenses; in which case the incoming water supply must be filtered and softened to avoid corrosion or blockage in the cooling system (see Appendix D). The positions of the features described are shown in Fig. 59. The specific requirements for alignment,[4] changing specimens, apertures, cathodes, plates, etc., for each type of instrument are usually included in the manufacturer's manuals; thus it will not be necessary to go into these here.

3.5.2 *The Electron Gun*

Intense electron sources, with brightness greater than 2×10^4 amp/cm²/unit solid angle,[11] are usually made from tungsten pointed filaments.† Cold cathode sources are not generally employed (except in the Trüb-Täuber instrument) because the voltage fluctuates with the gas pressure in the tube. The temperature dependence of the electron emission I_s, in amperes per square centimeter, of a thermionic emitter is given by

$$I_s = AT^2 e^{(-b/T)} \tag{74}$$

where A and b are empirical constants, with b proportional to the work function[4] (see section 1.5) and T is the absolute temperature. The constant A is relatively unimportant compared with the exponential term. From this it can be seen that the emission is conveniently controlled by the temperature of the filament. With tungsten filaments (melting point 3370° C) the usual operating temperature is ~2200° C. With use, the filaments become thinner because of evaporation and oxidation and eventually break near the tip. The filament is kept pointed to maintain a high specific emission, and although this may lead to a short life (100 to 1000 h), they are readily interchangeable in the control grid. The electrons generated are accelerated up to −100 kv through the anode which is kept at ground potential as shown in Fig. 62. The high voltage is obtained by stepping up the mains ac voltage, rectifying to dc and stabilizing to better than 1 volt fluctuation. The H.V. supply, power transformer, and

† However, there is little doubt that the pointed cathode will soon be used (see reference 12).

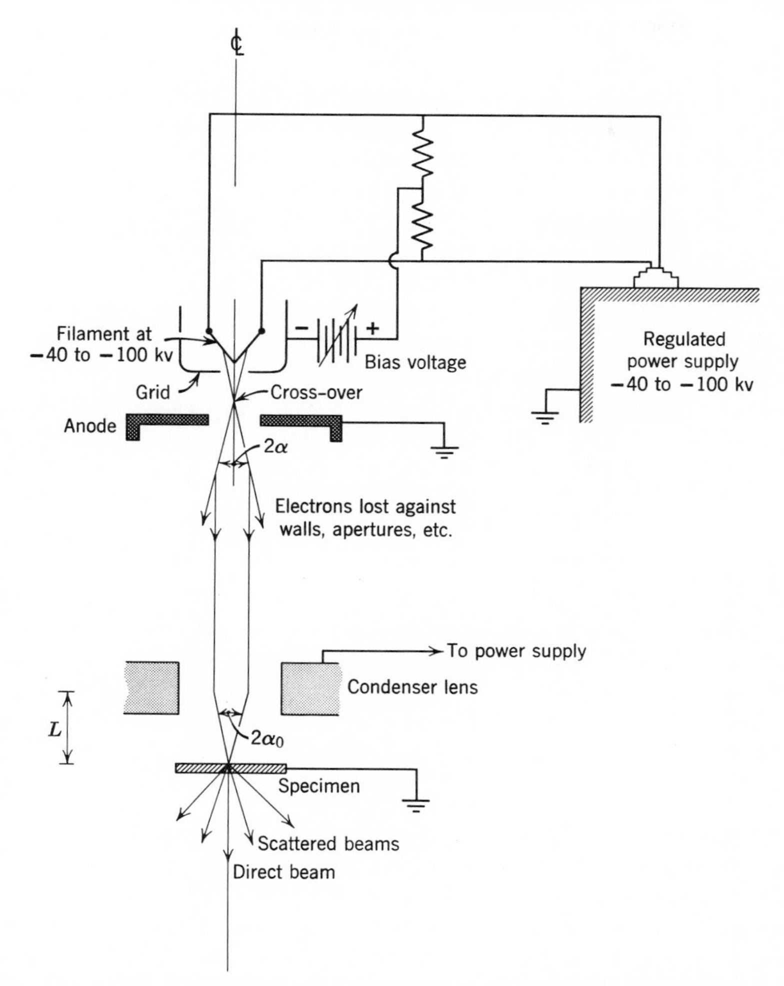

FIG. 62. Schematic illustration showing operation of electron gun and illumination system for a single condenser lens. The gun may be tilted for reflection microscopy or dark field work.

lens current supplies are housed in a cabinet which is usually placed outside the microscope room (see Appendix D).

The beam is narrowed by the first condenser lens (Figs. 62 and 63) so that the anodes are often damaged by the outer electrons and therefore must be kept highly polished. The control grid surrounds the filament and is kept slightly more negative than the cathode. To reduce the beam current the bias is made more negative. The function of the bias is to

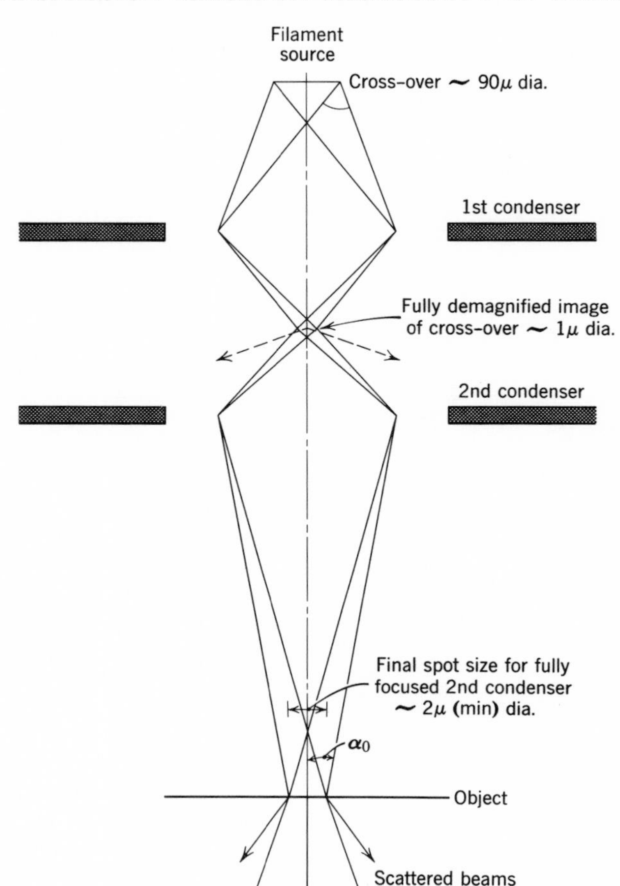

FIG. 63. The double condenser lens system.

provide a limiting aperture for the electrons so that only the central zone of the grid permits passage of electrons (see Fig. 62) since they cannot enter regions of negative potential. Changing the grid voltage, i.e., the bias, has the same effect as changing the size of the grid aperture. The beam suppressing action of the grid is largely supported by the space charge of the beam itself, and the negative charge of the slow electrons in front of the cathode forms a potential barrier which causes slow electrons to return to the cathode.

In most modern microscopes the electron gun is of the self-biased type, and the negative potential across the grid-filament gap is produced by the flow of beam current through a resistor rather than by a battery as indicated in Fig. 62. The potential is of the order of a few hundred volts. For example, for a beam current of 100 μa and a bias resistor of 1×10^6 ohms, the bias potential is ~ 100 volts. The height of the filament tip above the opening in the grid cap may be adjusted to give a small intense source unless the gun is so designed that this gap is always fixed. In general, the filament is 1 to 2 mm above the grid opening.

The filaments are heated electrically; with increasing filament current I_f there is beyond a certain temperature a rapid rise in beam current I_b to a flat maximum, whereafter I_b is practically independent of I_f. Increasing I_f tends to increase I_b, but this in turn would increase the negative bias. When this condition is reached, the filament is said to be saturated. In practice this is attained by gradually increasing the filament current and observing the beam current meter until there is no further change in I_b. During this time the image of the source (unmagnified) on the fluorescent screen will be seen to change as shown in Fig. 64. The position of the anode is determined by the operating voltage, and, as a result of the negative bias, the field about the circular opening in the grid acts as an electrostatic lens to produce a cross-over of small diameter between the grid and the anode as shown in Fig. 62. The image obtained is thus of the cross-over and not of the filament. The diameter d of the cross-over is determined by the accelerating voltage (approximately $d \propto$ K.E. of electrons/V in volts) and is usually less than 0.1 mm.

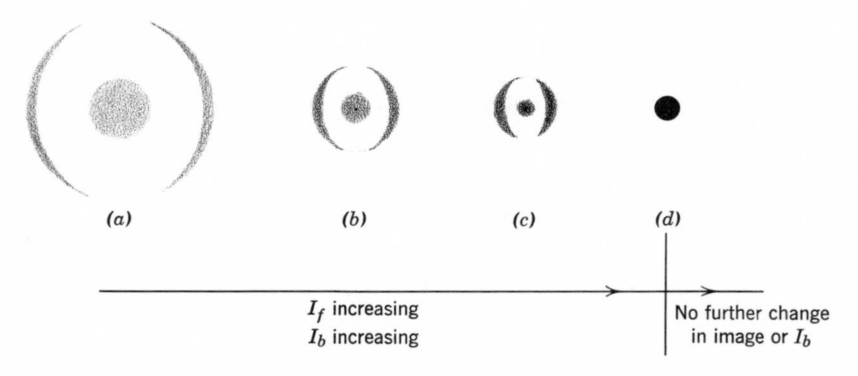

FIG. 64. Schematic representation of emission pattern from a self-biased gun as the filament current is increased to saturation. (*a*) to (*c*) undersaturated, (*d*) saturated. (After Hall.[4] By permission from *Introduction to Electron Microscopy* by C. E. Hall. Copyright 1953, McGraw-Hill Book Co.)

3.5.3 *The Condenser Lenses*

Figure 62 shows the principle of the illumination system of a microscope utilizing a single condenser lens. The strength of the lens determines the position of the image of the cross-over and thus governs the diameter of the beam and the aperture of illumination α_0. Thus if the radius of the cross-over (i.e., beam source) focused by the condenser is x, and f is the focal length of the lens (assumed to be thin), it can be shown[3, 4] that the aperture

$$\alpha_0 = \frac{x}{\{L(L/f - 2)\}} \tag{75}$$

where L is the object distance from the lens (Fig. 62). Using the conventional single condenser lens system, the diameter of the spot focused on the specimen is about 5×10^{-3} cm. Since this is unnecessarily large at high magnifications, a second condenser lens has been introduced to reduce the diameter of the focused spot of illumination to a few microns (see Figs. 59 and 63).

Since for a thin magnetic lens f is inversely proportional to I_c^2, the aperture of illumination and the beam diameter are a simple function of the condenser lens current I_c. The purpose of the first condenser lens in a double condenser system is thus to demagnify the beam source. The maximum demagnification corresponds to imaging the cross-over at a diameter of about 1 μ near the back focal plane of the first lens as shown in Fig. 63. This corresponds to the minimum spot size of the beam. The second condenser lens then projects this source onto the object with a magnification of about two times. Thus the final illuminating spot on the specimen is only about 2 μ in diameter. This is a considerable improvement over the single condenser system, since it is possible to obtain a smaller spot of higher intensity. Furthermore, defocusing the second condenser spreads the beam, thereby decreasing the illuminating aperture α_0, so that it is possible to approximate to axial illumination, giving rise to a smaller beam divergence and consequently sharper diffraction patterns and images (section 1.13.2).

The same effect is obtained when physical apertures are inserted below the second condenser lens (Fig. 59), since the smaller the condenser aperture, the smaller is the value of α_0. Consequently, however, the intensity I of the beam decreases. For a Siemens Elmiskop 1 electron microscope, if the initial beam diameter at the objective is about 10 μ†, we

† In practice, it is convenient to choose a beam diameter so that at the final screen, when the second condenser is focused, the illuminating spot just fills the screen. Thus the first condenser lens current may be selected on the basis of the total magnification.

get the following values for α_0 and I for different sized condenser apertures:[13]

Aperture diameter	α_0 (radian)	I_b (amp/cm^2)
no aperture	9.0×10^{-3}	20
600 μ	4.8×10^{-3}	5.7
400 μ	3.2×10^{-3}	2.5
200 μ	1.6×10^{-3}	0.6

This shows that the minimum value of the beam divergence is $\sim 10^{-3}$ radian.

The advantages of the double condenser lens system may be summarized as follows:

1. Using the double condenser avoids thermal overloading of the specimen.
2. A high intensity beam is produced, thereby obtaining better transmission through metal foils.
3. Only the examined area is illuminated so that nearby areas are not contaminated by the beam. Deposition of carbonaceous material severely impairs resolution.
4. Many metal foils charge up in the beam when large areas are illuminated, thus making focusing impossible. Reducing the beam diameter usually eliminates this.
5. A fine focused beam can cause high thermal stress concentrations in the illuminated area. This can result in the movement of dislocations in thin metal foils. (See Chapter 5.)
6. As we saw in Chapter 1 (section 1.13) the size of the diffraction spot is governed almost entirely by the beam divergence, so that better focusing of diffraction patterns can be achieved by defocusing the second condenser. Thus sharper diffraction spots may be obtained for an equivalent defocusing by using double condensers rather than a single condenser, since this gives less beam divergence owing to a reduced source image. This also leads to higher image resolutions.

3.5.4 *The Objective Lens*

This is the most important part of the microscope, and consists of iron-encased coils with a nonmagnetic gap. Pole pieces of high permeability material are attached to this casing to restrict the effective gap to a very small region around the beam and to provide a short focal length (down to 1 mm). To minimize astigmatism the pole pieces are machined to less than 0.2 μ variation in diameter and the lens currents stabilized to minimize chromatic aberration. The focal length is varied by changing the current in the lens, thereby adjusting the primary magnification. The rotation in the image is proportional to the field strength divided by the axial electron velocity, i.e., H/u or I/u, whereas the focal length[1] is proportional to (u/H^2) or (u/I^2) where I is the current through the lens. Usually the focal lengths are about 2 mm for high-resolution work, but sometimes much longer focal lengths are required when using accessory equipment such as reflecting stages, hot and cold stages, etc., since in these cases the specimen position is too high above the objective to allow focusing using the normal pole piece. Longer focal lengths are obtained by using pole pieces of larger bore. To maintain high contrast variable apertures can be inserted in the back focal plane (Fig. 59) and these must be kept clean and circular at all times.

3.5.5 *Intermediate and Projector Lenses*

These are similar to objective lenses but have longer focal lengths and are used singly or together to give a two- or three-stage magnification of the image formed by the objective. Unlike the objective, however, projector lenses are required to image relatively large areas with rays of relatively small aperture. This is more clearly seen by reference to Figs. 65, 66, and 67. In Fig. 65, a single-stage magnification is provided by means of the objective only. Figures 66 and 67 represent two-stage magnification, and Fig. 68 represents three-stage magnification using all the lenses. Since the projector lens is normally provided with four variable pole pieces, magnifications of $\times 200$ to $\times 150{,}000$ are readily obtainable.

Under certain conditions distorted images may result from projector lenses. Usually the image is perfectly sharp, but for a square grid the image is distorted into barrel or pincushion shapes.[3, 4] This distortion is attributable mainly to spherical aberration (section 3.2.1), and arises from the fact that the aperture of beams arriving at the intermediate or projector lens (with intermediate off) is so small that each beam can be considered as a single ray. Spherical aberration does not, therefore, noticeably affect the sharpness of the image, but their location at the image plane is markedly affected so that the magnification varies through the plane. Obviously, this defect is most noticeable for small lens currents. However, the use of intermediate and projector lenses together, so that the spherical aberration in one lens is compensated by that in the other, will completely correct this defect. For maximum compensation the projector lens must be exactly at the focal point of the intermediate lens.

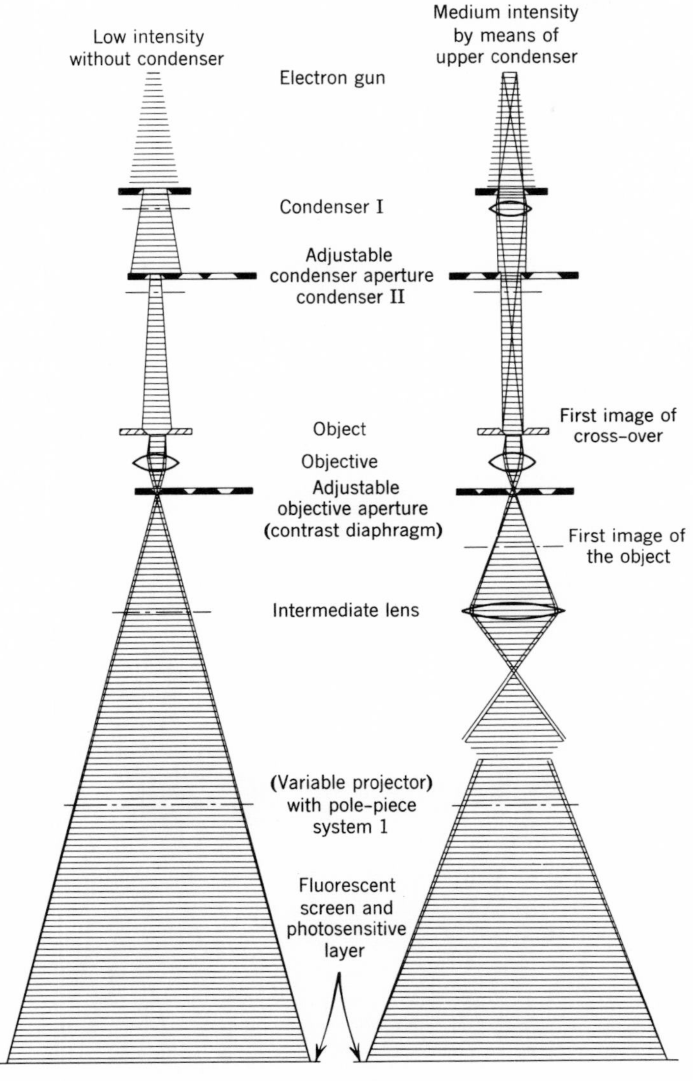

FIG. 65. Single stage magnification with objective lens only, up to approximately 200×. Low intensity without condenser. (Courtesy Siemens and Halske Ltd.)

FIG. 66. Two-stage magnification, using objective and intermediate lenses, up to approximately 2000×. Medium intensity (single condenser). (Courtesy Siemens and Halske Ltd.)

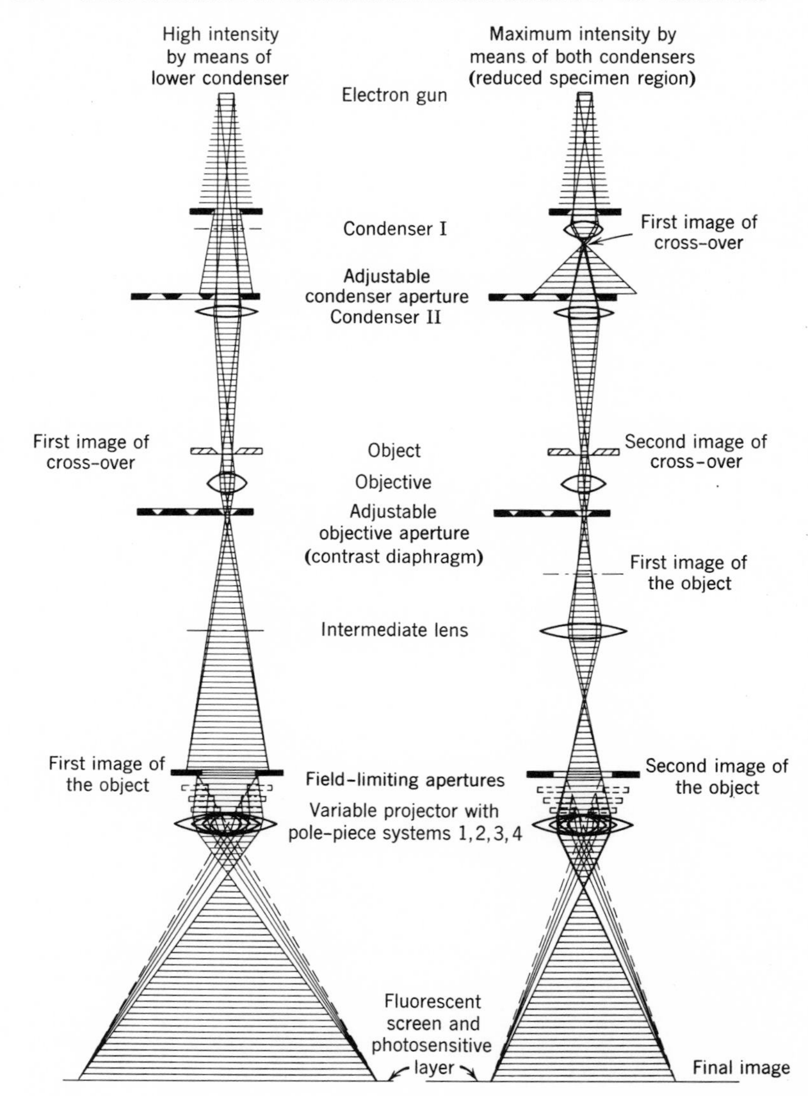

FIG. 67. Two-stage magnification using objective and projector lenses up to approximately 20,000×. High intensity (double condenser). (Courtesy Siemens and Halske Ltd.)

FIG. 68. Three-stage magnification with objective, intermediate, and projector lenses, up to approximately 160,000×. Maximum intensity (double condenser, see also Fig. 63). (Courtesy Siemens and Halske Ltd.)

3.6 Calibration of Microscopes

In order to obtain a continuous range of magnification without changing the projector pole pieces, e.g., from 100 to 1000, 1000 to 10,000, 10,000 to 100,000, etc., it is usual to operate with a fixed projector current and to vary the magnification by changing only the current in the intermediate lens. The magnification can then be calibrated† by using standard gratings of accurately known mesh size and by taking photographs at all the possible combinations of intermediate and projector lens settings and for different accelerating voltages. For relatively low magnifications it is convenient to use a 1000 mesh silver screen since this enables field distortion measurements to be made. For higher magnifications carbon replicas of ruled diffraction gratings may be used and with these it should be possible to maintain an accuracy of about 1 to 2%. These are obtainable up to 50,000 lines per inch and are shadow cast with a heavy metal to give high contrast.‡ This type of specimen is preferred to spherical latex particles since the replica is stable and not subject to line spacing changes. Latex particles tend to change their size and shape because of carbon contamination.

Probably the best system to adopt is to calibrate the microscope as accurately as possible at one well-chosen system of lens current settings, e.g., for a total magnification of $\times 20{,}000$. All other measurements can then be referred to this setting. Figure 69 is an example of a series of calibrations for a Hitachi H.U. 10 electron microscope using ruled diffraction grating replicas. This is obtained for two different projector pole pieces for 50 kv and 100 kv operation keeping the projector lens current fixed and varying only the intermediate lens current. Photographs are recorded for increasing intervals of 10 ma in the intermediate lens current. In this way a complete set of calibration charts may be obtained for all the accelerating voltages available with the instrument and for each projector pole piece. In the same way the rotation of the image with changing lens current can be determined.

When the intermediate lens is uncorrected for rotation, it is more convenient to use a specimen of molybdenum trioxide on a carbon support film. This is obtained by vaporizing MoO_3 onto a crucible lid and floating off the pieces in water before collecting on a specimen grid containing the

† There is no completely satisfactory method for accurate determination of the magnifications because of the peculiarities of electron lenses, e.g., the focal length varies with beam potential, coil currents, hysteresis effects, etc. To maintain reasonable accuracy, therefore, calibrations should be checked from time to time.

‡ In the United States these materials are obtainable from E. F. Fullam, Inc., Schenectady, New York.

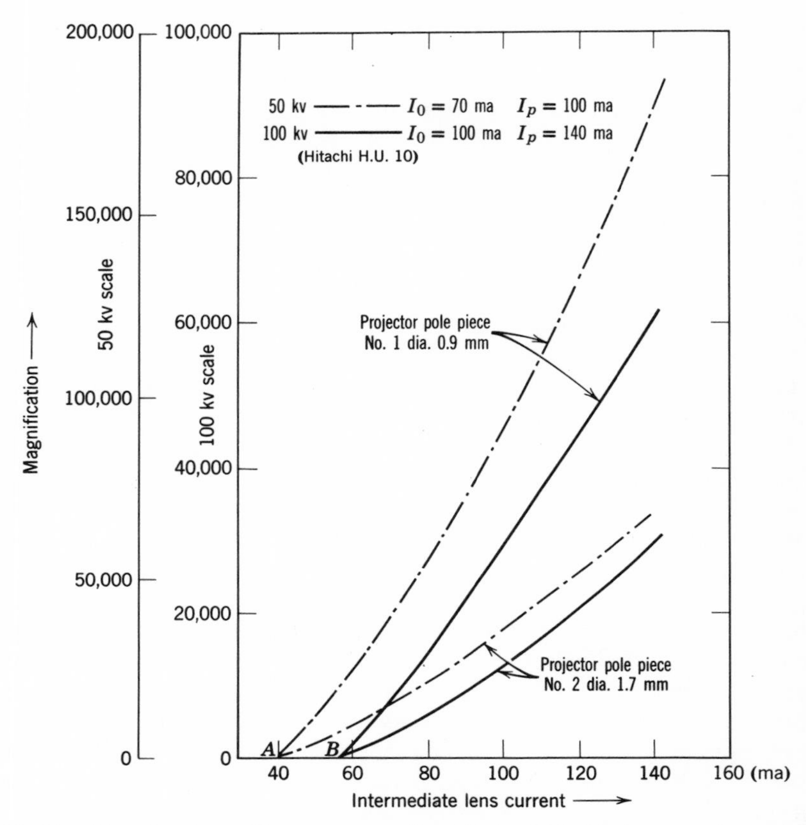

FIG. 69. Magnification calibration curves for Hitachi H.U. 10 electron microscope. Three-stage system for 100 and 50 kv operation. *A*, *B* are intermediate currents for selected area diffraction (cross-over). I_0 and I_p are the objective and projector lens current settings.

support film. The MoO_3 crystals grow as thin, long, flat needles (the structure is pseudo-orthorhombic) with straight edges running parallel to $\langle 100 \rangle$. By photographing a single crystal at different settings of the intermediate lens (projector lens current fixed) and at each stage superimposing the diffraction pattern on the micrograph, the rotation is then easily measured on the photographic plate. An example is shown in Fig. 70. A typical calibration for a Siemens Elmiskop 1 microscope using this technique is shown in Fig. 71. It should be pointed out that these calibrations will at times be in error owing to small differences in the objective current associated with slight variations in position of the speci-

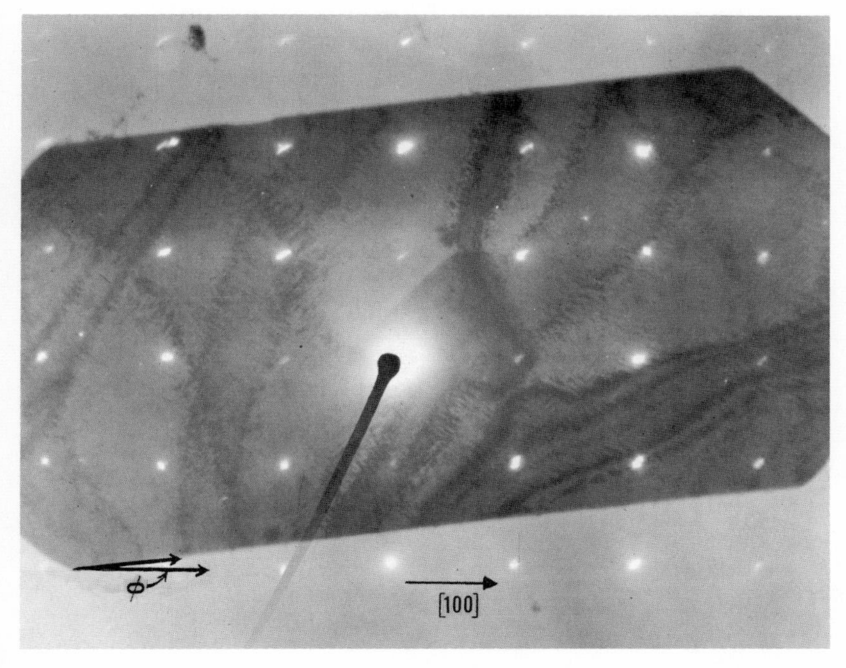

FIG. 70. Single crystal of MoO_3 with its selected area diffraction pattern superimposed (100 kv). The rotation $\phi(=7°)$ is the angle between the edge of the crystal and the [100] rows of spots. (×45,000.)

men, particularly if it is not lying perfectly flat on the grid. However, by plotting the changes in magnification and rotation due to small changes in objective current, this error can be corrected.

3.7 Selected Area Diffraction

Intermediate apertures may be inserted into the intermediate image plane (see Fig. 72) so as to select any required portion of the image. The position of the intermediate image plane depends on the energizing current in the intermediate lens. This current has to be set to a fixed value so that when the objective lens is focused, the intermediate image and selecting aperture are coincident, i.e., when both the image and the edge of the aperture are in sharp focus. As we saw earlier in section 2.2 (Fig. 29), by reducing the strength of the intermediate lens, the back focal plane of the objective is focused on the final screen when the image of the objective

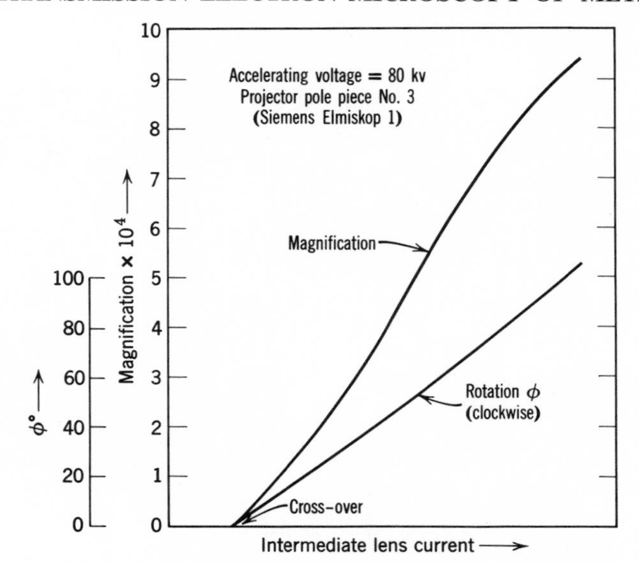

FIG. 71. Magnification and rotation calibration for Siemens Elmiskop 1 electron microscope for 80 kv operation using third projector pole piece.

aperture will be sharp. In this plane the transmission electron diffraction pattern is formed, and the complete pattern is obtained by removing the objective aperture. This is illustrated by the ray diagrams in Fig. 72. The pattern may be sharpened by underfocusing the second condenser.†

When there is a relative rotation between the pattern and the micrograph (due to the change in strength of the intermediate lens) the rotation can be evaluated from a previous calibration using MoO_3, e.g., from Figs. 70 and 71. As can be seen from Fig. 70 the rotation can be obtained directly from the photographs since the edge of the crystal (parallel to [100]) can be seen to deviate from the [100] direction on the diffraction pattern, by the angle of rotation.

In Chapter 1 we saw that diffraction from crystals gives prominent cross-grating patterns (e.g., Figs. 16 and 70) corresponding to a plane in the reciprocal lattice of the structure being examined. It is possible to recognize several distinct zones of reflection in these lattices, and since the crystals or foils are normal to the electron beam, it is possible to determine the normal to the foil from the orientation (determined by indexing

† It is an advantage to place a beam stop over the central spot (see Figs. 14, 70), since, because of its high intensity, it may be difficult to detect any low intensity diffracted beams. This may be removed for photography if an accurate location of the pattern center is to be made.

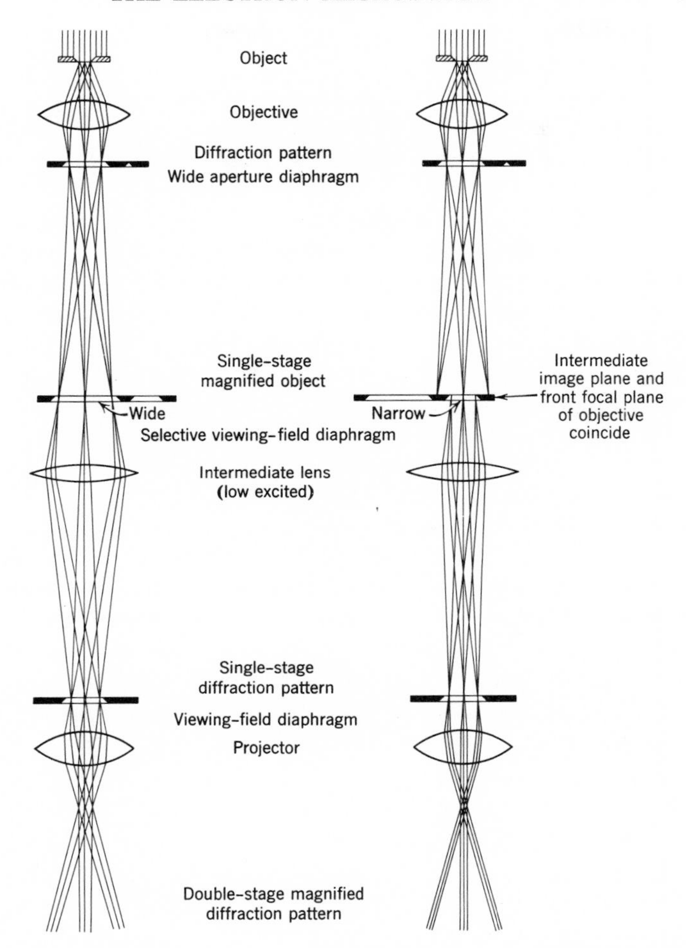

FIG. 72. Selected area diffraction of small area of specimen, to the left without additional aperture in the plane of the intermediate image; to the right with selective aperture in the plane of the intermediate image. (Courtesy Siemens and Halske Ltd.)

the spots). This gives the direction of the beam incident to the crystal. The data are most conveniently represented by graphical interpretation using stereograms,† and an example is given in Appendix A.

The interpretation of diffraction patterns is also facilitated by building models of the reciprocal lattices of the materials under investigation (e.g.,

† A full account of stereographic techniques is given in Chapter 2 of C. S. Barrett's *Structure of Metals* (McGraw-Hill), 1953 (2nd edition).

see Fig. 14*b* for the FCC reciprocal lattice). Orienting the model into the zone axis of the specimen will readily show what reflecting planes are to be expected and also the directions of these planes in the structure. With experience most diffraction patterns can be identified and indexed merely by inspection provided the structure of the specimen is known.

The facility of selected area diffraction thus provides valuable additional information regarding the micrograph. It is to be recommended, therefore, that for each micrograph the corresponding diffraction pattern should also be photographed. When taking selected area diffraction patterns, the following technique is recommended. With a 25 μ objective aperture inserted in the back focal plane, take a photograph of the complete area of the specimen by raising the screen and exposing it for 2 to 5 sec. Lower the screen and select the required area by inserting a suitably sized field-limiting aperture, making sure that this is properly centered in position. Now re-expose this area on the same plate as before. On the next plate, expose the diffraction pattern from this selected area after removing the objective aperture. When the plates are removed from the microscope and developed, the image of the field-limiting aperture will be clearly visible on the first plate, so that the diffraction pattern can be accurately compared to the area selected (i.e., the area inside the image of the aperture). This technique is particularly useful for analyzing specimens containing more than one phase, since in this way each type of precipitate can be separately recognized and identified.

Owing to spherical aberration in the objective lens, two defects may arise in the selected area diffraction pattern. First, at the edges of the field limiting aperture, regions of the specimen do not contribute uniformly to the diffraction pattern and may give rise to asymmetry. Second, areas outside the region defined by the aperture contribute to the pattern and may cause the appearance of extra reflections.† These defects become more important the smaller the area being examined and so great care must be exercised when analyzing the diffraction pattern.

3.8 Contrast

Although we have previously discussed in detail the mechanism of contrast in crystalline materials (Chapter 2), so far we have not considered the effect of the beam voltage. Neglecting diffraction effects for the

† This effect has been recently discussed independently by A. W. Agar and R. Phillips, *Brit. J. App. Phys.* 1960, **11**, pp. 185, 504.

moment and reconsidering Fig. 4 (Chapter 1) we see that because of relativistic effects the scattering curves are much flatter for high beam voltages than for low ones. Although this means that the use of very high accelerating voltages (~1000 kv) does not seem to be advantageous as far as transparency thickness is concerned, it has been reported by the Japanese workers (see *Proc. International Conference Electron Microscopy*, Springer-Verlag, Berlin, 1960) that a considerable improvement in transparency is obtained by using 300 kv electrons. One of the difficulties of very high voltage operation is maintaining stabilization; nevertheless, this technique looks promising from the point of view of examining thick specimens even though the contrast effects may be complicated.

It can be seen from Fig. 4 that the total scattering cross section varies inversely with the voltage, which suggests that lower beam voltages may be used, thereby increasing contrast—remembering that contrast is a result of the subtraction of part of the incident intensity from the imaging beam. At the same time, however, owing to the increased wavelength of electrons, the resolution becomes poorer. In noncrystalline specimens, e.g., replicas, contrast results from variations in the transparency thickness of the specimen (section 4.1.4), and since resolutions not greater than about 20 Å can be expected from such films, the use of lower beam voltages may be an advantage.[4] With metal specimens, however, because of their large scattering cross sections, high beam voltages are necessary to obtain good transmission. In this case, high contrast is obtained by using small objective apertures, e.g., a 30 μ objective aperture will stop off all the Bragg reflections from thin metal foils. Thus it is unlikely that low voltage microscopy will become important in thin foil work.

The Bragg diffraction contrast mechanism enables high-resolution images to be obtained with both bright field and dark field illumination. The dark field image is best obtained by tilting the electron gun through 2 to 3° (corresponding to the Bragg angles) to exclude the direct beam and to allow a Bragg reflection to pass through a centered 30 μ objective aperture. The projector lens then magnifies the image as shown in Fig. 73. This technique enables us to decide whether the contrast arises from a diffraction mechanism and which reflection is operating to produce the contrast.

Superimposed upon the image is the background intensity resulting from the inelastically scattered electrons. Although these do not contribute to the recognizable image, they can introduce chromatic aberration, particularly in metals of low atomic number, where the ratio of the inelastic to elastic scattering is large (see Fig. 3). This defect becomes important when the highest resolutions are to be obtained.

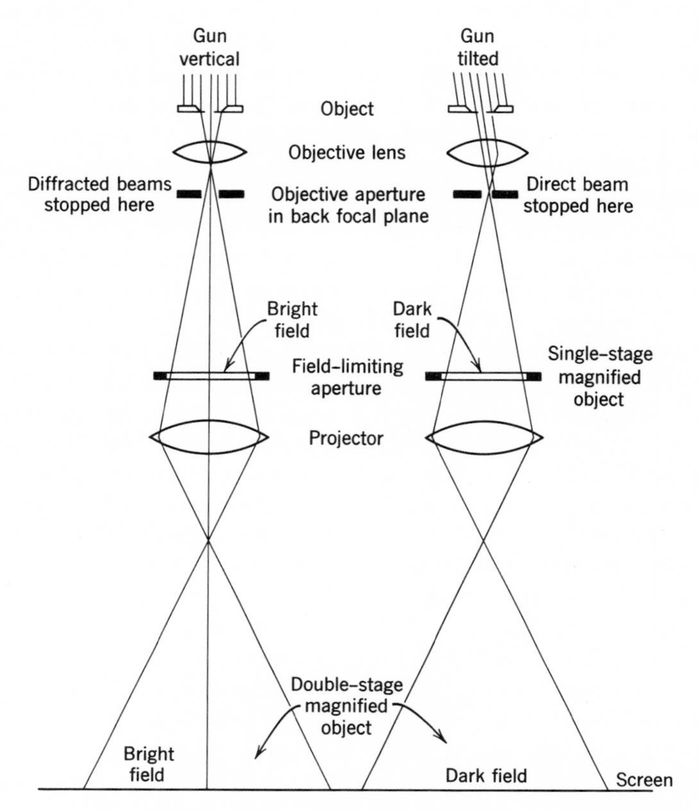

FIG. 73. Illustrating arrangement for bright field imaging (to the left) and dark field imaging with tilted illumination (to the right). (Courtesy Siemens and Halske Ltd.)

3.9 The Specimen Holder

Since contrast in crystalline substances arises mainly from Bragg diffraction, it is necessary to be able to tilt the specimen in order to bring it into a strong diffracting position, particularly for revealing defects in crystals. A tilting device is thus essential for all experiments with thin foils and for correct interpretation of contrast effects (see also section 5.2.14). As we saw in section 2.3.6, tilting the specimen will reveal features not visible at first sight (Fig. 40). It is to be hoped that future microscopes will be provided with goniometer stages so as to be able to tilt the specimen about any given axis. At the present time tilting is available only by swinging the specimen about a vertical axis, usually by means of the stereo-drive mechanism which is normally supplied with the instrument.†

The normal use of the stereoscopic device is of course, to record stereoscopic views of the specimen.‡ Because of the large depth of field good focusing is possible even though the object may be considerably extended in one direction. The technique is to record images of the same area of the specimen at two positions equally inclined on opposite sides of the microscope axis, the specimen having been previously mounted in the stereo-holder. For this to be efficient, the maximum tilt of the specimen should be $\sim 10°$. When viewed through a stereopticon, after allowing for any rotation, the two pictures produce the visual effect of a three-dimensional object. In some respects stereoscopic pairs can be valuable aids to interpretation, e.g., in deciding between "hills" or "valleys" on the specimen surface, and they are useful for work with surface replicas or biological specimens. Actually the existence of specimen tilting devices in commercial electron microscopes really stems from stereoscopic requirements rather than from contrast considerations. This is not surprising considering that stereoscopic techniques were widely used by workers in the medical and biological fields long before the advent of thin metal foils.

3.10 Cleaning the Microscope

It is useful here to point out some of the more convenient methods for cleaning the parts of electron microscopes in contact with the beam, as this forms a routine, but essential, part of all electron microscope work. Cleaning is necessary to avoid charge-up effects and to reduce to a minimum spherical aberration and astigmatism. The techniques are as follows:

3.10.1 *The Gun*

The ceramic insulator around the H.V. supply to the cathode becomes contaminated with brown deposits after prolonged use owing to the decomposition of carbonaceous matter in the impure vacuum. This may be removed with a chamois leather moistened with benzene, provided the

† Tilting devices for use in the Metropolitan-Vickers E.M.3 microscope have been described by G. R. Booker and J. Norbury (*J. Sci. Instr.* 1959, **36,** p. 368), R. E. Burge and H. R. Munden (*ibid.* 1960, **37**, p. 199), and for the Hitachi H.U. 10 by M. C. Huffstutler and G. Thomas (*Rev. Sci. Instr.* 1961, **32**, p. 86).

‡ The stereo-mechanism usually consists of a push-rod, which enters the microscope at a short distance above the objective pole piece and engages a spring-loaded specimen holder. Thus by screwing the push-rod inwards the specimen is tilted about a vertical axis. Upon unloading, the spring returns the specimen to its normal position.

contamination is not allowed to build up over a long period of time. If this is not removed, discharging will occur in the cathode chamber ("flashing"), resulting in beam drift and instability. More tenacious deposits may be removed (taking great care not to damage the insulating shield) by rubbing lightly with a smooth cloth impregnated with polishing wax, which is soluble in an organic solvent such as benzene or alcohol, so that residual traces can be finally removed. After this the insulator should always be polished lightly using a dry chamois leather.

After removing the grid from the cathode, the grid cylinder may be cleaned by first dipping in 5% NH_4OH (which removes tungsten oxide deposits), washing in water and a dilute soap solution, drying off with alcohol, and polishing with a clean, dry, chamois leather. The anodes may be cleaned in a similar manner. All other metal parts in the cathode and anode chambers should be cleaned with a chamois leather moistened with benzene followed by polishing with a dry chamois.

3.10.2. *Lens Pole Pieces*

The condenser and objective pole pieces do not usually become contaminated because of the protection provided by the apertures. However, dust, etc., can be removed by wrapping a piece of lens tissue around a cocktail stick, dipping this into alcohol, and gently wiping the inside areas with the tissue. The fixed apertures can be removed from the pole pieces and cleaned, by wrapping a piece of cotton wool impregnated with a soluble polishing medium around a cocktail stick sharpened to a fine point, and polishing the aperture hole and surrounding parts. Any residual polish on the apertures may then be dissolved in a suitable organic solvent. The aperture in the grid cylinder, the intermediate and projector pole pieces and the whole of the specimen holder may also be cleaned in this way. However, great care must be exercised so as not to damage the pole pieces during cleaning.

3.10.3 *Adjustable Apertures*

If these are made of molybdenum, they can be cleaned by heating them to white heat in a vacuum system (such as an evaporator). For this purpose the apertures are previously mounted in a molybdenum boat. When the apertures are made of platinum, they can be cleaned by placing them in a platinum basket and heating to white heat over a bunsen flame. If this does not remove all the contamination, it is necessary to use more severe methods. For example, the following sequence usually works well: (1) boil the apertures with solid sodium bisulfite until white fumes of SO_3

evolve; (2) dissolve the residues from (1) in H_2SO_4 and boil again; (3) remove apertures and wash in boiling water; (4) then boil in HNO_3; (5) wash in boiling water; and (6) finally clean in alcohol and flame dry. To ensure the apertures being perfectly clean they should be examined in an optical bench microscope before being returned to the electron microscope.

Some of the sources of trouble due to dirt occur in the slots in the lens pole pieces through which the apertures are inserted. These should be carefully cleaned using lens tissue wrapped round a cocktail stick, making sure that no pieces of tissue, dust, hair, etc., are left behind before returning the pole pieces to the instrument.

For all high-resolution work the parts of the microscope should be disassembled and cleaned in the manner described before each program of investigation. The gun should be cleaned daily and the apertures weekly. The time and trouble taken over routine cleaning are well worth the effort and will ensure constancy of high-resolution operation at all times. After cleaning and reassembly of the column, pumping for a few hours using only the rotary pumps will allow complete out-gassing of any residual volatiles.

3.11 Microscope Accessories

In present-day research problems it is an advantage to be able to carry out dynamic experiments on specimens, e.g., plastic deformation, heating, and cooling. Many of the modern microscopes are equipped with heating and cooling devices, and here again it is most useful to be able to tilt the specimen to maintain contrast during expansion and contraction. An example of a heating stage with this facility is shown in Fig. 74, and may be used in the Hitachi H.U. 10 and 11 electron microscopes. A 6-volt battery or mains dc supply provides the necessary electrical heating to the furnace *A*, in which the specimen may be mounted prior to operation. Tilting about two horizontal axes may be done using the controls *C* and *D*, and the specimen may be traversed using controls *E* and *F*. A calibrated thermocouple at *G* enables the temperature to be read off directly. The available cold stages also work on the same principle, e.g., see Watanabe et al.[15] Experimental hot stages have been made by Whelan[16, 17] and Pashley[18] for use in the Siemens Elmiskop 1 microscope, but as yet these cannot be tilted while in use. Some of the applications of high temperature work are described in Chapter 5.

The most effective deformation stage at present available is that designed by Wilsdorf[19, 20] for use in the Phillips desk model microscope (E.M.100B).

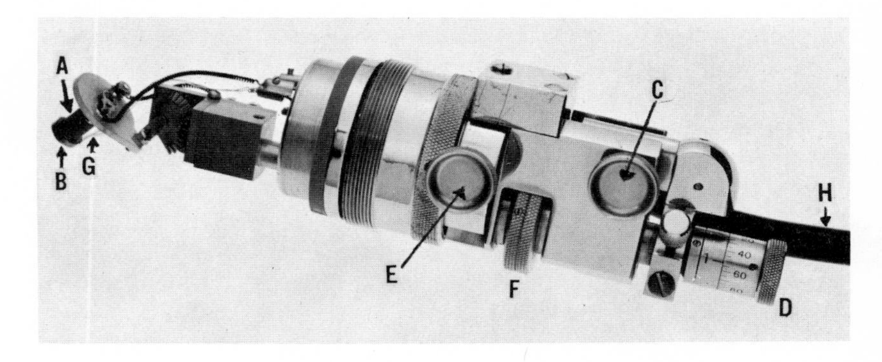

FIG. 74. High temperature stage for Hitachi H.U. 10 and 11 electron microscopes. *A*, furnace; *B*, specimen holder; *G*, thermocouple; *E*, horizontal adjustment; *F*, adjustment to move furnace along direction normal to *E*; *C*, tilting control; *D*, raise or lower control for furnace; *H*, heating current leads. (Courtesy Hitachi Ltd.)

The standard specimen holder is adapted as a straining device as shown in Fig. 75 so that when a rod *R* is inserted through the specimen holder and a movable shoe (*MS*) is attached to the end of it, a specimen mounted on the shoe can be pulled in tension by unscrewing the rod. Straining devices based on opening out a set of plates in a stereo-holder have been used in Siemens microscopes by Weichan,[21] Pashley,[22] and Fisher,[23] and a simple straining device for the Metropolitan-Vickers E.M. 3A has been described by Forsythe and Wilson.[24] Microscope manufacturers now provide straining devices as a standard accessory, but it usually requires the ingenuity of the research worker to design his own deformation stage to fit the space limitations in the specimen chamber of the particular microscope being used. Facilities for deforming specimens at high temperatures should also become available.

Many electron microscopes are convertible into reflection microscopes and high-resolution diffraction cameras. Although reflection microscopy suffers from a resolution limit of about 200 Å,† investigations of the surfaces of bulk specimens too thick for transmission are possible, and because of the strong scattering from projections on the specimen surface, the image contrast is exceptionally high. Figure 76 shows an example of this—the shadows are cast from an extrusion in aluminum after fatigue deformation.[25] Reflection microscopy thus affords a convenient method

† By adjusting the angles of illumination and viewing so that the diffracted electrons contribute to the image, Halliday and Newman (*Brit. J. Appl. Phys.* 1960, **11**, p. 158) have succeeded in improving the resolution in reflection microscopy to about 80 Å.

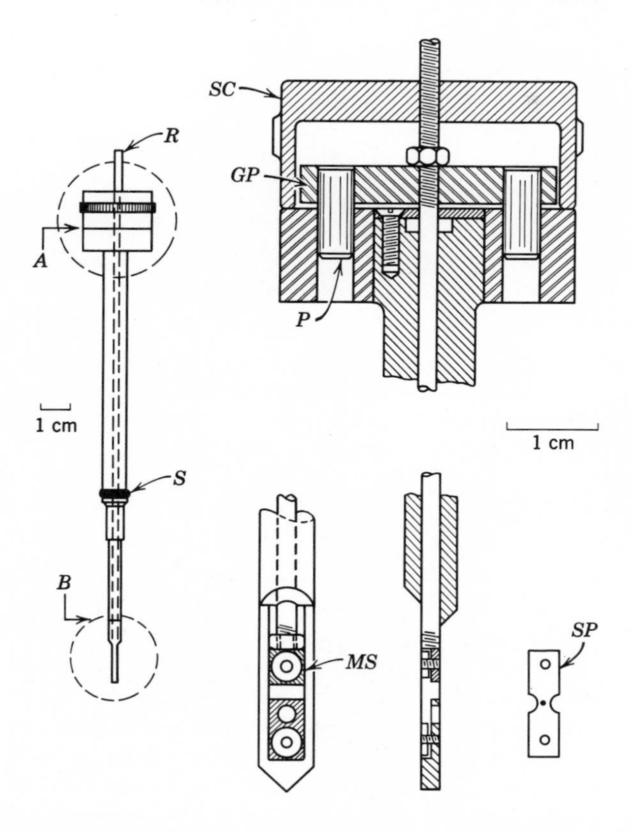

FIG. 75. Sketch of straining device for Phillips E.M. 100B electron microscope Cross-sectional view of movement controls *A* and tip *B* with clamps. *S*, vacuum seal; *SC*, screw cap; *GP*, guiding plate; *P*, guiding pin; *MS*, movable shoe; *SP*, specimen. (After Wilsdorf.[1])

of studying surface relief effects in metals without using involved methods of preparing specimens.†

From the foregoing we can conclude that electron microscopes are extremely versatile in performance, giving high resolution and, with accessories, providing a means of looking into the dynamic behavior of crystals under the influence of stress and/or temperature.

† See also references 136 to 142, Chapter 5.

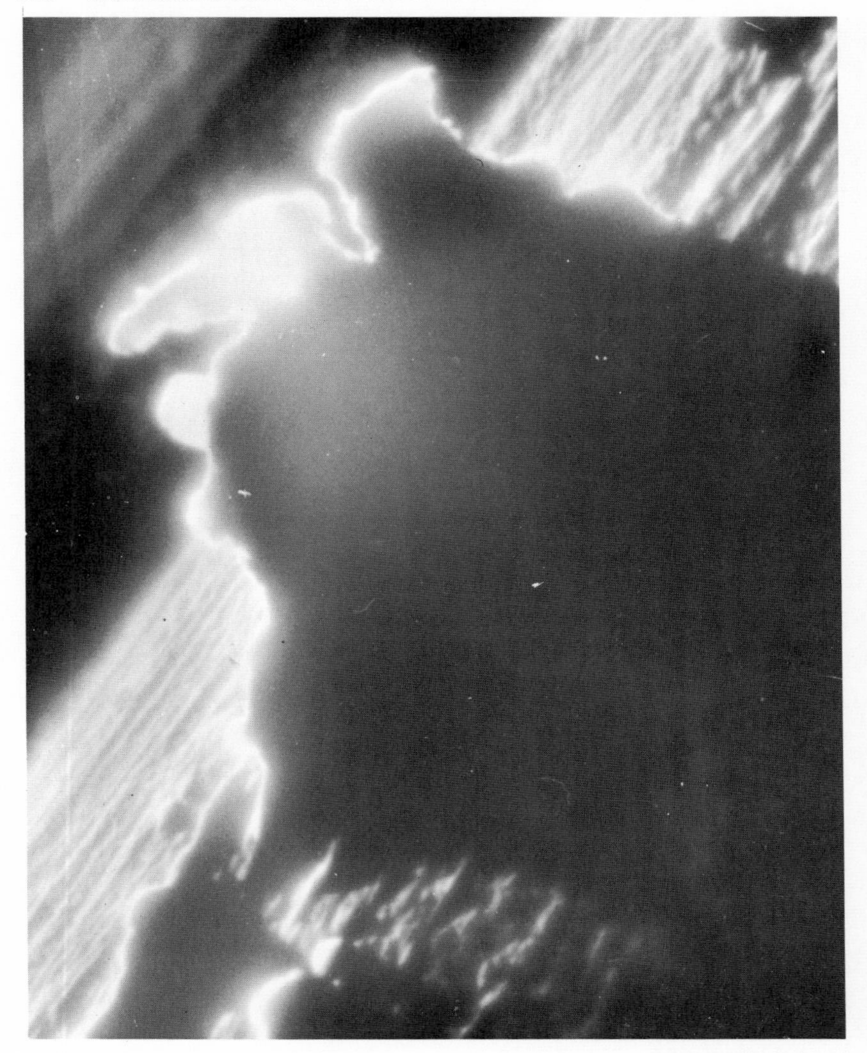

FIG. 76. Reflection electron micrograph of an extrusion from a slip band in fatigued aluminum; linear magnification ×3500.

3.12 Ciné Techniques

Ciné techniques provide admirable means of recording dynamic processes which may be taking place in the specimen while it is under observation.[26] Applications to plastic deformation[27] and annealing experiments[17, 28] have already been carried out. The availability of a double condenser lens system for fine focus illumination gives enough intensity at the final fluorescent screen to enable ciné recordings to be made at instrumental magnifications up to ×80,000. For convenience, however, it is usual to operate at an instrumental magnification of ×20,000 when the brightness is quite adequate for visual observation of the image during filming.

In order to place the ciné camera as close as possible to the viewing window on the microscope, it is advisable to use a short focal length lens (e.g., a 1-in. *f*/0.95 Anferieux lens). The camera may then be supported on a tripod or on a frame sitting on the microscope table. The fastest available ciné negatives should be employed, using speeds of 16 frames per second. The fluorescent screen should be clean and of the highest possible intensity. It is useful to place a 1 mm mark at the center of the screen in order to calibrate the magnification easily when projecting the final, processed film. During exposure it is necessary to tilt the screen so as to bring it into a plane normal to the axis of the ciné camera. This is done by means of the handle normally used to raise the screen for exposing plates.

3.13 Operational Requirements

The main features of the electron microscope have now been described, and for these to function properly in order to obtain the highest quality electron micrographs, certain mechanical and electrical requirements must first be fulfilled. These include accurate alignment of the various sections with respect to the beam, preservation of mechanical, electrical, and thermal stability, attainment of high vacuum, and routine cleaning and maintenance.[2, 4] Although these requirements are generally satisfied by following the instructions in the manufacturers' handbooks the following hints may be found useful.†

Before switching on, it is advisable to ensure that all the variable apertures and the specimen are removed from the path of the beam and that the largest bore projector pole piece is in position. Alignment of the gun with the beam-adjusting device may then be easily made so that with all lenses off, the beam exactly concides with the axis of the instrument. By increasing the filament current, the intensity build-up to the saturation level should occur with no sweep of the beam. To avoid damaging the screen by the intense beam the filament current may then be turned down to obtain a moderately bright beam. The first condenser lens may now

† See Fig. 59 for positions of the features mentioned in this section.

be switched on and the current set to give the required spot size, after which the second condenser may be turned on. The condenser system is then aligned by mechanical or magnetic controls so that when focusing or defocusing the second condenser, the beam spreads (or diminishes) concentrically without any sweep of the illumination. The variable condenser aperture is then inserted, and when aligned its image should also change concentrically as the current in the second condenser is altered. Centering may be achieved by moving the aperture into the direction of deviation from concentricity. The illumination system is now ready to be aligned with the objective, intermediate, and projector lenses as each is switched on and successively centered. After this procedure the filament current may again be raised to saturation and the microscope is then ready for use.

During operation, if the instrument is well aligned, it is only necessary to use the condenser controls to keep the beam centered. The brightness of the image is varied by changing the current in the second condenser lens. To center the objective aperture it is advisable to set the instrument for selected area diffraction with the specimen in position. In this way the aperture is quickly placed around the zero-order beam (bright field) or around the first order beam (dark field)† when viewing the diffraction pattern on the fluorescent screen. The screens (usually made of fluorescent material such as zinc sulfide coated on a backing plate with a fixative) may deteriorate with time owing to exposure to the beam. These are fairly easily recoated, e.g., with a suspension of fine particles of zinc or cadmium sulfide in a 0.1% solution of collodion in acetone, so that replacement is not a serious problem. For direct recording of the image using ciné cameras, screens of the highest intensity must be employed. Photographic plates must also be of high intensity, fine-grained, and yield good contrast without loss of detail so that the exposure time may be as short as possible.

During use, the variable condenser and objective apertures become contaminated; thus constant attention to cleanliness and preservation of perfectly circular apertures is necessary for reducing astigmatism and spherical aberration to a minimum. It is advisable to insert the smallest condenser and objective apertures prior to photographing the image to increase both resolution and contrast even though lower brightness may result. These can then be quickly replaced by larger apertures for general scanning of the specimen. Consequently, it is recommended that the variable condenser aperture holder should contain 600 μ, 400 μ, and 200 μ

† For high-resolution work it is better to tilt the illumination system to bring a diffraction spot to the center of the screen than to move the objective aperture, since in the latter case spherical aberration is introduced by an off-centered aperture.

apertures and the variable objective aperture holder 100 μ, 50 μ, and 25 μ apertures. Similarly, the intermediate apertures (field-limiting apertures for selected area diffraction) should also have a size range to facilitate selecting large and small fields of view. For the highest resolutions, when it is necessary to work at high magnifications so that the intensity of illumination is low, visual focusing of the image is difficult; in fact, for dimensions less than 10 Å, even impossible. In this case it is necessary to take a through-focus series of photographs, selecting the best one after developing the plates. This operation must be carried out as quickly as possible because contamination of both specimen and apertures during exposure will adversely affect the resolution.†

Before we can make all these observations and carry out refined experiments, it is necessary to be able to prepare thin specimens for transmission examination; herein lies the art of the electron microscopist. He must be prepared to devote much time and labor to perfecting preparation techniques, to be able to obtain clean surfaces which represent in the final thin specimen the conditions prevailing in the bulk sample. One of the chief difficulties is removing oxidation and chemical products from the surfaces of the specimens. Much attention has recently been paid to thinning techniques for metals. In the following chapter we shall discuss these, indicating the advantages and disadvantages in each case.

References

1. V. E. Cosslett, *Electron Optics* (Clarendon Press, Oxford), 1950.
2. V. E. Cosslett, *Practical Electron Microscopy* (Butterworths Scientific Publications, London), 1951.
3. V. K. Zworykin, G. A. Morton, E. G. Ramberg, J. Hillier, and A. W. Vance, *Electron Optics and the Electron Microscope* (John Wiley and Sons, New York), 1945.
4. C. E. Hall, *Introduction to Electron Microscopy* (McGraw-Hill Book Co., New York), 1953.
5. J. W. Menter, *Advances in Physics* 1958, **7**, p. 299.
6. M. Born and E. Wolf, *Principles of Optics* (Pergamon Press, London), 1959.
7. H. Boersch, *Z. Phys.* 1943, **44**, p. 202.
8. M. E. Haine and T. Mulvey, *J. Sci. Instr.* 1954, **31**, p. 326; *ibid.* 1957, **34**, p. 9.
9. R. Rebsch, *Ann. Physik* 1938, **31**, p. 551.
10. R. B. Nicholson and J. Nutting, *Phil. Mag.* 1958, **3**, p. 531.

† Voltage alignment is essential for high-resolution work. To adjust this an ac voltage may be superimposed on the accelerating voltage so that the final image oscillates concentrically with a period equal to that of the ac voltage. The center of the concentric circles is then brought to the center of the screen by means of the lateral adjusting screws. The ac voltage is switched off during examination of specimens.

11. B. v. Borries, *Proc. 3rd Int. Conf. Electron Microscopy* 1954 (Roy. Mic. Soc., London, 1956), p. 4; S. Leisegang, *ibid.* p. 184; M. E. Haine, *Adv. Electronics* 1954, **6**, p. 295.
12. M. Dreschler, V. E. Cosslett, and W. C. Nixon, *Proc. 4th Int. Conf. Electron Microscopy* 1958 (Springer-Verlag, Berlin, 1960), p. 13.
13. M. J. Whelan, *Ph.D. Thesis*, Cambridge University, 1956.
14. J. W. Menter, *Proc. Roy. Soc.* A 1956, **236**, p. 119.
15. M. Watanabe, I. Okazaki, G. Honjo, and K. Mihama, *Proc. 4th Int. Conf. Electron Microscopy* 1958 (Springer-Verlag, Berlin, 1960), p. 90.
16. M. J. Whelan, *ibid.*, p. 96.
17. M. J. Whelan and J. Silcox, *Phil. Mag.* 1960, **5**, p. 1.
18. D. W. Pashley and A. E. B. Presland, *J. Inst. Metals* 1959, **87**, p. 419.
19. H. G. F. Wilsdorf, *Rev. Sci. Instruments* 1958, **4**, p. 323.
20. H. G. F. Wilsdorf, L. Cinquina, and C. J. Varker, *Proc. 4th Int. Conf. Electron Microscopy* 1958 (Springer-Verlag, Berlin, 1960), p. 559.
21. C. Weichan, *Z. Wiss. Mikroskop* 1955, **62**, p. 147.
22. D. W. Pashley, *Proc. 4th Int. Conf. Electron Microscopy*, 1958 (Springer-Verlag Berlin, 1960), p. 563; also *Proc. Roy. Soc.* A 1960, **255**, p. 218.
23. R. M. Fisher, *Rev. Sci. Instr.* 1959, **30**, p. 925.
24. P. J. E. Forsythe and R. N. Wilson, *J. Sci. Instr.* 1960, **37**, p. 37.
25. G. Thomas, N. P. Sandler, and I. Cornet, *J. Inst. Metals* 1961, **89**, p. 253.
26. R. W. Horne, *Sond. Aus. Research* 1959, **3**, p. 150.
27. P. B. Hirsch, M. J. Whelan, and R. W. Horne, *Phil. Mag.* 1956, **1**, p. 677.
28. G. Thomas and M. J. Whelan, *ibid.* 1961, in press.

4

Preparation of Specimens for Transmission Electron Microscopy

4.1 Replica Methods

Although the main purpose of this book is to bring the reader up to date with modern methods of examining metals in direct transmission, it is important to mention some of the replica techniques which have stood the metallurgist in such good stead over the past fifteen years. The advent of thinning techniques has not necessarily meant the eclipse of replicas, but, of course, the advantage of being able to see directly into the interior of metals and alloys means that more people are using and will want to use thin foils rather than replicas. Not all laboratories, however, may be equipped with high-resolution, double condenser lens electron microscopes which are the most suitable for examining metal foils in transmission. In this instance, therefore, surface replicas may still be the only available means of studying materials at resolutions better than can be obtained with the light microscope. It must also be kept in mind that even in the electron microscope the best replicas will only resolve 20 Å or so and then perhaps unfaithfully, and will only show surface structures. Foils thinned down from bulk material, however, have no such restrictions. Foils also have the added advantage of showing up the nature and distribution of defects as well as being suitable for electron diffraction. Nevertheless, of all the replica methods now available, at least for ferrous alloys, the carbon extraction replica technique is still to be strongly reckoned with, for as we shall see, much added information about the chemical nature of constituents may be obtained using fluorescent analysis after examination by electron microscopy and diffraction.

We shall now consider very briefly some of the more widely used replica techniques[1] before concentrating in detail on methods for preparing thin sections of metals and alloys.

4.1.1 *Plastic Replicas*[2-7]

For routine metallographic work requiring a resolution of not greater than 200 Å the dry-stripped replica method is simple, rapid, reproducible, and does not involve destruction of the sample. The method involves careful metallographic polishing and etching, cleaning the surface, and dropping a plastic solution (e.g., Formvar, Collodion, Parlodion,† or polystyrene and silica) onto the surface. The replica which represents a "negative" of the specimen surface (see Fig. 84*a*), can then be dry-stripped using cellulose tape. The resolution limit of these replicas is determined by the size of their molecular structure, and they decompose by polymerizing when the intensity of the electron beam is high. They are also difficult to strip from deeply etched surfaces and tear rather easily. In general, the contrast with plastic replicas is not good, but this can be improved by shadowing with heavy metals such as gold, platinum, and chromium. This is done by evaporation onto the replica at a predetermined angle (see Fig. 77). If it is assumed that the metal evaporates uniformly in all directions, then if M is the total weight of metal evaporated, ρ is the density of the metal, θ is the angle, and R the filament-specimen distance, then the thickness of the deposit is approximately[8]

$$T = \frac{M \sin \theta}{\rho 4 \pi R^2} \tag{76}$$

The development of carbon replicas with a resultant increase in resolution has now almost entirely replaced the use of plastic replicas.

4.1.2 *Carbon Replicas*

These are universally applicable to all metals, alloys, and indeed any kind of solid surface, and are now widely used for work with ceramics. The two-stage preshadowed technique first developed by Bradley[9] and Smith and Nutting[10] was soon replaced by the direct method of evaporating carbon onto the surface of the metal specimen.[11] The technique is carried out as follows:

The polished and etched sample is placed in an evaporating unit, capable of reducing the pressure to 10^{-4} mm Hg, with two carbon rods (one blunt,

† These are trade names for the following substances: *Formvar*, polyvinyl formal or polyvinyl acetal resin—solution of 1–2% in dioxane. *Collodion*, pyroxylin, cellulose nitrate—4% solution in 3 parts ethyl ether and 1 part ethanol. *Parlodion*, pyroxylin, cellulose nitrate—2% solution in amyl acetate.

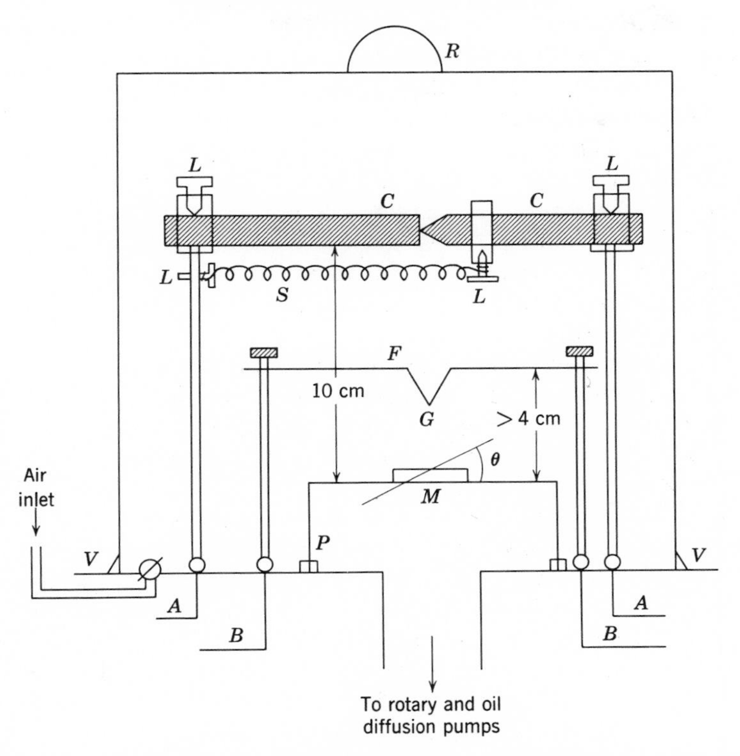

FIG. 77. Schematic sketch of evaporating unit. *A*, electrical leads to carbon rods (for carbon replication); *B*, electrical leads to filament (for metal evaporation); *C*, dense carbon rods; *F*, filament (tungsten wire); *G*, tungsten basket (or molybdenum boat); *S*, spring to maintain contact between carbon rods during evaporation; *M*, specimen which can be tilted through $\theta°$ (e.g. self-shadowing); *P*, specimen stand, rotatable; *R*, glass bell jar; *L*, clamps; *V* vacuum seal.

the other sharp) just touching each other, placed about 10 cm above the sample (Fig. 77). At low pressures evaporation of carbon for a few seconds by resistance heating at the contact point produces a replica only a few hundred angstroms thick, with a resolution of about 20 Å. If the current setting to the carbon rods is previously adjusted to give an input of 50 amp, switching on for about 3 to 4 sec produces a film about 200 Å thick. To judge the thickness of the film it is useful to place a clean glass slide with a drop of oil on it on a piece of white paper alongside the specimen. Since carbon will not form a film on the oil droplet, the whiteness of the paper continues to show through the droplet during evaporation, whereas the other areas become progressively more deeply tinted. When this tint has become light gray, evaporating may be stopped. A similar procedure may be adopted for making carbon support films, in this case the "specimen" is a clean glass slide. After evaporation the slide is removed from the evaporator and squares scratched on the carbon film of size comparable to the dimensions of the microscope grids. These are then floated off in water. Since carbon films are very hydrophobic, the addition of a few drops of alcohol or wetting agent to the water prevents the films from rolling up. The films can then be caught on microscope grids and used to collect molybdenum trioxide, MgO, or ZnO smoke, or other crystals for test specimens (e.g., magnification calibration, diffraction calibration, resolution tests, etc.).

Stripping the carbon film from a metal specimen may be carried out as follows: the film is scored into squares while on the specimen. It is then removed by electropolishing in an electrolyte normally used for that purpose. For steels, for example, a solution of 10 parts glacial acetic acid to 1 part 60% perchloric acid may be used.† The films can be removed from the electrolyte using microscope grids, and are washed in 50% nitric acid solution in water, followed by final washing in distilled water for 10 min. They can then be caught on microscope grids ready for examination. Contrast in these replicas (normally referred to as direct carbon replicas in which the replica is of uniform thickness and follows the contours of the etched sample) is determined by the local variations in the number of electrons transmitted through the film (see section 4.1.4). The contrast may be enhanced by self shadowing, i.e., by evaporating carbon at an acute angle to the specimen surface, or by shadowing with heavy metals. Bradley[12] has recently described a method for improving contrast by the simultaneous evaporation of carbon and platinum using a single composite source (rods of carbon with 80 wt. % platinum). Evaporation is carried out as for plain carbon films.

An example of a carbon replica taken from a tempered carbon steel is shown in Fig. 78. The effect of self-shadowing can be seen by the dark contrast around the cementite particles since these were left in relief after the polishing treatment. Elsewhere the thickness of the carbon film is perfectly uniform.

Most of the work using carbon replica techniques has been applied to the study of ferrous materials, with the result that an enormous number of papers have appeared in the literature; in fact, too many to be discussed here. Some examples may be found in references 3 to 7, 10 to 14.

With radioactive metals, handling is a major problem and it seems that replica methods are the best way of studying the metallographic features

† A list of electropolishing solutions suitable for stripping replicas from other metals is given in Table III of this chapter.

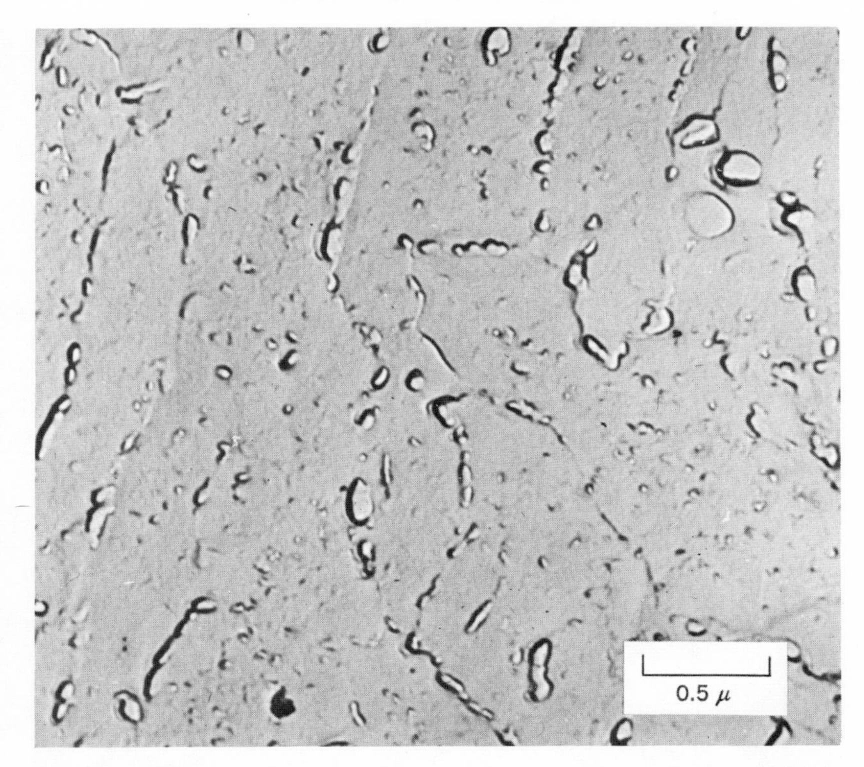

FIG. 78. Direct carbon replica of 1.2% carbon steel, quenched and tempered, showing positions of Fe_3C particles (× 30,000). (After Smith and Nutting.[10])

of such materials, since it is impossible to examine active metals (particularly if α, β, γ active) directly in the electron microscope. A convenient two-stage formvar-carbon replica method has been devised by Coiley et al.[15] in which the formvar film is dropped onto the active metal, which is arranged in a tilted position to allow drainage of the formvar. The whole unit is shielded inside a lead box. This part of the method as well as stripping with cellulose tape is done by remote handling using tongs. The tape is then removed, examined for any activity, and a carbon film evaporated onto it and stripped in the usual way.

Thin foil techniques are only possible with radioactive materials in which the radioactivity decays rapidly to a safe level (e.g., uranium), but there are indications that a remote, direct carbon replication method may be used, at least for β, γ active metals.[15]

A great advance in the technique of carbon replication was the extraction method first devised by Fisher.[16] This method is applicable to multiphase solid systems in which the matrix is chemically anodic to any precipitates contained in it; it has been widely employed in metallographic investigations of steels. Thus etching through the normal (direct) carbon replica results in dissolution of the matrix, leaving precipitate particles attached to the film (see Fig. 84*c*). Extracted particles may then be identified by electron diffraction and chemically analyzed by X-ray fluorescent analysis. Figure 79 shows the microstructure of a normalized 0.6% carbon steel with Fe_3C particles extracted in the replica.

With this technique it is sometimes helpful if the specimen is lightly electropolished immediately after the evaporation of carbon since this tends to loosen the film. Extraction stripping by etching is then carried out {e.g., with steels by using a 5% solution of nital (nitric acid in ethanol)}

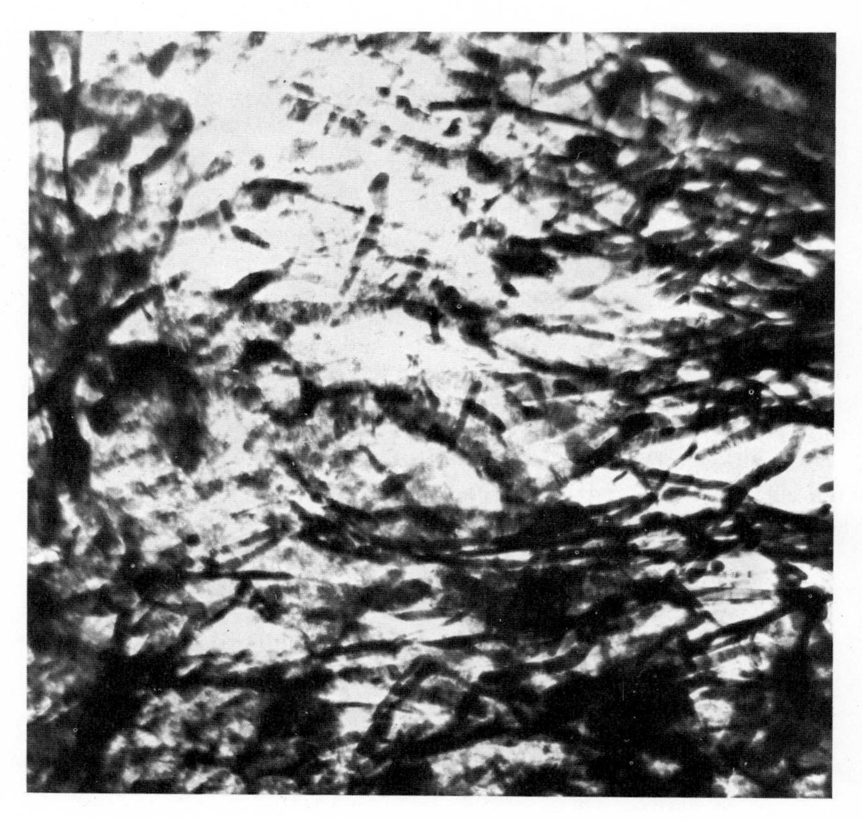

FIG. 79. Carbon extraction replica from pearlitic 0.6% carbon steel. Extracts are Fe_3C particles; some distortion of these is evident (×25,000).

followed by immersion in distilled water where the surface tension forces complete the stripping. The removal of particles by means of the extraction replica technique can also be done in bromine or iodine solutions taking the usual safety precautions with these chemicals. After washing in distilled water, the replicas can be caught on grids for examination. As with other replicas contrast may be enhanced by metal shadow casting. Fisher[16] has shown that most of the extracted particles retain the same relative positions and orientations they had in the metal.

An example of an extraction replica taken from a creep resisting alloy steel is shown in Fig. 80. In this, small precipitates of NbC (~150 Å dia.) and mixed carbides $M_{23}C_6$ (~500 Å), where M denotes any metal (e.g., Cr, Mo, W, Fe, etc.) identified by selected area electron diffraction, are seen along what must have been a dislocation network in the steel. This micrograph also serves to illustrate how the extracted carbides considerably enhance the contrast, since they strongly scatter electrons elastically. These scattered beams fall outside the objective aperture so that in the image the extracts appear black, whereas the electron beam passing through the carbon film is only scattered incoherently—the major portion being transmitted—giving light contrast in the image (see section 4.1.4). Besides the added contrast obtained from Bragg scattering, the structure of the particles can be determined from the diffraction patterns using the selected area technique. With this method good progress is being made in understanding the behavior of complex alloy steels subjected to conditions of high thermal and mechanical stress, e.g., the creep properties are determined by the nature and distribution of precipitated carbides.[17] Recently an added advantage has arisen by being able to analyze chemically the extracts using fluorescent analysis,[18] thus giving a complete picture of morphology (from micrographs), structure (from diffraction patterns), and composition (by fluorescent analysis). The carbon replica technique is still a powerful method for metallographic research and has been widely used in studies of ferrous materials.[13-18]

4.1.3 *Oxide Replicas*

Alumina Replicas. Aluminum and its alloys can be successfully replicated by using its own tenacious and coherent surface oxide layer. This layer may be formed under controlled conditions by anodizing in a suitable solution (e.g., 12% solution in water of disodium hydrogen phosphate containing 0.5% H_2SO_4) after the specimen has been electropolished.[19, 20] The replicas may be stripped by electropolishing again. Electropolishing of aluminum and its alloys may be conveniently carried out in an electrolyte of 1 part 60% perchloric acid to 4 parts ethanol (see also Table III)

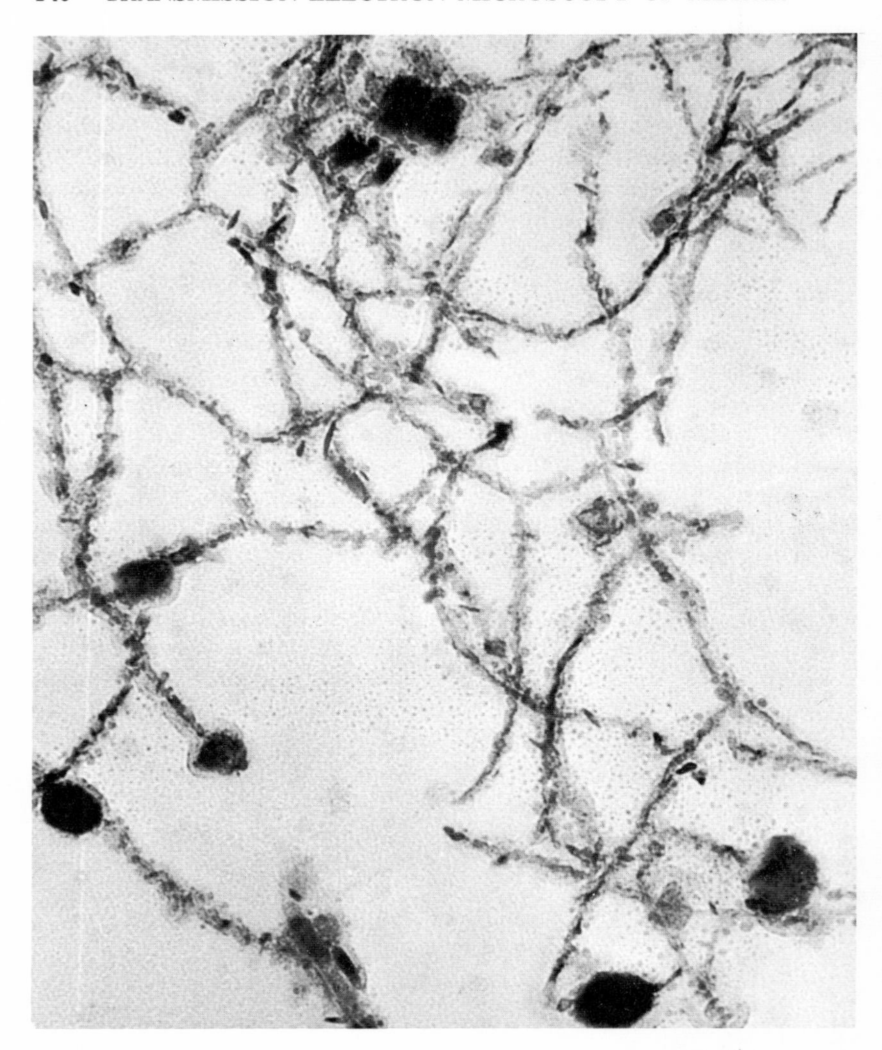

FIG. 80. Carbon extraction replica from a creep resisting alloy steel, showing small particles (extracted) of niobium carbide. These are identified by selected area diffraction (×90,000). (Courtesy of J. M. Arrowsmith and J. Nutting.)

which must be kept at a temperature less than 30° C; otherwise the solution becomes inflammable. The specimen is made the anode, and only that part to be polished is exposed to the solution, the remaining parts being previously covered with a suitable lacquer (see later). Polishing is carried out at 0.5 amp and 20 to 30 volts (depending on the alloy), using a hollow cylinder of aluminum as cathode. Anodizing is done by immersing the specimen in the anodizing solution and raising the voltage gradually from 0 to 30 volts, using a lead cathode. Since the thickness of the oxide film is about 14 Å per anodizing volt, accurate control of thickness is possible. The time of anodizing is unimportant—the current flow will stop when anodizing is complete because the cell is then polarized. If etching occurs, the specimen should be washed in 10% HCl and reanodized. The films are then stripped, after scoring the specimen into squares, by reelectropolishing. The replicas should be removed in a blackened dish in which they are more visible than in clear glass.

In alloys there is no need to etch to reveal second phases since they oxidize at a different rate from the matrix, giving rise to variations in film thickness so that good contrast in the image is obtained (Fig. 84*b*). An example of this is shown in Fig. 81, which shows the Widmanstatten {100} orientations of θ' particles in an aged Al–4% Cu alloy. Precipitate structures smaller than 100 Å are not faithfully reproduced by this technique,[24] but surface displacements of 20 to 50 Å (e.g., after plastic deformation) are well resolved.[20] Figure 82 shows an example of slip lines, duplex slip,

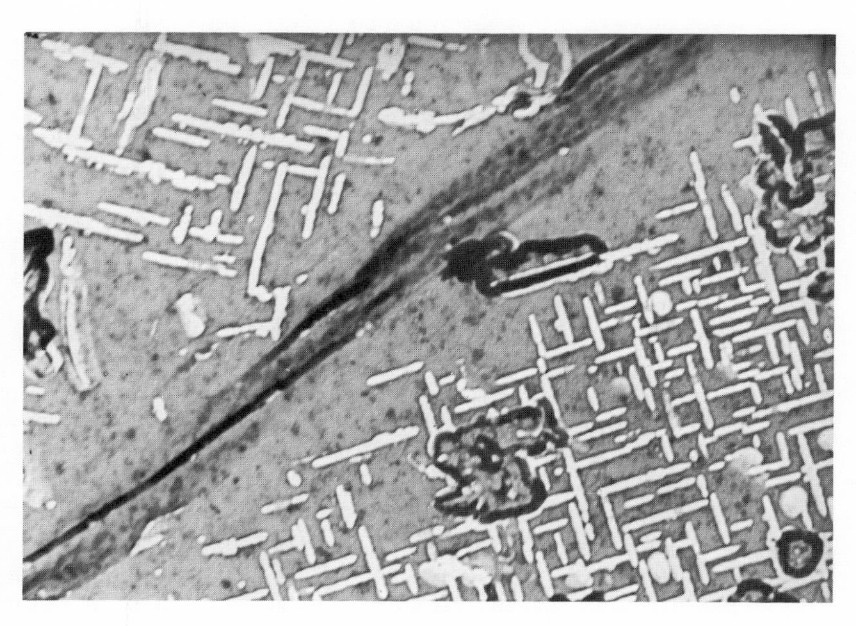

FIG. 81. Aluminum oxide replica of Al–4% Cu alloy aged to produce θ'–$CuAl_2$ (white regions). A grain boundary runs diagonally across the micrograph (×18,000).

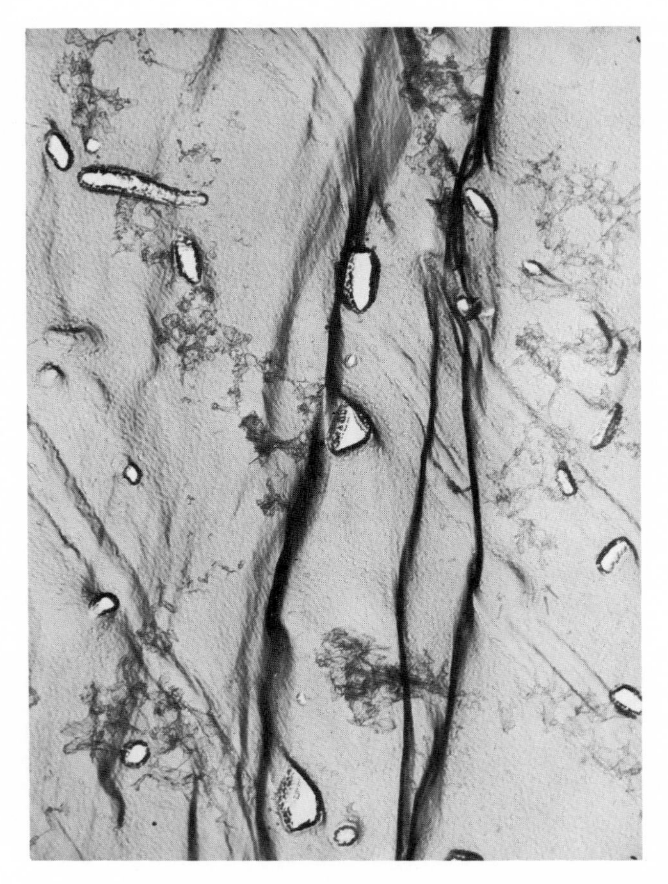

FIG. 82. Aluminum oxide replica of Al–4% Cu alloy deformed 10% in tension after aging for 20 hours at 320° C. Notice slip lines, duplex slip, and θ–$CuAl_2$ particles (white areas) ($\times$ 15,400).

and the distribution of $CuAl_2$ particles in a fractured Al–4% Cu alloy. Aluminum oxide replicas have been widely used for investigations of age-hardening problems,[19–23] and plastic deformation in light alloys,[20–24] but apart from the surface effects of plastic deformation this method is now being replaced entirely by thin metal foil techniques.

Silicon Monoxide. Silicon monoxide replicas give better resolution than alumina replicas, and are extremely strong and stable to electron bombardment. These were used by Wilsdorf and Kuhlmann-Wilsdorf[25] to investigate slip lines on deformed metals.[26–27] The method is to evaporate silicon monoxide (placed in a tungsten basket, Fig. 77) under a low pressure onto the surface of the prepared sample. After removing the specimen from the evaporator the film may be stripped chemically or by electropolishing as described in the previous sections. An example, shown in Fig. 83, is taken from deformed α-brass. Fine details down to 20 Å may be resolved using this technique. All the methods for stripping carbon and oxide replicas involve chemical etching or electropolishing (i.e., wet methods) so that the surfaces of specimens are destroyed by the stripping process. Some distortion of the replica may also occur during stripping.

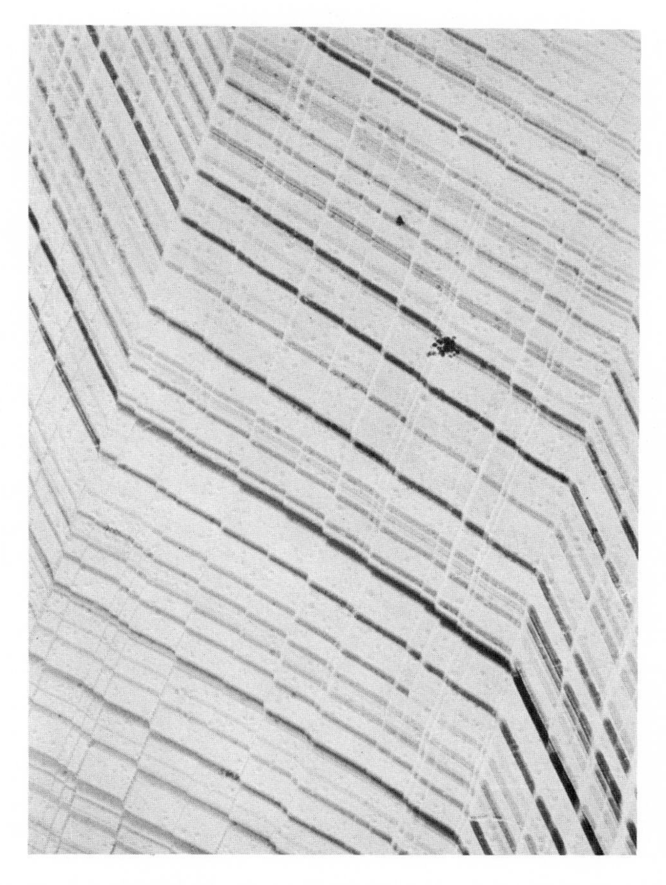

FIG. 83. Silicon monoxide replica of deformed α-brass. Notice change in direction of slip bands across the twin (× 15,400). (Courtesy of P. R. Swann.)

4.1.4 *Contrast from Replicas*

With noncrystalline films contrast in electron microscope images arises mainly from variations in the scattering cross section. Dark contrast in the image is thus produced by a deficiency of electrons from regions of high scattering power, i.e., low transparency thickness, or regions containing atoms of higher atomic number.† In films containing no extracted crystalline particles the scattering is almost entirely incoherent (see section 1.4). However, when extracted crystalline particles are present, e.g., in carbon extraction replicas, and some oxide replicas, they scatter electrons coherently (Bragg diffraction) so that these beams fall outside the objective aperture and black contrast is produced. We can conveniently represent contrast from replicas by referring to Fig. 84*a* to *c*. In direct carbon replicas, since they are of uniform thickness, the electron intensity in the image is the same for all horizontal surfaces. Hence, in

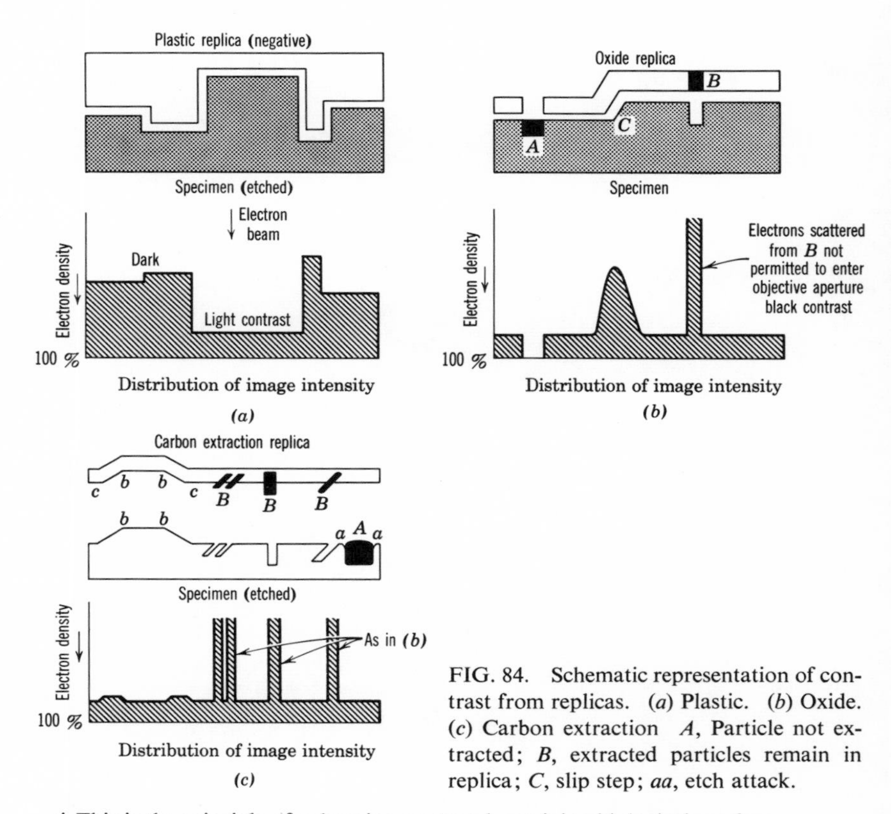

FIG. 84. Schematic representation of contrast from replicas. (*a*) Plastic. (*b*) Oxide. (*c*) Carbon extraction *A*, Particle not extracted; *B*, extracted particles remain in replica; *C*, slip step; *aa*, etch attack.

† This is the principle of enhancing contrast by staining biological specimens.

the region *bb* Fig. 84*c*, because the electron beam goes through the same thickness of replica as at *cc*, there is no indication in the image that the regions at *b* are in relief. With plastic replicas, as in Fig. 84*a*, their thickness is related to the relief of the specimen surface, so that the interpretation of surface relief is much easier than with ordinary carbon films. However, because plastic replicas are mechanically weak and, being insulators, tend to be destroyed by the electron beam, it is much more satisfactory to use metal shadowed or self-shadowed carbon replicas, e.g., as in Fig. 78. The enhancement of contrast with extraction replicas is immediately apparent from Fig. 84*c*.

In conclusion we can sum up by saying that replicas are still very useful for surface studies and for resolutions of about 20 Å. The replicas may not always be truly representative of the surfaces from which they are stripped. For example, carbon films often stretch out during wet stripping so that when removed from deformed samples it is not always possible to see fine slip.[28] In extraction replicas some mechanical deformation of extracted particles may occur during stripping, so that the morphology of precipitation in the original sample may not always be faithfully reproduced; this usually arises with large particles, e.g., carbides in the pearlite structure of steel (Fig. 79). Nevertheless, the carbon extraction replica technique is still an excellent method for metallographic analysis of ferrous materials.

4.1.5 *Fractography*

The microscopic examination of the fracture surfaces of metals and alloys is useful in providing information on the mechanisms of fracture and is a valuable technique for routine investigations of service failures. The electron microscope is especially useful for fracture examination because of the high resolution and great depth of focus available. Low[29] has recently reviewed microstructural aspects of cleavage fracture and has considered results obtained from electron optical studies for ferrous materials using replica techniques. When the fracture surfaces are not too rough, replication is conveniently carried out by the direct carbon technique previously described. In other cases, however, it may be necessary to employ a two-stage replication method, e.g., that devised by Scott and Turkalo.[30] In this, a primary replica is made by pressing a surface of a cellulose acetate strip ($\sim$0.005 in. thick) wet with acetone against the fracture surface of the specimen. After drying, it is then stripped from the surface and preshadowed at 45° using chromium evaporation. Carbon is then evaporated in the normal way. The cellulose acetate may be dissolved in acetone after the composite is cut and mounted on microscope grids. The same technique has been used by Guard and Turkalo[31] for fracture studies of the intermediate compounds NiAl and Ni_3Al, and by Davies and Martin[32] for uranium. Direct carbon replicas have been used by Crussard, et al.[33-34] for investigating fracture in steels; for light alloys the aluminum oxide replication method is suitable. In materials containing dispersed phases, the micrographs show characteristic patterns of closed loops around particles and curved lines, indicating fracture has commenced near the particles. In some cases fracture of the particle has been observed.[20] A typical micrograph (oxide replica) of a fractured surface from a two-phase alloy is shown in Fig. 85.

Fracture studies of metals in the form of thin foils have been carried out directly by deformation experiments done inside the microscope. How-

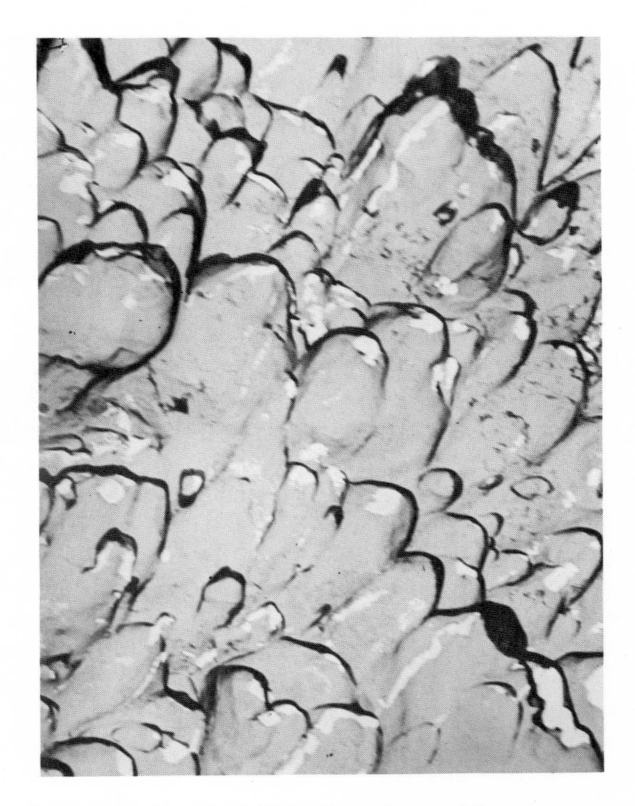

FIG. 85. Aluminum oxide replica of Al–7% Mg fractured in an impact machine after aging for 50 days at 150° C. White areas represent Mg_3Al_2 particles; cusps (black) are produced during fracture (× 11,600).

ever, we cannot strictly classify this technique under the general term "fractography." Nevertheless, it is now possible to observe directly the nucleation and propagation of cracks.[35–37] Since the mechanical properties of thin films and whiskers are vastly different from those of bulk material, these experiments are of direct interest only to thin film behavior. We shall return to this again in Chapter 5.

4.2. Thin Metal Foils

We have seen from the earlier chapters that metal samples become transparent to electrons when they are of the order of a few thousand angstroms thick. To correlate the properties of bulk samples with the structures that can be observed with thin metal foils, it is necessary to be able to thin specimens from relatively thick material without modifying or destroying the structure in any way.[38] The two possibilities of doing this are: (1) to perform experiments on bulk material and reduce the thickness without mechanical means or (2) to reduce the thickness to relatively thin sections (say 100 μ) and then use this as "bulk" material for experimental purposes, finally thinning to 1000 Å, again without mechanical deformation. Many investigations have been carried out using thin metal samples obtained by evaporation and other methods, but since these are only a few hundred angstroms in thickness, they are not truly representative of bulk material. Nevertheless, much useful information on structure has been obtained by this technique. We can thus classify thin foil techniques into two broad groups: (1) prepared from bulk samples, and (2) formed directly in thin sections.

4.3 Preparation from Bulk Samples $\sim$100 μ Thick

It is difficult to define precisely the term "bulk sample" as the starting point in thin foil techniques, but for most purposes it is sufficient to say that a polycrystalline bulk sample must contain at least three grains across its thinnest section, which must be thicker than 2 μ. The reason for this is twofold. First, in diffusion processes the surface is the most active part of the metal, and since diffusion distances are of the order of 1 μ, samples 1 μ thick have their surfaces so close together that practically all diffusion reactions occur at the surfaces. Thus studies of quenching, aging, etc., must be done on samples thicker than 2 μ. Second, transformations without diffusion (e.g., shear processes) have constraints imposed on them by the thickness of the material. In thin foils, however, there is practically no constraining effect on the foil when it is only 1000 Å or so thick, so that transformation mechanisms may be different from those in thick samples In the same way surface restrictions or relaxations may affect structures such as dislocation arrangements, i.e., the arrangement in the bulk material may be relaxed during thinning. This latter point will be discussed more fully later on. Thus to be as close as possible to actual bulk material it is convenient to start off by using specimens about 100 μ in thickness as typical of the bulk. The first stages in preparation involve reducing the material into sheet about 100 μ thick. With most metals this is conveniently done by mechanical work, e.g., cold rolling or beating, which may involve intermediate stages of annealing. The specimens so obtained may then be thoroughly annealed to remove the effects of the cold work and as a result will usually have a preferred orientation which will depend on the processing schedule used (e.g., after rolling, most FCC metals retain $\langle 110 \rangle$ textures). Any experiments such as deformation, age hardening, etc., can then be carried out on these samples, and foils suitable for transmission can be prepared in any of the following ways.

4.3.1 *Chemical Etching*

Hirsch et al.[39] thinned beaten gold samples by etching in a dilute solution of potassium cyanide, and although the foils produced were uneven in thickness, the heavily cold-worked structure resulting from beating was preserved. The examination of dislocation structures in aluminum was successfully carried out by Hirsch, Horne, and Whelan,[40] who obtained large uniform thin areas by etching their specimens (0.25 μ thick) in 0.5% hydrofluoric acid. The etching method is limited to single-phase metals and alloys because it is difficult to prevent preferential attack of one or other of the phases in multiphase materials. Unless the proper choice of etching reagent and conditions is obtained, etching substructures may be formed,[41] and this will confuse the true structure of the specimen.

4.3.2 *Ion Bombardment*

Castaing[42, 43] successfully obtained thin foils of aluminum and aluminum-copper alloys by bombarding both sides of a thin disc (1 to 2 μ thick) with a beam of ions. A schematic illustration of his apparatus is shown in Fig. 86. The regulating knob *A* squeezes against a vacuum seal *F* on the air leak valve stem *V*. A small scratch is made on this stem to introduce a very slight air leak to permit ionization when the H.V. (up to 10 kv) is applied to the anodes *C*. The ions produced are focused at *D* (magnesium discs with seven holes 1 mm dia. placed concentrically

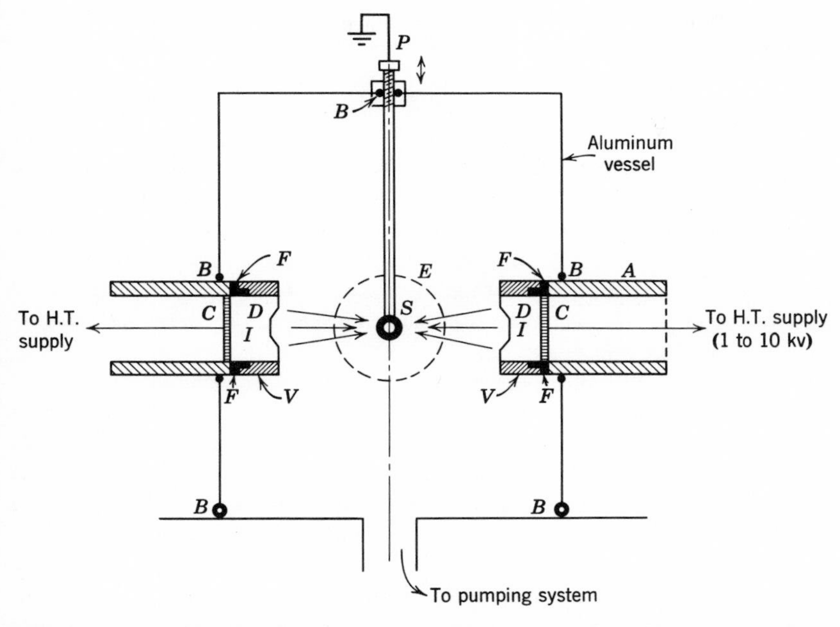

FIG. 86. Ion bombardment apparatus (schematic). (After Castaing.[43]) *A*, insulated knob to control air (or gas) leak; *B*, vacuum seals; *C*, anode; *D*, focusing discs; I, ionization chamber; E, viewing window; P, adjustable specimen support; S, specimen; V, leak valve stem.

about their center) so as to cover completely the 2 mm area on both sides of the specimen *S* which hangs vertically in the center of the chamber on the center line through the anodes. A low pressure may be obtained by connecting the system to an oil diffusion and rotary pump. During bombardment the specimen can be viewed through windows (*E*) placed at right angles to the position of the anodes. In using ions to remove metal atoms from a specimen, the energy of the ions must be adjusted to give random atom removal. If the energy of the ions is too high, heating and possible damage to the structure occur, whereas if too low, etching effects are produced. Critical adjustment of the accelerating potential is thus an essential requirement; for aluminum 3 kv is suitable. Attempts to use this method to thin specimens of stainless steel[44] and α-brass[43] have so far proved unsuccessful.

Since the rate of removal of metal by bombardment is about 10 μ in 24 h, control in the final stages is difficult. Although clean, uniformly thin foils can be produced, the method is complicated and tedious. It has been found that electropolishing techniques provide the most convenient and successful way of finally thinning specimens.

4.3.3 *Electropolishing*

The first real success with direct transmission through thin samples was achieved as early as 1949 by Heidenreich.[45] His technique consisted of electropolishing aluminum and aluminum-copper specimens 3 mm dia. and 125 μ thick in a bath of 1 part methyl alcohol, 1 part concentrated HNO_3, and 1 cc HCl to 50 cc of the mixture. The electropolishing caused perforation over the specimen, and the areas near the holes were transparent to the electron beam.

The main difficulty in producing thin foils is to achieve uniform removal of metal from all parts of the specimen surface. To do this, the current density at all points on the specimen must be the same. At low current densities the edges of the specimen polish preferentially, whereas at high current densities preferential polishing occurs at the center. These effects are probably due to differences in uniformity of the anode layer. The decision when to stop electropolishing is usually attained by experience as it is difficult to measure foil thicknesses of 200 to 2000 Å. The criterion usually employed is that as soon as a few tiny holes appear, the areas near the holes are thin enough for transmission. If electropolishing is carried on after the appearance of the holes, then because the current-density distribution is disturbed, the holes become rounded off and the thin regions lost. Thus electropolishing by the perforation method tends to produce wedge-shaped foils. It is interesting to consider in a little detail the conditions existing during electropolishing.

When specimens are placed vertically as the anode in an electropolishing solution with a flat vertical cathode, it is found that the current density is a maximum at the edges of the specimen and at the liquid surface-metal interface. Electropolishing solutions, which are composed of an oxidizing reagent and a solvent for dissolution of the complex anode products formed during electropolishing, can be classified into two groups: (1) solutions forming viscous layers which flow down under gravity, and (2) solutions which evolve gas bubbles at the anode.

Solutions of the first type will give preferential attack at the top of the specimen, since the viscous layer is flowing away from this region and is increasing in thickness as it moves down the specimen. Solutions of the second type will form a blanket of gas bubbles which is thin at the bottom of the specimen and thick at the top; consequently, metal is removed more quickly from the bottom of the specimen. A horizontal anode eliminates the effect of gravity on the viscous layer and on the bubbles, but now the conditions are different on the top and bottom surfaces of the specimen. In this case, it will be difficult to attain the correct polishing conditions on both sides of the anode, particularly if the current-density range for polishing is small for the electrolyte. Some advantage is gained when using vertical anodes if they are surrounded by a cylindrical cathode closed at the bottom end.

The experimental arrangement for electropolishing is carried out quite simply by a voltage dividing circuit as shown in Fig. 87*a*. When there is no dc supply available, the mains ac may be rectified to dc using a resistance—capacitance filter to reduce any ripple to less than 10%, and the voltage changed with a variable voltage divider as shown in Fig. 87*b*.

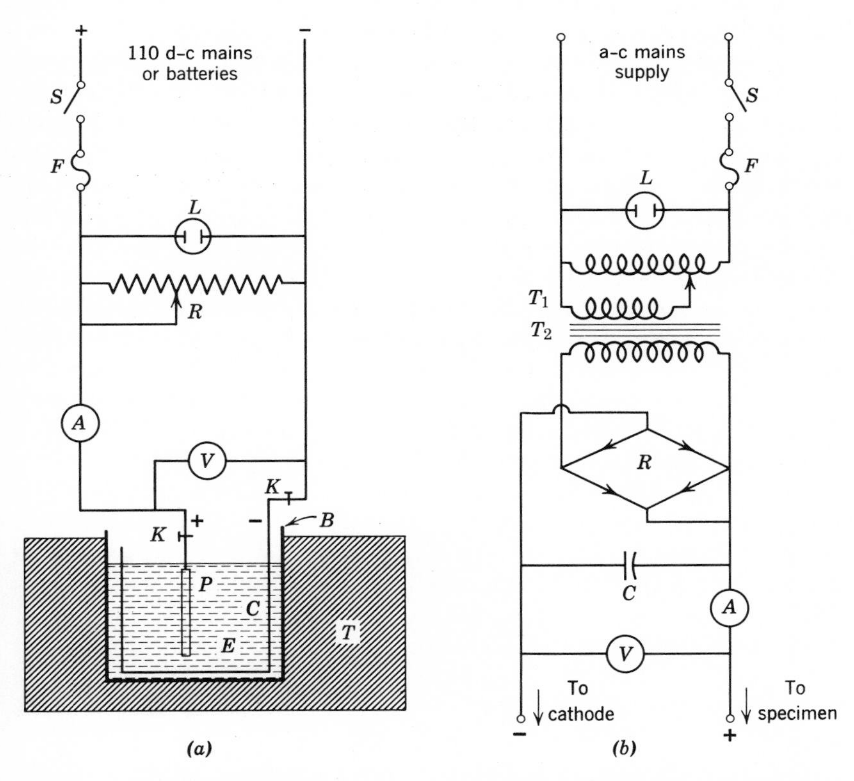

FIG. 87 (*a*) Electropolishing circuit for d-c mains or battery supply. *A*, ammeter; *B*, beaker for electrolyte *E*; *C*, cathode; *F*, fuse; *K*, contact slips; *L*, light bulb; *P*, specimen; *R*, variable resistance; *S*, switch; *T*, controlled temperature bath; *V*, voltmeter. (*b*) Variable d-c source from an a-c supply. *C*, capacitor; *R*, rectifier; T_1, variable transformer (voltage divider); T_2, transformer.

To establish the right conditions for electropolishing it is necessary to obtain the voltage current relationship for the electrolyte and specimen material being used. This is done by placing a test specimen as anode, as in Fig. 87*a*, exposing the same area as is to be used in later work, and plotting graphically the voltage-current curve for increments of 5 volts (allowing a few minutes between each increment to allow the bath to settle down). The plateau of the curve then gives the voltage range for a constant current necessary for electropolishing. This is illustrated schematically in Fig. 88. For the first few minutes of electropolishing, there is usually a large surge in current which then diminishes with time as the anodic layer is being formed. To ensure uniform removal, it is wise to agitate the solution as this helps to dislodge the film mechanically. If a viscous anode layer forms during polishing, it is advisable to use the highest possible voltage to effect rapid dissolution (Fig. 88).

Elimination of preferential polishing at specimen edges may be achieved by coating them with a nonconducting lacquer.† In this case, electropolishing produces preferential thinning at the edge of the lacquer coating, leaving the center of the specimen thicker than the outside. To overcome this difficulty Bollmann[44,46] used pointed cathodes blanked off with lacquer, leaving only the tips free, mounted close to the center of the specimen as shown in Fig. 89. Electropolishing is done until a hole appears at the center (Fig. 89*b*). After this, the electrodes are moved as far apart as possible and polishing continued until the stage shown in Fig. 89*c* is obtained. Specimens may then be cut from the thin edges.

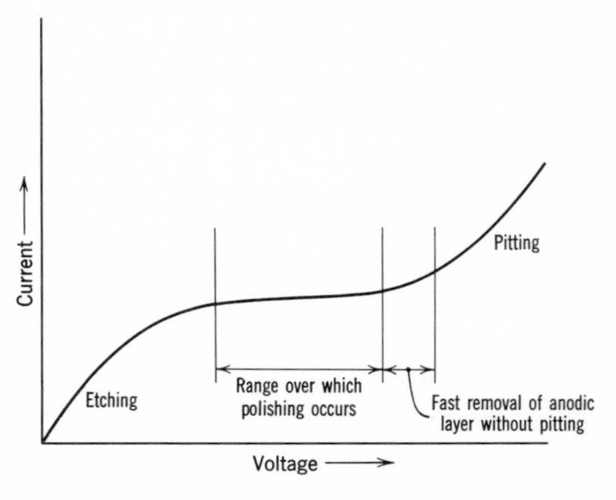

FIG. 88. Schematic representation of voltage-current characteristics for electropolishing solutions.

† Suitable lacquers are as follows: United States—Micromask or microstop, manufactured by Michigan Chrome and Chemical Co., Detroit, Michigan. Britain—Lacomit, manufactured by W. Canning and Co., Ltd., Birmingham, England. These are all soluble in acetone.

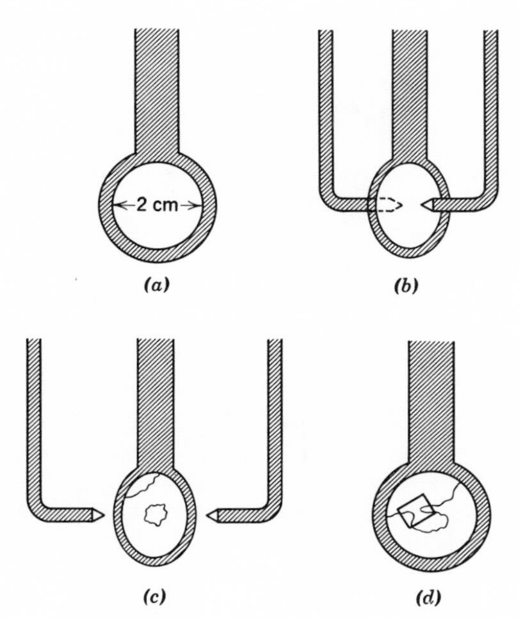

FIG. 89. The Bollmann method for preparing thin foils. (*a*) Initial specimen. (*b*) Point cathodes arranged close to specimen to give central perforation. (*c*) Cathodes moved apart after central perforation. (*d*) Area from which foils are cut.

For cutting specimens a sharp steel scalpel may be used by pressing hard onto the specimen which is placed on a flat piece of plastic, e.g., lucite, and not by dragging the knife across it as in normal cutting operations. This prevents tearing and damaging the foils and also reduces the number of bend contours, due to buckling, when the specimen is under observation. It is always advisable to relacquer the edges of the specimen before final perforation is achieved, as this will stop preferential thinning there, and will also prevent the polished area from being detached from the main body of the specimen.

Kelly[38] has avoided the necessity for removing the specimen to adjust the position of the cathodes between the stages shown in Fig. 89*b* and *c* by using two sets of cathodes on the specimen holder (Fig. 90). In this way one set of pointed electrodes placed close to the specimen may be switched into the negative of the power supply for central perforation and then the outer flat electrodes switched in for the final stages of perforation (after switching off the inner set). The Bollmann technique is limited to solutions of low throwing power and usually produces wedge-shaped foils. However, it is a simple method and is used widely for thin foil preparation.

The preferential polishing of a vertical specimen due to the formation of a heavy viscous layer has been utilized by Nicholson, Thomas, and

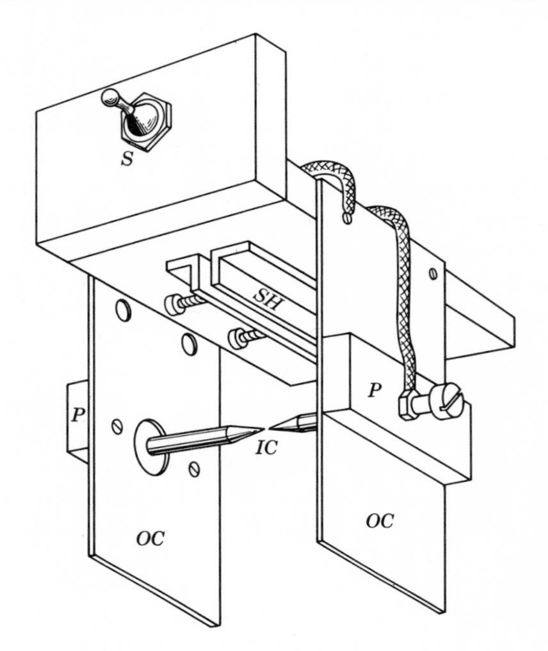

FIG. 90. Refined Bollmann apparatus. *S*, switch; *SH*, specimen holder; *P*, lucite insulating blocks; *IC*, inner point cathodes; *OC*, outer flat cathodes. (After Kelly and Nutting.)

Nutting[47] for use with light alloys, and by Tomlinson[48] for making thin foils of Ni, Al, Cu, Fe, Co, and Mg. This technique, which has become known as the "window" method, is illustrated in Fig. 91. The specimens are in the form of sheets 2 in. × $\frac{1}{2}$ in. and 100 μ thick (Fig. 91*a*) and are

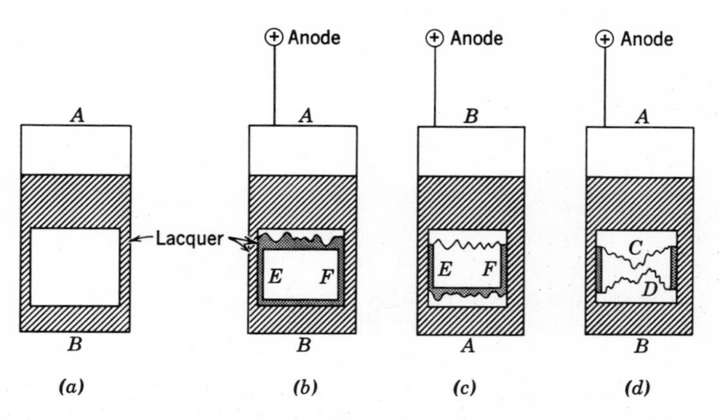

FIG. 91. The window method for preparing thin foils (schematic). (*a*) Initial specimen both faces lacquered as shown. (*b*) After first perforation. (*c*) Reversed until second perforation. (*d*) Reversed; specimens cut from edges *C*, *D*.

electropolished until perforation occurs at the top (Fig. 91*b*). The specimen is then turned upside down, relacquered to cover the perforated edge and the sides as shown, and repolished until perforation again occurs at the top (Fig. 91*c*). The specimen is then again turned upside down and polishing continued until only a small area of the specimen remains (Fig. 91*d*). Foils may then be cut from the thin edges *C* and *D*. To prevent break-through at the sides *EF*, it is advisable to relacquer there between each reversal. The foils are more uniform in thickness over larger areas if the current is switched on and off rapidly near the end of the polishing.[48] The complete explanation for this is not yet known, but at least one of the benefits is that when the current is off, time is allowed for dissolution of the anode layer by the electrolyte. Similarly, it is an advantage to keep the electrolyte flowing continuously past the specimen provided the cathode is suitably shaped to avoid turbulence. An arrangement for doing this is shown in Fig. 92 where the cathodes are flat, stainless steel plates rounded off at the top and bottom. Circulation of the electrolyte is maintained by a centrifugal pump. To avoid the labor of

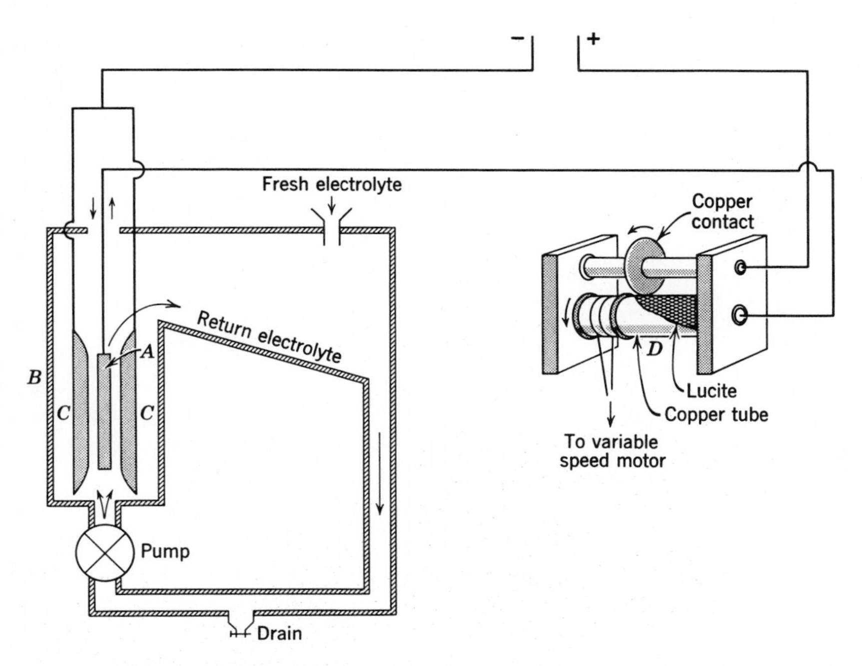

FIG. 92. Sketch of electropolishing apparatus for use with viscous electrolytes. *A*, specimen; *B*, transparent container; *C*, cathodes shaped to minimize turbulence; *D*, drum to make and break contact. (Courtesy of P. R. Swann.)

manually switching on and off near the end of the thinning operation the specimen is electrically connected to a drum (Fig. 92). This is made up of two sections—one-half being insulating (e.g., with lucite) and the other half being conducting (e.g., with copper). The rate of switching on and off can then be controlled by the speed of rotation of the drum; thus the whole operation is automatic. The switching sequence is on $\frac{1}{3}$ sec and off $\frac{2}{3}$ sec. This arrangement has been found to be particularly convenient for polishing silver specimens using a solution of 6 gm KCN in 100 cc of water and polishing at 20 volts.†

Polishing should always be carried out as slowly as possible, and this may be helped by using low temperatures, e.g., by immersing the electrolytic bath in a suitable cooling medium (Fig. 87*a*). This is particularly helpful with metals and alloys which oxidize rapidly (e.g., copper and its alloys) and in which the anodic layer is reactive. In the latter case etching occurs if the current is switched off when the specimen is still in the electrolyte. To avoid this the specimen must be removed from the bath with the current still on, immediately washing the surface with a jet of cold alcohol. This technique has been developed by Swann and Nutting[49] for making thin foils of copper, Cu–Zn, Cu–Sn, Cu–Ge, Cu–Al, and Cu–Si alloys. The electrolyte is suspended in a container placed over liquid nitrogen to keep the bath temperature below $-20°$ C. Electropolishing is carried out by using the window method followed by the Bollmann method, and final polishing is done with the window method again. Just before finally removing the specimen, some liquid air is poured onto the top of the electrolyte to form a cold layer through which the specimen can be withdraw (with the current still on). Washing is carried out immediately by immersing the specimen in methyl alcohol covered with a layer of liquid nitrogen.

Another technique employing the viscous polishing layer has been developed by Brandon and Nutting[28] for use with iron specimens. The method, known as the "figure of eight" technique, is similar to the window method, but corresponding to the stage of Fig. 92*b* the surface is lacquered to give a figure of eight outline (Fig. 93*d*). Final polishing is carried out as indicated in Fig. 93.

To avoid the preferential rounding off of the edges of holes after perforation, Mirand and Saulnier[50] backed their specimens with a sheet of the same material; thus, when a hole appears, attack continues on the backing material (Fig. 94). Polishing is continued until fragments drop from the specimen into the electrolyte, where they can be collected for examination. This technique works well for titanium alloys,[50, 51] but the difficulty in extending it to other alloys is in finding an appropriate

† P. R. Swann (private communication).

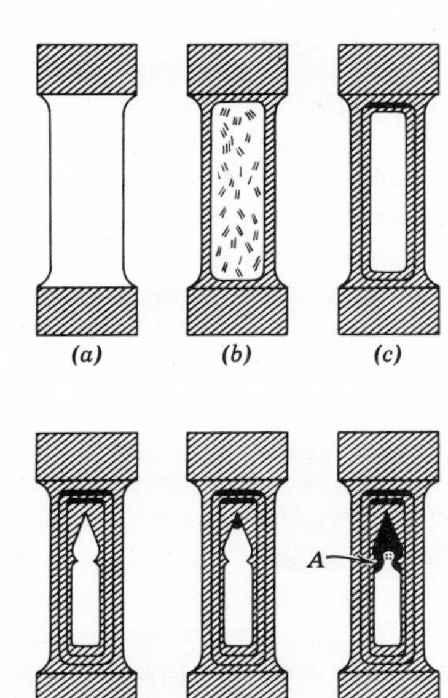

FIG. 93. Polishing sequence for "figure-of-eight" technique. (*a*) Initial specimen. (*b*) Specimen coated with lacquer prior to initial electropolish. (*c*) Specimen lacquered after first perforation. (*d*) "Figure-of-eight" outline. (*e*) Perforation beginning in final electropolish. (*f*) Area from which foils are cut (*A*). (After Brandon and Nutting.[28])

electrolyte which will not attack the fragments while lying in the solution. The electrolyte may also seep between the specimen and the backing piece, thus etching the back of the specimen.

The production of exact geometrical shapes and the preservation of certain profiles have been discussed by Michel[52] who gives an indication of

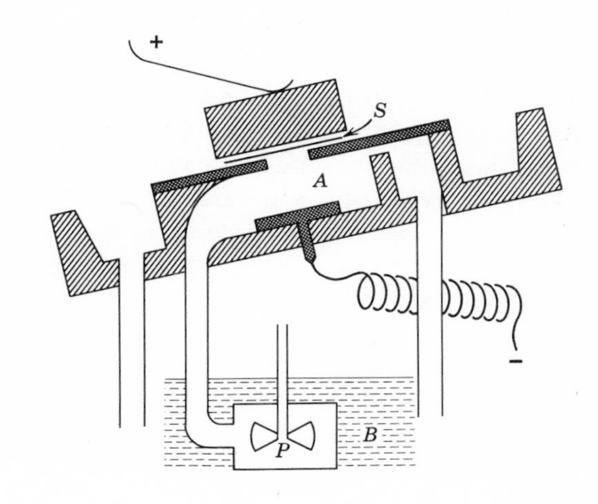

FIG. 94. Apparatus for collecting fragments of foils from backed specimens. Electrolyte is circulated through the chamber *A* by means of paddle *P*. Fragments collected at *B*. *S*, specimen. (After Mirand and Saulnier.[50])

the methods to be adopted to attain uniform thinning. Fisher[53] has shown that to give a uniform current density over the anode surface, the anode should be pear-shaped, e.g., by placing suitable washers on either side of a flat sheet (uniform-field method). Perforation then occurs simultaneously at a number of points, when the current must immediately be switched off to prevent rounding off the holes.

Having discussed the various techniques for electropolishing, we can now apply them to particular cases. This is illustrated in Table III, which shows the methods that have given consistent results for a wide range of metals and alloys.† If a new material is to be investigated, the only approach is to find an electrolyte which will polish slowly and evenly without etching or preferentially attacking second phases, grain boundaries, etc. There have been numerous papers published on electrolytic polishing, and the works of Jacquet[54] and Tegart[55] are particularly useful in helping to find suitable electrolytes. Ceramic materials may also be thinned by electropolishing.

Finally, it may be useful here to set out a step-by-step procedure for preparing a foil using the rolling method to obtain strips initially 100 μ thick and the window method for final thinning. (1) Roll down with mechanical rolls to maximum reduction in area. If there is any edge-cracking, the material should be well annealed before attempting further rolling. Final rolling is best accomplished by hand rolling using 1 to 3 in. diameter rolls. The strips should now be $\sim 100\ \mu$ thick. (2) Anneal overnight to remove effects due to rolling. (3) Heat treat or perform the desired experiments. (4) Clean surfaces with grade 000 emery paper to remove oxide scale, etc. Cut specimens 2 in. by $\frac{1}{2}$ in. from strip. (5) Swab specimen with acetone or alcohol. (6) Paint on window with lacquer (e.g., Microstop, Lacomit, etc.). (7) Clean window with alcohol after lacquer has dried (5 to 10 min). (8) Determine the plateau on the voltage-current curve for the chosen electrolyte using as anode a piece of the same material as the specimen. This gives the range of voltage over which good polishing conditions exist. (9) Electropolish specimen (with some initial agitation to remove oxide layers) until first hole just appears. Remove specimen, wash, dry, and relacquer. (10) Turn foil upside down and electropolish again until a hole appears at the opposite end to that where the first hole formed. (11) Remove specimen, wash, dry, and relacquer. (12) Turn specimen upside down again and repolish until about half the foil has dissolved away. During this stage reduce the voltage to allow slow, even electropolishing. (13) When jagged edges have formed and the foil has a "limp" appearance, stop polishing, carefully rinse in alcohol

† Numerous examples of the results obtained with these techniques are to be found in the illustrations to Chapter 5.

or distilled water, then cut off appropriate pieces with a scalpel, keeping the specimen immersed in alcohol. (14) The specimens may then be mounted on 100 mesh copper grids so that the jagged edge of the foil lies across the center of the grid. Sometimes it is necessary to stick the foils firmly to the grids. This can be done by first dipping the grid into a 2% solution of polybutene in xylol, then allowing the xylol to evaporate, and pressing the foil firmly onto the now sticky grid. It is also possible to hold the foil firmly in the microscope specimen holder by sandwiching it between two grids. However, this tends to reduce the open area of the grids unless they can be exactly matched in position. If the specimen holder is designed so that the grids sit in a cap, which is either pushed or screwed onto the holder barrel, it is useful to cut a thin slot along one side of the cap almost to the base. It is then easy to hold the grid with a pair of finely pointed tweezers and place it in the cap with the tweezers sliding down the slot until the specimen is properly positioned. In this way, the grid can be prevented from turning upside down, so that the specimen is not altogether lost during the mounting operation. To have the maximum area of foil available for examination the grids can be dispensed with altogether. The specimen can then be stuck to the bottom of the microscope specimen holder (again using polybutene). In this way, there will be no problems of having grid bars obstructing the thin areas of the foil from the field of view, but there is a danger of losing the foils inside the microscope.

It is always advisable to examine the foils in the microscope as soon as possible after they have been prepared, because very often they deteriorate owing to oxidation, and sometimes with alloys, surface precipitation occurs. Oxidation can be avoided by storing the foils in a vacuum desiccator or covering them with lacquer. This can then be removed with solvent just before examination. Foils can also be stored in glycerol.

4.3.4 *Preparation from Massive Bulk Samples*

The electropolishing methods described above are limited to materials that can be obtained in the form of bulk specimens $<200\ \mu$ thick. Although this is possible with ductile metals, it is virtually impossible to roll hard materials (such as alloy steels) down to thicknesses of $\sim 100\ \mu$. Consequently, a jet-machining method has been developed by Kelly and Nutting,[38, 56] whereby the starting bulk sample may be any thickness before it is jet-machined down to 100 μ thick so as to be in a suitable form for final thinning by electropolishing. The method is suitable for all thick specimens provided a suitable electrolyte can be found.

The apparatus used for this technique is shown schematically in Fig. 95*a*

Table III Electopolishing Techn

Metal or Alloy††	Condition	Initial Form of Specimen	Technique	Electrolyte Compositior
Ag	All heat treatments	Rolled sheet, 25–200 μ thick	"Window" method	Per liter of: 67.5 KCN, 15 g R elle salt, 14.5 orthophospho acid, 15 g pot sium ferrocya 2.5 cc NH_4OH
Ag–Al	,,	,,	,,	20% perchloric 80% absolute al
Ag–Al	,,	,,	,,	70% absolute alc 20% perchlori acid, 10% gly
Ag–Zn	,,	,,	,,	6 gm KCN, 100 water
Al Al–Cu Al–Ag Al–Ag Al–Zn Al–Mg	... As quenched ,, Overaged As quenched or aged All conditions	Rolled sheet, 25–200 μ thick	,,	20% perchloric (60%), 80% absolute al
Al Al–Cu Al–Ag Al–Zn–Mg Al–Zn–Mg–Cu Al–Zn	... Aged ,, ,, ,, ,,	Rolled sheet, 25–100 μ thick	Bollmann method or window method	817 cc orthophos phoric acid (d = 1.57), 13 sulfuric acid, 156 g chromic oxide, 40 cc water
Al–Cu	All treatments	Rolled sheet, 25–80 μ thick	Electropolished with specimen backed by sheet of same material. Collect fragments	Any electrolyte gives a good p and will not e without the ac of current
Al Al–Cu	As quenched or aged	Discs 3 mm dia., 1–2 μ thick	Ion bombardment	...
Al	All treatments	Beaten foil, 0–5 μ thick	Etching	Dilute hydrofluo acid

roducing Metal Foils†

	Polishing Conditions			Remarks	Ref.
node	Voltage	C.D. amp/cm^2	Temp. °C		
nless l	4–5	0.5	< 30	Wash in water, keep under ethyl alcohol	65
,,	10–15	0.2	< 30	,,	66
,	15–50	...	,,	,,	See 44 Chap. 5
,,	6–8	...	room	,,	,,
eet or nder	15–20 finish at 10–12	0.2	< 30	Clean Al–Cu and Al–Ag as quenched in 50% nitric acid, Al–Zn in ethyl alcohol, and Al and Al–Ag aged in moving water. Specimen may be electromachined from 1 mm thick to 50 μ with nitric acid electrolyte	47 48 63
ints or nder	10–12	0.05	70	Clean Al–Cu and Al–Ag in cold phosphochromic acid solution. Specimen may be electromachined from ~1 mm thick to 50 μ with nitric acid electrolyte	47 63
..	...	...	...	Specimen may be electromachined from ~1 mm thick to 50 μ with nitric acid electrolyte	50 51
..	...	...	...	Specimen electropolished from rolled sheet to 1–2 μ	42 43
..	...	...	...	...	40

Table III Electropolishing Techni

Metal or Alloy††	Condition	Initial Form of Specimen	Technique	Electrolyte Composition
Al Al–Cu	All treatments	Rolled sheet, 125 μ thick	Electropolishing of small discs	1 part methyl alc 1 part nitric ac 1 cc hydrochlo acid to 50 cc o mixture
Au Au–2% Cu Au–2% Ag	,,	Beaten foil, 0.1 μ thick	Etching	Dilute potassium cyanide
Au	,,	Sheet, 75 μ thick	"Window" method	17 g potassium cyanide, 3.75 g potassium ferrocyanide, 3.75 g potassium sodium tartrate 3.5 cc phosphoric acid, 1 cc ammonia, 250 cc water
Au	,,	Sheet, 50–200 μ thick	"Uniform-field" electropolishing	As above
Co	,,	Sheet, 25–250 μ thick	"Window" method	77% glacial acetic acid, 23% perchloric acid
Cu Cu–Zn	,,	Rolled sheet, 25–250 μ thick	,,	33% nitric acid, 67% methyl alcoh
Cu Cu–Zn ‡ Cu–Al Cu–Ge § Cu–Sn ¶ Cu–Si	For all heat treatments	Rolled sheet, 25–200 μ thick	"Window" method in addition to Bollmann method	As above
Fe Low-C steels	,, As quenched or lightly tempered	Rolled sheet, ~50 μ thick Electromachined sheet, ~50 μ thick	"Figure-of-eight" method	10 parts glacial a acid, 1 part per chloric acid (60

Producing Metal Foils†—*continued*

Polishing Conditions				Remarks	Ref.
:athode	Voltage	C.D. amp/cm²	Temp. °C		
sheet	...	...	...	Wash specimen in water, then methyl alcohol. Final wash in 1:1 methyl alcohol and acetone	45
inless steel	...	...	...	...	39
,,	...	...	...	Wash in distilled water	64
,,	...	...	...	...	53
,,	22	0.75	< 30	Wash in methyl alcohol	48
sheet	4–8	0.5–0.6	< 30	,,	48
sheets and Cu points or stainless steel	5 6 9 8 9 6	0.5	< −20	Solution must be cooled to give correct C.D. Before removing specimen pour liquid nitrogen onto solution to form layer and withdraw specimen through this layer. Wash in cold methyl alcohol	49
inless teel	12	0.1	< 30	Wash and cut foils in methyl alcohol. Electromachining from 1 mm thick to 50 μ done with hydrochloric acid electrolyte	28 56

Table III Electropolishing Techniqu

Metal or Alloy††	Condition	Initial Form of Specimen	Technique	Electrolyte Composition
Fe	All treatments	Rolled sheet, 25–200 μ thick	"Window" method	20 parts glacial ace acid, 1 part perchloric acid
Fe	,,	Sheet, 50–200 μ thick	"Uniform-field" electropolishing	10 parts glacial acetic acid, 1 par perchloric acid
Fe Plain C steels Low-alloy steels Weld metal	Quenched, tempered, or normalized Early stages of tempering	Electromachined sheet, ~50 μ thick	"Figure-of-eight" method	135 cc glacial aceti acid, 25 g chromic oxide, 7 water‖
Fe Plain C steels Uranium steels 20% Ni steel	... Quenched, tempered, or normalized As quenched to γ or mixture of γ+ martensite	,,	Bollmann method	,,
Stainles steel Si steel	All treatments	Rolled sheet, 25–250 μ thick	,,	60% orthophosphor acid, 40% sulfuri acid
Stainless steel	,,	Sheet, 50–200 μ thick	"Uniform-field" electropolishing	,,
Si iron 50% Fe–50% Co	,,	Sheet, 50–200 μ thick	"Uniform-field" electropolishing	135 cc glacial acetic acid, 27 g chromi acid, 7 cc water
Mg	,,	Sheet, 25–250 μ thick	"Window" method	33% nitric acid, 67% methyl alcoho
Mg	All treatments	Rolled sheet, 25–250 μ thick	,,	20% perchloric acic 80% ethyl alcohol

r Producing Metal Foils†—*continued*

Polishing Conditions				Remarks	Ref.
Cathode	Voltage	C.D. amp/cm²	Temp. °C		
ainless steel	35–45	0.7	< 30	Wash in methyl alcohol	48
,,	100	...	< 30	...	53
,,	25–30	0.1–0.2	< 30	Wash in water, then methyl alcohol, and cut specimen under methyl alcohol. Electromachining from ~ 1 mm to 50 μ with hydrochloric acid electrolyte	28 56
ainless steel points	,,	,,	,,	,,	56
,,	9 20 falling to 9	1.5 3.5	60	Specimen may be electromachined from ~1 mm to 50 μ with hydrochloric acid electrolyte	44 46 48
,,	9	1.5	60	...	53
,,	25–30	0.1–0.2	< 30	...	53
ainless steel	9	0.5	< 30	Wash in methyl alcohol	48
g	10	0.2–0.5	0	Wash in orthophosphoric acid. Keep under alcohol	67

Table III Electropolishing Techniqu

Metal or Alloy††	Condition	Initial Form of Specimen	Technique	Electrolyte Composition
Mo	All treatments	Rolled sheet, 25–200 μ thick	,,	870 cc H_2SO_4, 30 cc H_2O§§
Nb	,,	,,	,,	40% HF+HNO_3, 60% H_2O
Ni	,,	Sheet, 25–250 μ thick	,,	23% perchloric aci 77% glacial acet acid
Ti Ti–Cu Ti–Al, V Ti–Al, Mn	,,	Rolled sheet, 20–80 μ thick	Electropolish with specimen backed by sheet of same material. Collect fragments	Any electrolyte wh gives a good pol and will not etch without the actic of current
U	Quenched, irradiated, annealed	Rolled sheet	"Window" method	33% orthophosphc acid, 33% ethyl alcohol, 33% glycerol
U	,,	,,	,,	7.5% perchloric ac 92.5% glacial acetic acid
U	,,	,,	,,	20% orthophosphoric acid, 40% H_2SO_4, 40% wa

† Some precautions should be observed when using certain of the electrolytes listed i Table III, particularly for all perchloric acid solutions. The perchloric acid-ethyl alcohol mixtures are inflammable above 35° C and such solutions, even during their preparation, shoul always be kept well cooled. Similar remarks apply to perchloric-acetic acid mixtures. Th usual care must be taken when handling cyanide and HF solutions because of the hazards c poisoning, burning, and possible necrosis of the bone.

‡ Pink deposits of Cu indicates solution must be changed. Proportion of HNO_3 should b increased to 40% (max.) as Zn content is raised.

Producing Metal Foils†—*continued*

Polishing Conditions				Remarks	Ref.
:athode	Voltage	C.D. amp/cm^2	Temp. °C		
or Mo :ylinder	10–21	0.02 0.06	< 30	Start at 21 v, finish at 10 v. Wash and keep under glass-distilled water	67
sheet or stainless steel	10–15	...	< 30	Wash in distilled water	93 94
inless teel	20–30	0.7	< 30	Wash in methyl alcohol	48
,,	...	...	...	...	50 51
sheet	10–20	0.2–0.5	...	Wash in glass-distilled water or methyl alcohol	68
inless teel	20–30	...	...	,,	69
,,	10–20	...	...	,,	55

§ A black film forms on the specimen. This can be removed by vibrating the specimen.

¶ Remove flaky film by immersing finished foil in distilled water prior to washing in alcohol.

‖ This solution is also suitable for preparing foils from Au–Cu and Ni–Mn alloys of composi-ons corresponding to those which form superlattice structures.

†† The greater part of Table III is due to Drs. P. M. Kelly and J. Nutting, courtesy of ıstitute of Metals.

§§ This solution is also suitable for Nb, Ta, and W. Alternatively use 5% H_2SO_4, 1.25% F in methanol, current density 4.4 amp/cm^2 at 50–70 volts, polishing at O° C.

and a completely assembled model in Fig. 95*b*. A glass jet ~1 mm in diameter connected to a stainless steel tube cathode is moved horizontally backwards and forwards at constant velocity thirty times a minute while the specimen, mounted perpendicularly to the jet at a distance of 1 to 2 mm, is moved vertically up and down once every 6 min also at constant velocity. The cams imparting these motions have a 1-in. throw, so that a square area on the specimen of 1-in. side is covered by the jet. The accuracy of the machining is better than 2% of the metal removed. With this and using a HCl electrolyte at 2 amp and 50 volts, for steel specimens, the rate of metal removal is ~250 μ/amp/h. Since it is possible to eliminate the effects of mechanical cutting and grinding by electromachining without destroying or modifying the structure of the interior of the metal, thin foils can be prepared from as large a specimen as required. This technique is used to thin down to about 100 μ thick, after which any of the methods described previously for electropolishing may be used to obtain the thin foils. For steels it has been found[56] that the Bollmann technique works best.

A similar jet technique has been used by Washburn[57] to obtain foils ~1000 Å thick directly from cleaved MgO crystals ~$\frac{1}{4}$ mm thick. The

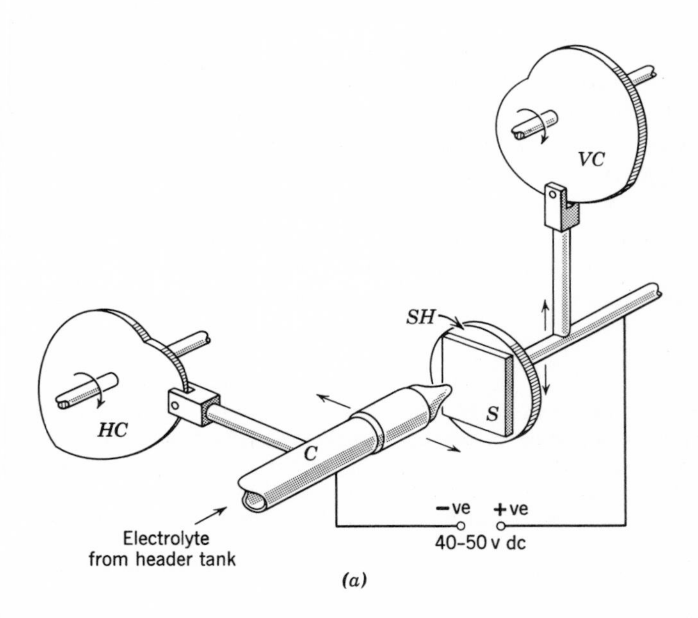

FIG. 95(*a*). Sketch of electrolytic jet-machining apparatus. *HC*, horizontal traverse cam; *VC*, vertical traverse cam; *C*, cathode; *S*, specimen; *SH*, specimen holder. (After Kelly and Nutting.[38])

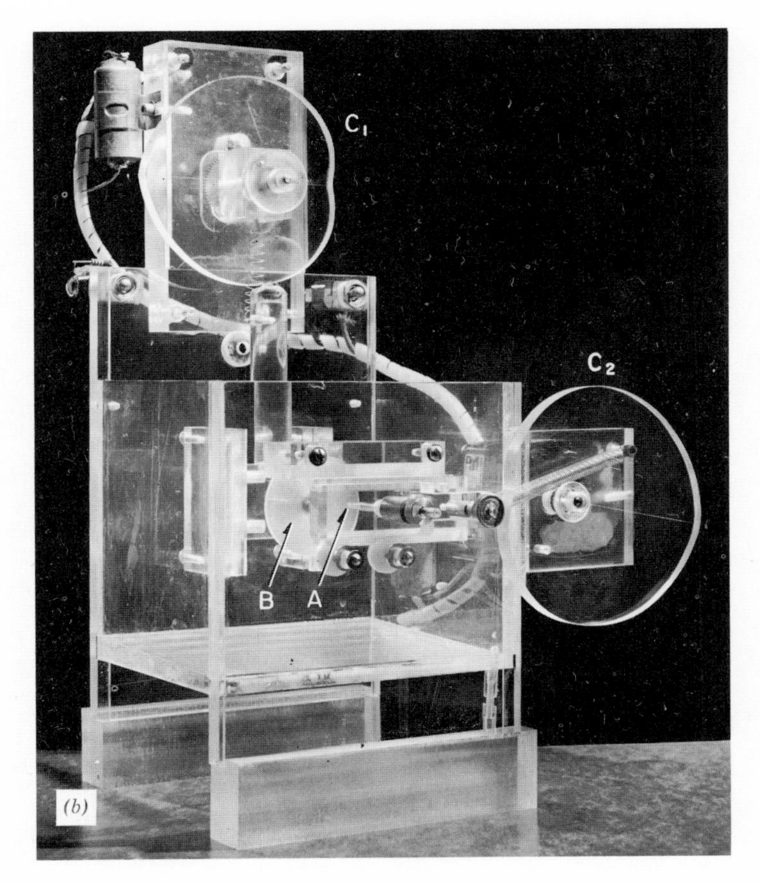

FIG. 95(*b*). Assembled jet-machining apparatus. C_1, vertical traverse cam; C_2 horizontal traverse cam; *A*, jet; *B*, specimen holder.

method is illustrated in Fig. 96. The etchant used is concentrated orthophosphoric acid which is heated to 100° C and sprayed onto the specimen as shown. The height of the specimen above the jet is such that breakaway of the liquid is just prevented. The specimen is rotated about a vertical axis to produce a circular thin area with a slightly thicker center. Within a few seconds of the first appearance of a hole good thin edges are obtained. After washing thoroughly and drying by rinsing in methyl alcohol and then ethyl ether, pieces may be broken off with a needle. New thin areas from the original sample can then be obtained by dipping into hot 85% orthophosphoric acid.

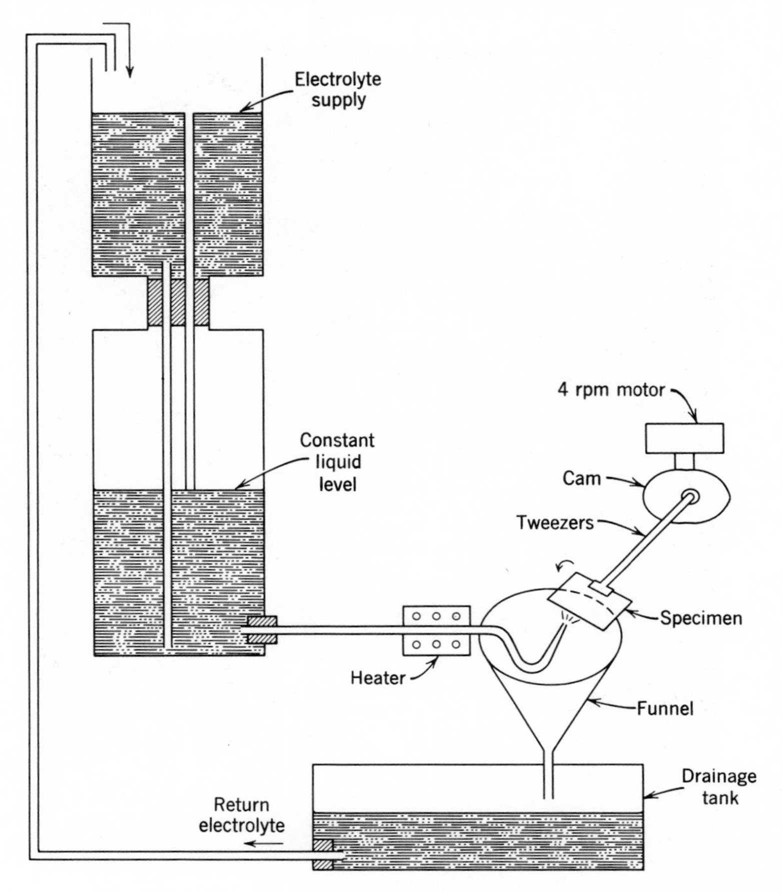

FIG. 96. Sketch of apparatus for preparing thin foils of MgO. (After Washburn et al.[57])

Since MgO is transparent to light, the thinned area may be viewed under an optical microscope so that selective sampling of various areas can be done. The thin areas will show extinction fringes as shown in Fig. 97, and these thin regions can be selected and carefully mounted on a microscope grid under an optical microscope prior to examination in the electron microscope. In this way it is possible to examine particular regions in the crystal e.g., slip bands at *A*. (See also section 5.2.13.)

Thin foils of the alkali halides can be prepared using the window method and etching in tap water. Unfortunately, the nonconducting properties

FIG. 97. Light optical photograph of MgO, after thinning, mounted on a 75 mesh microscope grid. The region showing fringes will be suitable for transmission electron microscopy (× 65).

of these materials (e.g., LiF) make electron microscopic examination extremely difficult, since the specimens decompose under prolonged electron bombardment.

In some cases it is possible to thin down bulk specimens by normal bright-dipping or chemical polishing techniques.[54] Usually, however, the attack is nonuniform, but by constantly reversing the specimen or rotating it in the solution satisfactory results may be obtained. With aluminum and its alloys, e.g., the following two solutions have been found to work well:† (1) 400 cc/liter HCl, 5 g/liter NiCl—solution kept at 30° C. (2) 200 g/liter NaOH—solution kept at 70° C. Both solutions remove metal uniformly from the surface at the rate of about 0.001 in. per minute. Overheating the specimens during chemical dissolution is avoided either by maintaining an efficient cooling system or washing the specimens in cold water every minute or so.

For copper and its alloys an acid solution consisting of 50% nitric acid,

† F. Keller (private communication).

25% orthophosphoric acid, and 25% acetic acid is suitable for chemical polishing. If the specimens are masked off with lacquer at a distance of $\frac{1}{16}$ in. from the edges and immersed in this solution, the polishing rate is approximately 0.00025 in. per min from each surface. The specimens should be inverted every 10 min or so, depending upon the initial thickness. When the center portion is 0.002 to 0.004 in. thick, it drops away owing to the accelerated attack at the lacquered edge. At this stage the center part should be uniformly thick. This portion may then be thinned using the usual electropolishing solutions (Table III).

A similar procedure may be adopted for silver and its alloys using an agitated 50% aqueous solution of nitric acid. In fact, some of the Ag–Al alloys of hexagonal structure (e.g., 33 at % Al) may be thinned directly from 1 mm thickness to give foils transparent enough for electron transmission without the necessity of final thinning by electropolishing. There is one disadvantage, however, namely, that for the Ag–Al alloys this solution will only chemically polish specimens in the basal plane orientation.

4.3.5 *Cleavage Techniques*

Many nonmetallic materials are found to cleave easily when struck sharply along the cleavage plane by a wedge-shaped tool. This fact has recently been utilized for preparing thin flakes of layer materials such as graphite, molybdenite, bismuth and antimony tellurides, mica, talc, and chlorite, which are suitable for examination by transmission electron microscopy.[58–62] Most of these materials possess some hexagonal structure and cleave easily along {0001} planes. These planes are also the glide planes so that movement and interaction of glide dislocations can readily be studied, particularly since the cleavage operation introduces cold work in the specimens. (See also Chapter 5.) The technique involves cleaving natural, single crystals, using a wedge-shaped tool to produce a clean surface, followed by successive cleaving using adhesive tape. Eventually, small, thin flakes are obtained which are stuck to the tape. These can be removed by immersing the tape in petroleum ether which dissolves away the gum, thus leaving the flakes floating on the surface. They are then collected on microscope grids, ready for examination.

4.4 Direct Formation of Thin Films

4.4.1 *Vacuum Evaporation*

It is well known that thin films of metals can be produced by evaporating the metal in the form of a thin wire in vacuo and condensing the vapor onto a substrate. The films can be removed by stripping from, or dissolution of, the substrate. If the substrate is a crystalline material and evaporation is carried out at a high temperature, the orientation of the film follows that of the surface of the substrate and thus grows epitaxially.[70] In this way single crystal thin films of many metals can be prepared. In most cases, it is more convenient to use a substrate of a metal of given orientation (by previous evaporation onto a substrate such as mica or rock-salt cleavage surfaces for cubic metals) in which the surface is atomically smooth. If evaporation is carried out on a cold, amorphous or polycrystalline substrate, polycrystalline deposits are formed.

Both metals and thermally stable compounds begin to evaporate rapidly in vacuo when their temperature has been sufficiently raised for their vapor pressure to have reached a value in excess of 10 μ mercury.[8] Some substances such as Mg, Cd, Zn, and ZnS can be evaporated from the solid state, but the majority of metals and dielectrics deposited as thin films evaporate from the liquid phase. The technique is to use a refractory metal heater in a standard evaporating unit such as that illustrated in Fig. 77. These heaters can be divided into two main groups:

1. Those made from wire such as straight helices or conical baskets (usually made of tungsten).
2. Those constructed from foil bent into the form of troughs (usually made of molybdenum). This type is not suited to the evaporation of metals which may readily wet the surface heater, because on melting the metal flows over the heater and considerably lowers its resistivity, so that the heater leads may burn out.

Holland[8] has compiled a table of melting points and temperatures at which the vapor pressure is equal to 10 μ Hg for a range of metals. This is reproduced in Table IV and is intended only as a guide, because with some sources vapor pressures greatly in excess of 10 μ Hg may be reached.

As a specific example of the formation of oriented single crystal deposits we shall describe the techniques used by Pashley et al.[35, 71, 78] Films of cubic metals are formed in (111) orientation by evaporating onto a cleaved mica substrate about 1 cm^2 in area, kept at 270° C in a standard evaporating unit. For films of (100) orientation the substrate used is a cleaved rock-salt single crystal. The metal film, such as silver, is then used as the substrate for further evaporation since the surface is atomically smooth. For example, oriented films of Au, Pt, Pd, and Rh may be formed by evaporation onto silver (on mica or rock salt), after which the films are detached by dissolving the silver in nitric acid. In general, the higher the temperature of the substrate the greater likelihood there is that orientation will occur. It should be noted that although a slow rate of evaporation

Table IV Vapor Sources. Techniques and Refractory Support Materials for Evaporating Metals together with some useful Vapor Pressure Data

Metal	*M.P. (°C)*	*vap. temp. (°C) V.P. = 10 microns Hg*	*Resistance heated sources in order of merit*†		*Other evap. techniques*		*Remarks*
			Filaments	*Boats*	*Source*	*Method of heating*	
Aluminium	660	996 (?) 1148 (?)	W, Ta, helical coil	—	—	—	Alloys freely with refractory metals and reacts with carbon and oxide crucibles.
Antimony (Sb_2)	630	678	Chromel, Ta, conical basket	Mo, Ta	Alumina crucible	External W-heater	Wets Chromel.
Arsenic	—	—	—	—	Alumina crucible	External W-heater	—
Barium	717	629	W, Ta, Mo, Cb, Ni, Fe, Chromel, conical basket	Ta, Mo	— —	— —	Freely wets without alloying with the heater metals quoted. Reacts with alumina.
Beryllium	1284	1246	Ta, W, Mo, conical basket	—	Carbon crucible BeO crucible	Electron bombardment High frequency	Wets heater metals quoted.
Bismuth	271	698	Chromel, Ta, W, Mo, Cb, conical basket	Fe, Mo, Ta	Alumina crucible	External W. heater	Wets chromel.
Boron	2000 –2080	1355	—	—	Carbon	Resistance heated	Deposits from carbon probably impure.
Calcium	810	605	W, conical basket	—	Alumina crucible	External W-heater	
Caesium	29	153	—	—	—	—	
Cadmium	321	264 (subl.)	Chromel, Cb, Ta, Mo, W, Ni, Fe, conical basket	Mo, Ta	Alumina crucible Fe-crucible	External W-heater Nichrome heater	Freely wets Chromel and Cb.
Carbon	3700 ±100	2681	—	—	Carbon rods pressed together forming high resistance contact †	Resistance heating	—
Cerium	785	1305	—	—	—	—	—
Chromium	1900	1205	W, conical basket	—	Electro-dep. Cr on W-helical coil	Sub-limation	—

† See Fig. 77.

Table IV Vapor Sources—*continued*

Metal	*M.P. (°C)*	*Evap. temp. (°C) V.P. = 10 microns Hg*	*Resistance heated sources in order of merit*		*Other evap. techniques*		*Remarks*
			Filaments	*Boats*	*Source*	*Method of heating*	
Cobalt	1478	1649	Cb, W	—	Alumina or BeO crucible Electro-plated W-spiral	Embedded W-heater Resistance heater	Alloys with W, Ta, Mo, Cb, Pt. Evaporant weight must not exceed 35 per cent of that of W spiral.
Columbium	2500	(V.P. at M.P. = 1μ Hg.)	W, helical coil	—	—	—	Refractory metal used as a source material.
Copper	1083	1273	Pt-helical coil; Cb, Mo, Ta, W, conical basket	Mo, Ta	Alumina crucible	Embedded W-heater	Copper alloys with Ni, Fe, Chromel. Does not wet readily Mo, W, Ta.
Gallium	30	1093	—	—	BeO, SiO_2 Al_2O_3	—	Alloys with metals, oxides quoted resist attack up to 1000° C.
Germanium	959	1251	Ta, Mo, W, conical basket	Mo, Ta	Alumina crucible Carbon crucible	Embedded W-heater Resistance heater	Wets Ta and Mo.
Gold	1063	1465	W, Mo, conical basket	Mo	—	—	Wets but reacts with Ta, possible alloy formation. Partially wets W, Mo.
Indium	157	952	W, Fe, conical basket	Mo	—	—	—
Iridium	2454	2556	—	—	—	—	—
Iron	1535	1447	W, helical coil	—	Alumina or BeO crucible	Embedded W-heater	Alloys with W, Ta, Mo, Cb. Evaporant must not exceed 35 per cent of the weight of W-filament.
Lead	328	718	Fe, Ni, Chromel, conical basket	Mo —	Alumina or iron crucible Carbon (?)	External Nichrome heater Resistance heater	Does not wet W, Ta, Mo, and Cb.
Lithium	179	514	—	—	Mild steel (?) crucible	Nichrome heater	—
Magnesium	651	443 (subl.)	W, Ni, Fe, Ta, Mo, Cb, Chromel, conical basket	Mo, Ta	Iron crucible Carbon	External Nichrome heater Resistance heater	Does not melt when volatilized from open spirals and boats.
Manganese	1244	980	W, Ta, Mo, Cb, conical basket	—	Alumina crucible	Embedded W-heater	Freely wets heater metals quoted.

Table IV Vapor Sources—*continued*

Metal	*M.P. (°C)*	*Evap. temp. (°C) V.P. = 10 microns Hg*	*Resistance heated sources in order of merit*		*Other evap. techniques*		*Remarks*
			Filaments	*Boats*	*Source*	*Method of heating*	
Molybdenum	2622	2533	—	—	—	—	Refractory metal used for filament and boat type vapour sources. Evaporates rapidly if oxidized to form MoO_3.
Nickel	1455	1510	W, heavy gauge helical coil	—	Alumina or BeO crucible	Embedded W-heater	Alloys with Mo, Ta and W. Evaporant must not exceed 30 per cent of the weight of W-filament.
Palladium	1555	1566	W, helical coil	—	—	—	—
Platinum	1774	2090	Multi-strand W-filament with Pt-wire twisted together	—	Electro-deposited Pt on W spiral	Resistance heater	Alloys with Ta and partially with W. Platinum may be used as a source heater for metal oxides to prevent decomposition of charge, see page 448.
Rhodium	1967	2149	—	—	Electro-deposited Rh on W-spiral	Sublimation	Requires very low pressure for deposition of neutral transmitting films.
					Resistance heated Rh-foil	Sublimation	
Selenium	217	234	Chromel, Fe, Mo, Cb, conical basket	Mo, Ta	Alumina crucible	Nichrome external heater or radiant heater	Very volatile, may contaminate plant. Wets filament metals quoted.
Silicon	1410	1343	—	—	BeO crucible	Embedded W-heater	Difficult to prepare Si-films free from SiO contamination.
Silver	961	1047	Ta, Mo, Cb, Fe, Ni, Chromel helical coil or W conical basket	Mo, Ta	Electro-deposited Ag on Mo helical coil	Resistance heater	Ag does not wet W. Can be kept in basket by binding fine platinum wire on outside.

Table IV Vapor Sources—*continued*

Metal	*M.P. (°C)*	*Evap. temp. (°C) V.P. = 10 microns Hg*	*Resistance heated sources in order of merit*		*Other evap. techniques*		*Remarks*
			Filaments	*Boats*	*Source*	*Method of heating*	
Strontium	771	549	W, Ta, Mo, Cb, conical basket	—	Carbon	Resistance heated	Freely wets without alloying with all filament metals quoted.
Tantalum	2996	(V.P. = 1μ Hg at 2820)	—	—	—	—	Refractory metal used for source heaters.
Tellurium	452	(V.P. = 760 mm Hg at 1390)	W, Ta, Mo, Cb, Ni, Fe, Chromel conical basket	—	Alumina crucible	External Nichrome heater	Very volatile, may lead to plant contamination. Wets without alloying all metal heaters quoted.
Thallium	304	606	Ni, Fe, Cb, Ta. conical basket	—	Alumina crucible	External Nichrome heater	Freely wets metal heaters quoted without alloying. Partially wets W, Ta, but not Mo.
Thorium	1827	2196	W conical basket	—	—	—	Wets W-heater.
Tin	232	1189	Chromel, helical coil ; Mo, Ta conical basket	Mo, Ta	Alumina crucible Carbon	Embedded W-heater Resistance heater	Wets chromel and Mo.
Titanium	1727	1546	W, Ta, conical basket or helical coil	—	Carbon	Resistance heated	Ti reacts with W-spiral, deposit contains trace of W. Ti does not react with Ta but filament may burn out during premelting of Ti.
Tungsten	3382	3309	—	—	—	—	Refractory metal used for source heaters. Evaporates more readily if surface oxidized to form volatile WO_3 or WO_2.
Uranium	1132	1898	W conical basket	—	—	—	Forms oxidized deposits at lowest gas pressures.

Table IV Vapor Sources—*continued*

Metal	*M.P. (° C.)*	*Evap. temp. (° C) V.P. = 10 microns Hg*	*Resistance heated sources in order of merit*		*Other evap. techniques*		*Remarks*
			Filaments	*Boats*	*Source*	*Method of heating*	
Vanadium	1697	1888	W, Mo, conical basket	—	—	—	Alloys with Ta and partially with W. Wets but does not alloy with Mo.
Yttrium	1477	1649	—	—	—	—	—
Zinc	419	343 (subl.)	W, Ta, Mo, Cb, conical basket	—	Alumina or iron crucible Carbon	external Nichrome heater Resistance heater	Wets without alloy formation all filament metals quoted.
Zirconium	2127	2001	W, conical basket or helical coil	—	—	—	Requires low pressures <0·1 microns Hg to prevent film oxidation. Evap. characteristics similar to Ti.

From L. Holland, *The Vacuum Deposition of Thin Films*, 1956, by permission of John Wiley and Sons, New York.

favors the occurrence of orientation (i.e., epitaxy), a fast rate favors the formation of a continuous coherent film so that a compromise is necessary for a coherent oriented film to be formed. The mechanism of growth of such films is still not well understood; however, the observations show[35] that the deposits grow continuously from small nuclei to full coherence when the thickness has reached ~800 Å.

It is difficult to control the thickness of the evaporated layer purely by evaporating a weighed quantity of metal owing to losses inside the evaporating unit [see Equation (76), section 4.1.1]. The thickness may, however, be controlled by the evaporating time and voltage and current density of the glow discharge, but for routine purposes the most satisfactory method is to measure the optical (by multiple beam interferometry) or electrical (e.g., resistivity) characteristics of the deposit during growth.

There is no precise law relating the partial vapor pressures of alloy components to their vapor pressures in the pure state, and the published data for the vapor pressures of alloy systems is mainly restricted to binary alloys containing a highly volatile component, i.e., alloys which can be separated into their respective components by fractional distillation in vacuo. Alloys of the latter type can only be deposited as thin films in which the alloy components are uniformly distributed by one of the following techniques:

1. Rapid evaporation of a succession of finely divided metal alloy particles onto a hot substrate.
2. Simultaneous evaporation from two or more sources each containing an alloy component (see e.g., Takahashi et al.[74–77]).
3. The alternate evaporation of pure metal films to form a multilayer deposit which is then homogenized by annealing at a temperature which facilitates rapid diffusion of the metal atoms within the layers.

Copper-gold alloy films have been prepared by method (3) for investigations of the order-disorder phase changes which occur at certain compositions.[72] The technique is as follows: a single crystal gold film in (001) orientation is prepared by evaporation onto silver on rock salt. After the film is detached from its substrate (rock-salt is dissolved in water and the silver in nitric acid), the required amount of copper for the alloy is evaporated onto it. The composite film is homogenized by annealing at $\sim 350°$ C for 1 h. This treatment also allows ordering of the alloy to take place. This technique is suitable for most alloys, and by choosing a suitably oriented substrate, single crystal films of any required orientation can be formed.

An application of the evaporation technique has been made to the study of metal lattices by means of the moiré pattern which is formed in the images obtained from two thin overlapping crystals.[78] The technique is to evaporate the second metal onto the oriented first metal as described previously, maintaining a substrate temperature in the range 300 to 450° C. Under these conditions the second metal grows in parallel orientation on the first. This method has been completely successful with Ni, Cu, Rh, Pd, and Pt deposits on gold.[78] Some of the results obtained are described more fully in the following chapter.

Polycrystalline iron-alloy foils have been prepared by Pitsch to study martensitic reactions in these alloys.[79] For Fe–C and Fe–N alloys, iron is first evaporated onto collodion substrates at room temperature. After dissolution of the collodion, the iron films are mounted on a grid and heat-treated in a furnace (at about 1100° C) in gaseous mixtures of hydrogen and ammonia (for Fe–N alloys), or methane (for Fe–C alloys) to produce alloying. Heat treatment is then carried out directly from this furnace. Iron-nickel alloy foils are prepared by using an iron-nickel alloy as the evaporation source.

The properties of thin foils formed by evaporation are quite different from those of the bulk metal; e.g., gold foils deform elastically in tension up to strains of 1 to $1\frac{1}{2}\%$. This corresponds to a tensile strength of approximately 60 tons/in.^{2} compared with 15 tons/in.^{2} for hard-drawn bulk wire.[35] The electrical and magnetic properties of thin foils are also different from those of bulk materials,† so that although the results from experiments on thin foils cannot be strictly related to bulk behavior, it is important to study the properties of thin films in relation to their microstructure; for besides their intrinsic interest, they are also becoming of great importance commercially.

4.4.2 *Distillation from the Vapor*

Coleman and Sears[80] have described a technique whereby zinc crystals can be prepared by distillation in vacuo ($\sim 10^{-6}$ mm Hg) at about 475° C, followed by condensation in a growth vessel under a helium atmosphere. This method has been used by Price[81] to study the nucleation and growth of twins and dislocation loops produced by pyramidal glide in zinc. The crystals formed by the distillation technique are usually dislocation-free. They can be grown as whiskers 2 to 6 mm long and 5 μ to 1 mm wide and have their largest faces in the basal plane. In general, the thickness varies from 0.05 to 0.5 μ, and are thus transparent to a 100 kv electron beam over their entire length. Price has also obtained thin crystals of cadmium by a similar technique. In this case, the cadmium is deposited from the vapor onto a quartz fibre in an argon diffusion cell. At low supersaturations, the cadmium crystallizes out in the form of thin whiskers and platelets.

4.4.3 *Deposition from Solution*

Large, thin single crystal flakes of gold have been prepared by Suito and Uyeda[82] by reduction of a dilute solution of auric chloride. The thin crystals (100 to 200 Å thick and several microns across) showed trigonal and hexagonal habit and grew parallel to (111). The method can be used for preparing oriented single crystals of other metals, but again the value of such thin crystals in giving information about bulk properties is limited.

4.4.4. *Electrodeposition*

There are two methods available for producing thin electrodeposits for examination in the electron microscope. First (due to Weisenberger[82]) thin carbon support films are used as conducting electrodes onto which the metal is deposited. The second method (Weil and Read[84]) consists in electroplating nickel onto copper or zinc and stripping the thin electroplate from the cathode. The electroplating bath used is a purified Watts bath at a *pH* of 4.0 and at a temperature of 60° C. The current density is 0.1 to

† See, e.g., Neugebauer et al., *Structure and Properties of Thin Films*, John Wiley and Sons, New York, 1959.

1 amp/sq. cm. Stripping is carried out by a solution of 250 to 500 gm/liter of CrO_3 in 25 gm/liter of H_2SO_4 kept at 60° C. This technique has been used by Reimer[85] to study epitaxial growth of electrodeposits. Apart from the investigation of the structure of electrodeposits this technique has little application.

4.4.5 *Preparation from the Melt*

Takahashi et al.[75] have developed a technique for preparing foils from the melt. An elliptical loop of iron wire (1 cm major axis, 0.3 cm minor axis) coated with carbon or alumina powder is dipped into the molten metal and withdrawn at the rate of 2 cm/sec. For metals which oxidize readily in air this must be done in an inert atmosphere as illustrated in Fig. 98, but otherwise the method should be universally applicable. Results obtained with alloys of Sn–Pb, Al–Ag, Al–Sn, and Al–Cu[77, 83, 87] have shown that the microstructure of the thin parts of the foil was not typical of bulk samples but thicker sections (of approximately a few thousand angstroms) were. However, other materials may show a

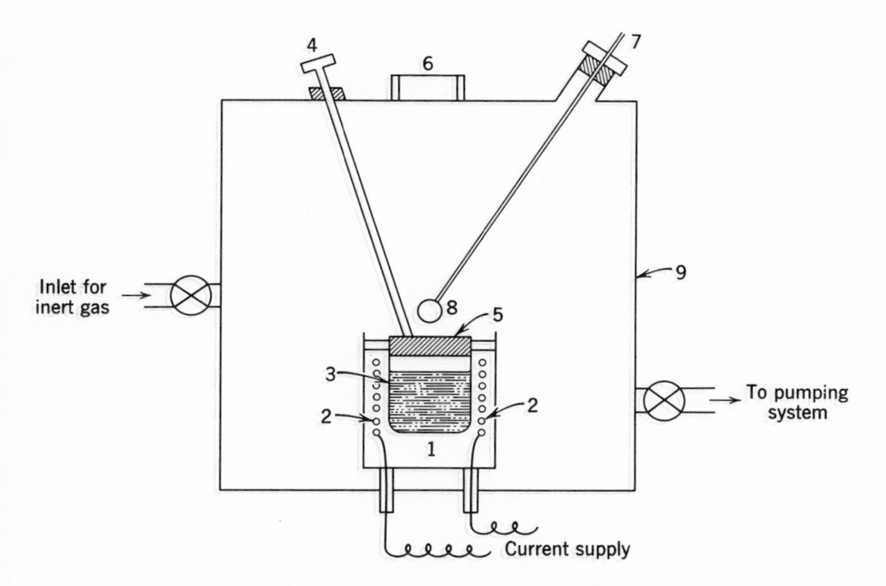

FIG. 98. Sketch of apparatus for preparing thin foils from the melt. 1. Alumina crucible. 2. Mo heating coils. 3. Molten metal. 4. Rod to withdraw cover 5. 6. Viewing window. 7. Rod to manipulate loop 8. 9. Vacuum tight vessel. (After Takahashi et al.[75])

different lower limit of thickness for truly bulk behavior, and so it is dangerous to extend results obtained from such foils to large sections.

An ingenious method for shock-chilling droplets of molten metal has been briefly described by Duwez et al.[88] A small liquid droplet (~25 mg) is kept molten in a graphite mold, and is then propelled by a shock-wave against a solid copper target. To maintain good contact between the metal and the target the latter is made as a cylinder and is rotated at high speed. The centrifugal force acting on the molten material insures good thermal contact. The metal spreads over the cylinder as a very thin film, so thin, in fact, that some areas are transparent to light. Alloys of Ag–Cu and Ga–Sb–Ge formed using this technique exhibit extremely interesting properties; thus this new field is well worth further exploration.

4.4.6 *Use of Ultramicrotomes*

Haanstra[89] has designed an ultramicrotome utilizing thermal expansion for advancing the specimen and the magneto-restriction of a nickel rod for withdrawal in order to cut biological sections 50 to 100 Å thick. Foils of soft metals were also cut using this apparatus, but they were heavily deformed. Reimer[85] used a diamond knife (described by Fernandez-Moran[90] for cutting biological sections) with an ultramicrotome and succeeded in obtaining thin sections of Al, Ni, Cu, Au, Fe, Pd, Pt, and Ag, but again the foils were heavily deformed. An example is shown in Fig. 99, in which a thin section of polycrystalline aluminum exhibits a flowed grain pattern typical of a heavily deformed metal. More recently Phillips[91] using a Leitz ultramicrotome † obtained sections 100 to 500 Å thick from samples of Al–Al_2O_3, Au–Ni, Pb–Sn, and Cu–Be alloys which had been previously mounted in araldite. The method is more likely to be useful in preparing foils of multiphased alloys (when the electropolishing technique is not successful) and hard metals, e.g., ceramics, steels, etc., and holds great promise as a rapid method for obtaining thin specimens.

4.5 Summary

Of all the methods discussed for thinning bulk specimens, the electropolishing one is the most widely applicable and gives consistent and reliable results. With alloys containing noncoherent second phases, difficulties may be encountered when preferential leaching-out of the second phase occurs. In this case, the use of a microtome seems the best solution.

To discover whether the thinning operation caused some readjustment

† Other instruments are also available commercially.

FIG. 99. Micrograph of thin section of aluminum obtained using an ultramicrotome. The grains have been deformed in the direction of cutting. (Courtesy of A. W. Agar and Aeon Laboratories.)

of the structure near the surface, a technique has been used by Hirsch et al.,[92] whereby thinning by electropolishing is carried out from one side only. That is, one surface of the specimen is completely covered with lacquer. The window method is then used in the normal way until perforation occurs. Thin specimens are cut from the piece and the lacquer is removed by dissolution. In this way the dislocation structure in slip bands were preserved, enabling a comparison to be made between the surface and interior of the same specimen (e.g., see Fig. 117). There is some evidence from this work that a rearrangement of structure does occur because of elastic strain relaxations produced during thinning, but the "one-side" technique does prevent this in most cases. Thus, we are fairly certain now that the examination by electron microscopy, of foils prepared from bulk samples by electropolishing, enables a direct and true correlation to be made between microstructure and bulk properties.

References

1. L. Reimer, *Preparation Methods for Electron Microscopy* (in German) (Springer-Verlag, Berlin), 1959.
2. V. J. Schaeffer and D. Harker, *J. Appl. Phys.* 1942, **13**, p. 427.
3. "Metallurgical Applications of the Electron Microscope," *Inst. of Metals*, Rep. and Mon., Ser. 8, 1950.
4. *Proc. Amer. Soc. Test. Materials*, 1950, **50**, pp. 444, 489.†
5. *Ibid.* 1952, **52**, p. 543–591.†
6. *Ibid.* 1957, **57**, p. 452–535.†

† These are also published separately as A.S.T.M Special Technical Publications, e.g., see also No. 155, 1953, No. 245, 1958 and No. 262, 1959.

7. J. Nutting, *Rev. Univ. des Mines* 1956, **12**, p. 1.
8. L. Holland, *Vacuum Deposition of Thin Films* (John Wiley and Sons, New York and Chapman and Hall, London), 1956.
9. D. E. Bradley, *J. Inst. Metals* 1954–55, **83**, p. 35.
10. E. Smith and J. Nutting, *Proc. 3rd Int. Conf. Electron Microscopy* 1954 (Roy. Mic. Soc., London, 1956), p. 206.
11. E. Smith and J. Nutting, *Brit. J. Appl. Phys.* 1959, **7**, p. 214; *J. Iron and Steel Inst.* 1959, **187**, p. 314.
12. D. E. Bradley, *Proc. 4th Int. Conf. Electron Microscopy* 1958 (Springer-Verlag, Berlin, 1960), p. 428.
13. "Ferrous Applications of the Electron Microscope," *Proc. 3rd Int. Conf. Electron Microscopy*, 1954 (Roy. Mic. Soc. London, 1956).
14. *Proc. 4th Int. Conf. Electron Microscopy* 1958, Section 6, pp. 574–686 (Springer-Verlag, Berlin, 1960).
15. J. A. Coiley, P. A. E. Clark, and B. F. Sharpe, *ibid.*, p. 440.
16. R. M. Fisher, *J. Appl. Phys.* 1953, **24**, p. 113, and *A.S.T.M. Special Tech. Publ.* No. 155, 1954, p. 49.
17. G. R. Booker and J. Norbury, *Brit. J. Appl. Phys.* 1957, **7**, pp. 109, 154; *ibid.* 1958, **8**, p. 361.
18. R. G. Baker and J. Nutting, *J. Iron and Steel Inst.* 1959, **192**, p. 257.
19. F. Keller, *Inst. of Metals Symp. on Met. Appl. of Electron Microscope*, Mon. and Rep., Series 8, 1950, p. 85.
20. G. Thomas and J. Nutting, *J. Inst. Metals* 1956, **85**, p. 1.
21. G. Thomas and J. Nutting, *Inst. of Metals Symp. on Phase Transformations* Rep. and Mon., Ser. 18, 1955, p. 57.
22. P. Lacombe and A. Berghezan, *Alluminio* 1949, **18**, p. 365.
23. R. Castaing, *Compt. rend.* 1949, **228**, pp. 1341, 2033.
24. G. Thomas and J. Nutting, *J. Inst. Metals* 1957–58, **86**, p. 7.
25. H. Wilsdorf and D. Kuhlmann-Wilsdorf, *Z. Angew Phys.* 1952, **4**, pp. 361, 409.

26. H. Wilsdorf and D. Kuhlmann-Wilsdorf, *Acta Met.* 1953, **1**, p. 394.
27. H. Wilsdorf and J. T. Fourie, *ibid.* 1956, **4**, p. 271.
28. D. G. Brandon and J. Nutting, *Acta Met.* 1959, **7**, p. 101.
29. J. R. Low, *Fracture*, ed. Averbach et al. (Tech. Press M.I.T.; John Wiley and Sons, New York; and Chapman and Hall, London), 1959, p. 68.
30. R. L. Scott and A. M. Turkalo, *Proc. A.S.T.M.* 1957, **57**, p. 536.
31. R. W. Guard and A. M. Turkalo, *Mech. Properties of Intermetallic Compounds*, ed. Westbrook (John Wiley and Sons, New York), 1960, p. 141. See also A. M. Turkalo, *Trans. A.I.M.E.* 1960, **218**, p. 24.
32. D. M. Davies and J. W. Martin, *European Congress Electron Microscopy*, Delft, Holland, 1960, De Nederlandse Vereniging voor Electronenmicroscopie, 1961, p. 482.
33. C. Crussard, R. Borione, J. Plateau, T. Morillon and F. Maratray, *J. Iron and Steel Inst.* 1956, **146**, p. 183.
34. J. Plateau, G. Henry, and C. Crussard, *Rév. Univ. des Mines* 1956, **12**, p. 543.
35. G. A. Bassett and D. W. Pashley, *J. Inst. Metals* 1959, **87**, p. 449 (see also D. W. Pashley, *Proc. Roy. Soc.* A 1960, **255**, p. 218).
36. P. J. E. Forsythe and R. N. Wilson, *J. Sci. Instr.* 1960, **37**, p. 37.
37. D. R. Brame and T. Evans, *Phil. Mag.* 1958, **3**, p. 971.
38. P. M. Kelly and J. Nutting, *J. Inst. Metals* 1959, **87**, p. 385.
39. P. B. Hirsch, A. Kelly, and J. W. Menter, *Proc. 3rd Int. Conf. Electron Microscopy* 1954, (Roy. Mic. Soc., London, 1956), p. 231.
40. P. B. Hirsch, R. W. Horne, and M. J. Whelan, *Phil. Mag.* 1956, **1**, p. 677.
41. R. Phillips and N. C. Welsh, *ibid.* 1958, **3**, p. 801.
42. R. Castaing, *Rev. Met.* 1955, **52**, p. 669.
43. R. Castaing, *Proc. 3rd Int. Conf. Electron Microscopy* 1954 (Roy. Mic. Soc., London, 1956), p. 379.
44. W. Bollmann, *Proc. Stockholm Conf. Electron Microscopy* 1957 (Almqvist and Wiksell, Stockholm), 1957, p. 316.
45. R. D. Heidenreich, *J. Appl. Phys.* 1949, **20**, p. 993.
46. W. Bollmann, *Phys. Rev.* 1956, **103**, p. 1588.
47. R. B. Nicholson, G. Thomas, and J. Nutting, *Brit. J. Appl. Phys.* 1958, **9**, p. 25.
48. H. M. Tomlinson, *Phil. Mag.* 1958, **3**, p. 867.
49. P. R. Swann and J. Nutting, *J. Inst. Metals* 1960, **88**, p. 478.
50. P. Mirand and A. Saulnier, *Compt. rend.* 1958, **246**, p. 1688.
51. A. Saulnier and P. Mirand, *Proc. 4th Int. Conf. Electron Microscopy* 1958 (Springer-Verlag, Berlin, 1960), p. 390.
52. P. Michel, *Sheet Metal Ind.* 1949, **26**, p. 2175.
53. R. M. Fisher, *A.S.T.M. Special Tech. Publication* No. 262, 1959, p. 104.
54. P. A. Jacquet, Inst. of Metals, *Metallurgical Reviews* 1956, **1**, pt. II, p. 156.
55. W. J. McG. Tegart, *Electrolytic and Chemical Polishing of Metals* 1956 (Pergamon Press, London).
56. P. M. Kelly and J. Nutting, *J. Iron and Steel Inst.* 1959, **192**, p. 246.
57. J. Washburn, A. Kelly, and G. K. Williamson, *Phil. Mag.* 1960, **5**, p. 192; also with G. W. Groves, *ibid.*, p. 991.
58. G. Geach and R. Phillips, *Proc. 4th Int. Conf. Electron Microscopy* 1958 (Springer-Verlag, Berlin, 1960), p. 138.
59. S. Amelinckx and P. Delavignette, *Nature* 1960, **185**, p. 603; also *Phil. Mag.* 1960, **5**, pp. 533, 729.
60. G. K. Williamson, *Proc. Roy. Soc.* A 1960, **257**, p. 457.

61. F. W. C. Boswell, *European Congress Electron Microscopy*, Delft, Holland, 1960, De Nederlandse Vereniging voor Electronenmicroscopie, 1961, p. 409.
62. D. W. Pashley and A. E. B. Presland, *ibid.* p. 417.
63. R. B. Nicholson, *Ph.D. Dissertation* 1960, Cambridge University.
64. J. Silcox and P. B. Hirsch, *Phil. Mag.* 1960, **4**, p. 72.
65. J. E. Bailey, *Ph.D. Dissertation* 1960, Cambridge University.
66. A. Howie, unpublished work.
67. G. Thomas, J. Nadeau, and R. Benson, *European Congress Electron Microscopy*, Delft, Holland, 1960, De Nederlandse Vereniging voor Electronenmicroscopie, 1961, p. 447.
68. J. Silcox, *Proc. 4th Int. Conf. Electron Microscopy* 1958 (Springer-Verlag, Berlin, 1960), p. 552.
69. P. A. Jacquet and R. Caillat, *Compt. rend.* 1949, **228**, p. 1224.
70. D. W. Pashley, *Advances in Physics* 1956, **5**, p. 173.
71. D. W. Pashley, *Phil. Mag.* 1959, **4**, p. 324.
72. A. Glossop and D. W. Pashley, *Proc. Roy. Soc.* A 1959, **250**, p. 132.
73. S. Ogawa and D. Watanabe, *Acta Cryst.* 1954, **7**, p. 377; *ibid.* 1957, **10**, p. 860.
74. J. J. Trillat and N. Takahashi, *Compt. rend.* 1952, **235**, p. 1306.
75. N. Takahashi and J. J. Trillat, *ibid.* 1953, **237**, p. 1246.
76. N. Takahashi and K. Mihama, *Acta Met.* 1957, **5**, p. 159.
77. N. Takahashi and K. Ashinuma, *J. Inst. Metals* 1958–59, **87**, p. 19.
78. G. A. Bassett, J. W. Menter, and D. W. Pashley, *Proc. Roy. Soc.* A 1958, **246**, p. 345.
79. W. Pitsch, *Phil. Mag.* 1959, **4**, p. 577.
80. R. V. Coleman and G. W. Sears, *Acta Met.* 1957, **5**, p. 131.
81. P. B. Price, *Phil. Mag.* 1960, **5**, pp. 473, 873.
82. E. Suito and N. Uyeda, *Proc. 3rd Int. Conf. Electron Microscopy* 1954 (Roy. Mic. Soc., London, 1956), p. 223.
83. E. Weisenberger, *Z. Wiss. Mikroskop.* 1955, **62**, p. 163.
84. R. Weil and H. J. Read, *J. Appl. Phys.* 1950, **21**, p. 1068.
85. L. Reimer, *Z. Metallkunde* 1959, **50**, p. 37; *Naturw.*, **46**, p. 68.
86. N. Takahashi and K. Kazato, *Compt. rend.* 1956, **243**, p. 1408.
87. N. Takahashi and K. Ashinuma, *ibid.* 1958, **246**, p. 3430.
88. P. Duwez, R. H. Willens and W. Clement, *J. Appl. Phys.* 1960, **31**, pp. 1136, 1137, 1500.
89. H. B. Haanstra, *Phillips Tech. Rev.* 1955, **17**, p. 178.
90. H. Fernandez-Moran, *Industrial Diamond Rev.* 1956, **16**, p. 128.
91. V. A. Phillips, *European Congress Electron Microscopy*, Delft, Holland, 1960, De Nederlandse Vereniging voor Electronenmicroscopie, 1961, p. 485.
92. P. B. Hirsch, P. G. Partridge and R. L. Segall, *Phil. Mag.* 1959, **4**, p. 721.
93. M. Cottin and M. Haissinsky, *J. Chim. Phys.* 1950, **47**, p. 731.
94. A. Fourdeux and A. Berghezan, *J. Inst. Metals* 1960, **89**, p. 31.

5

Applications of Thin Foil Techniques

5.1 Introduction

It was shown in Chapter 1 that metals become transparent to electrons in the thickness range up to a few thousand angstroms, and because of certain diffraction conditions contrast is obtained in the image. Any lattice displacements (from imperfections, second phases, etc.) produce a phase-contrast effect, making it possible to reveal almost all the structural features thought to exist in metals. Since resolutions of close to 5 Å can be expected, it is possible to investigate details of very fine structures which are so important in governing the properties of materials. Furthermore, the use of selected area diffraction techniques makes it possible to identify the crystallographic features revealed by the micrographs. Thus, the possibilities of applying thin foil techniques to studies of the structure and properties of materials by transmission electron microscopy are endless. Over the past few years a considerable amount of research has been reported covering the fields of defects in crystals, plastic deformation, phase transformations, etc., and in the following paragraphs some of these results are discussed. The case of dislocations is considered in some detail because of their importance with respect to the mechanical properties of metals, and the major part of research using thin foil techniques has so far been concerned with the study of these defects. In this account we shall be concerned mainly with results obtained by examining thin foils prepared from bulk metal since this technique may be readily applied to most metals and alloys. The major number of illustrations to the following sections are thus of thin foils prepared by electropolishing methods. Although direct lattice resolution experiments with evaporated foils provide more details regarding the atomic displacements near dislocations, the instrumental and specimen techniques are stringent in order to obtain the necessary resolution (see section 2.2), and the properties of thin evaporated foils may be quite different from those of bulk specimens. In metals only indirect resolution is possible by means of moiré patterns and although dislocations may be observed in these patterns it is not possible to determine in which crystal the dislocation lies. However, some of the results obtained by these techniques are described in 5.2.12.

It may be argued that the dislocation distribution in thin foils is not representative of that in the bulk material because of (1) heating in the electron beam and (2) the relief of internal stresses due to the thinning down procedure. However, the temperature rise in metal specimens may be only 20° C when a fine focus electron beam of 10 to 20 μ diameter is used,† and moreover, the dislocations are generally arranged in rather complicated networks, which suggests that only relatively small rearrangements would occur during thinning. Thus, it is now fairly certain that what is seen in the electron microscope is typical of the bulk metal. Most of the work has so far been carried out on face-centered cubic metals, but there is increasing activity in research on other materials.

5.2 Observations of Defects in Pure Metals

5.2.1 *Dislocations*

Dislocations are line defects and in the neighborhood of the dislocation the atoms are displaced from their position in the ideally perfect lattice. When metals are plastically deformed, dislocations move under the applied stress so that the dislocation line represents a boundary between the unslipped and slipped areas of the slip planes. The study of dislocations has been intense over recent years since their properties govern the mode of plastic flow in crystals (see references 1 to 6), and many observations of dislocations have been carried out (for reviews see references 7 to 11).

In the electron microscope, dislocations can be revealed as a result of certain diffraction conditions arising from the displacement of the atoms from their ideal positions around the dislocation line, e.g., in Fig. 100, which shows an edge dislocation. With an edge, the displacements produce compressive strains above the dislocation line (associated with the extra half plane *AB*) and tensile strains at the bottom of the dislocation line. For screw dislocations (Fig. 101) the displacements produce a

† This is probably a minimum figure. By spreading the electron beam and raising the beam current it is possible for the temperature to rise to ~1000° C. In poor conductors (e.g., MgO), even with a 10 μ beam, the temperature may be as high as a few hundred degrees. Nevertheless the temperature of the specimen can be kept to a minimum by using a fine focus beam at the smallest possible filament saturation current.

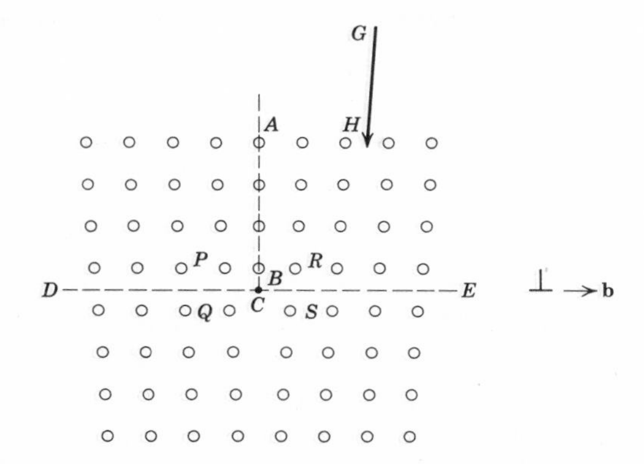

FIG. 100. The unit edge dislocation in a simple cubic lattice. The Burgers vector is normal to the dislocation line; the line is normal to the page. (After Cottrell.[1])

helical strain field. In general, a dislocation is characterized by the magnitude and direction of the slip movement associated with it, i.e., the Burgers vector. For an edge dislocation the Burgers vector is perpendicular to the dislocation line, and for a screw the Burgers vector is parallel to the dislocation line. Dislocations are usually neither pure edge nor pure screw, i.e., they are mixed dislocations having Burgers vectors at some angle to the line as shown in Fig. 102. Here only the segments at P and Q are pure edge and segments R and S pure screw, whereas along the arcs QR, RP, PS, SQ the dislocation is of mixed character.

Since a dislocation is the boundary line of a slipped surface, it must either be a closed loop or else end at the surface of the crystal. Inside the crystal a dislocation can only end at a point of intersection with other dislocations, i.e., at nodes. Two dislocations meeting at a point with Burgers vectors $\mathbf{b}_1$, $\mathbf{b}_2$ give rise to a resultant dislocation whose Burgers vector is $\mathbf{b}_3$, such that $\mathbf{b}_1 + \mathbf{b}_2 = \mathbf{b}_3$. Thus at nodes $\mathbf{b}_1 + \mathbf{b}_2 + \mathbf{b}_3 = 0$. Dislocations may move under a stress (glide) or by thermal activation (climb).

5.2.2 *Properties of Dislocations*

In face-centered cubic metals the slip planes are $\{111\}$ and the slip directions $\langle 110 \rangle$. The slip planes are arranged in a stacking sequence

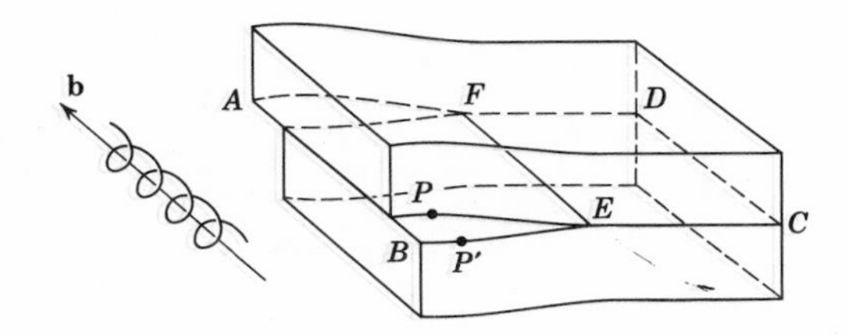

FIG. 101. A screw dislocation EF in a slip plane $ABCD$. The Burgers vector is parallel to EF. Describing a circuit round P to P' involves going through one turn of a helix. (After Cottrell.[1])

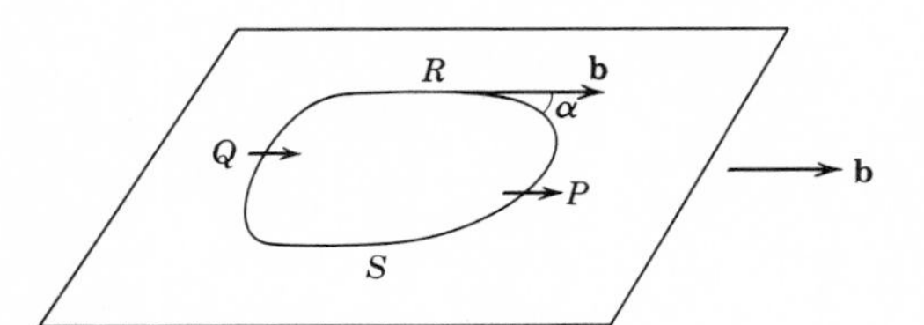

FIG. 102. A dislocation loop in a slip plane. Segments P and Q are edge, R and S are screw, and elsewhere, mixed. (After Cottrell.[1])

ABCABCABC, etc., as shown in Fig. 103. A dislocation line passing between planes A and B requires that the atoms at position B jump to an equivalent position B' at a distance b_1 (length of the Burgers vector). However, it is energetically easier for atom B to jump to position C ($\mathbf{b}_2$, Fig. 103) and then into position B' ($\mathbf{b}_3$, Fig. 103). Thus, the dislocation $\mathbf{b}_1$ splits into two partial dislocations by the reaction

$$\mathbf{b}_1 = \mathbf{b}_2 + \mathbf{b}_3, \quad \text{e.g.,} \ \tfrac{1}{2}a[10\bar{1}] = \tfrac{1}{6}a[2\bar{1}\bar{1}] + \tfrac{1}{6}a[11\bar{2}] \qquad (77)$$

The energy of a dislocation is roughly proportional to the square of the Burgers vector;† hence it is clear from the addition of the squares of the Burgers vectors that this dissociation is very favorable. Dislocations $\mathbf{b}_2$ and $\mathbf{b}_3$ are at 60° to each other and repel. As a result of the dissociation, the stacking sequence is changed to $ABCA\,|\,CABC$, i.e., a stacking fault has been introduced. It can be seen that this changes the structure locally to HCP (section 1.14.1). The stacking fault has surface energy and tends to pull the two dislocations $\mathbf{b}_2$ and $\mathbf{b}_3$ together. The separation d of these

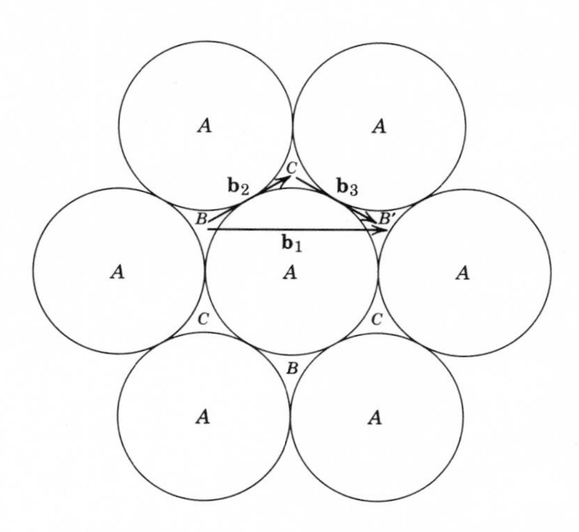

FIG. 103. Stacking sequence of (111) planes in the FCC structure, showing slip directions $\mathbf{b}_1 = \tfrac{1}{2}a[10\bar{1}]$, $\mathbf{b}_2 = \tfrac{1}{6}a[2\bar{1}\bar{1}]$, $\mathbf{b}_3 = \tfrac{1}{6}a[11\bar{2}]$. (After Cottrell.[1])

† Neglecting the core energy of the dislocation (references 1 to 3).

partial dislocations (called Shockley partials) is given by[2]

$$d = \frac{G\mathbf{b}_2\mathbf{b}_3(2 - \nu)}{8\pi\gamma(1 - \nu)} \tag{78}$$

Where G is the shear modulus, ν is Poisson's ratio, and γ the stacking fault energy. The width of the stacking fault thus depends on γ. We can therefore classify metals into three groups having (1) low, (2) intermediate, and (3) high stacking fault energies.[11] Group (1) includes Co, stainless steel, copper alloys; group (2), Cu, Ni, Pt, Au; and group (3) Al and most Al alloys. In these metals, cross-slip becomes easier as we move from group (1) to group (3), since in order for a dislocation to cross-slip onto another slip plane the partials must come together at a constriction; the wider the stacking fault ribbon the more difficult is this process. Dislocations interacting on different planes form networks; thus in face-centered cubic metals dislocations are expected to be arranged in two- or three-dimensional networks consisting of threefold nodes, and stacking faults are expected in metals of group (1).[1–5]

5.2.3 *Appearance of Dislocations in Electron Micrographs of Thin Foils Made from Bulk Specimens*

Stacking faults† have often been observed in stainless steel,[12–14] copper alloys,[15] and cobalt.[16] They are characterized in electron micrographs by the appearance of fringes running parallel to the intersection of the fault with the surfaces as seen in Fig. 104, and in which the ribbon of stacking fault is bounded by two partial dislocations which repel each other. We have seen in section 2.3 that a discontinuity on an inclined plane will produce fringes due to the variation in depth periodicity of the intensity oscillations along such a plane (e.g., Fig. 34). Hence fringe contrast is expected in images of stacking faults.[13] We shall consider this in more detail in section 5.2.14.

Because of the difficulty of cross-slip, dislocations in metals of low stacking fault energy will tend to pile up against any obstacles in their path, e.g., grain boundaries. An example of this is shown in Fig. 105. In the micrographs the dislocations appear as projections in the slip plane onto the top and bottom surfaces of the foil (Fig. 106); in the bright field image, because of Bragg contrast, they appear as black lines whereas the contrast is reversed in the dark field image.

Under a suitable stress these dislocations can be made to move, and slip traces are produced at the foil surfaces. Originally,[13] dislocation movement in thin foils was produced by what was thought to be thermal stresses induced by the electron beam. Now, however, it seems that

† Stacking faults have also been observed in silicon.

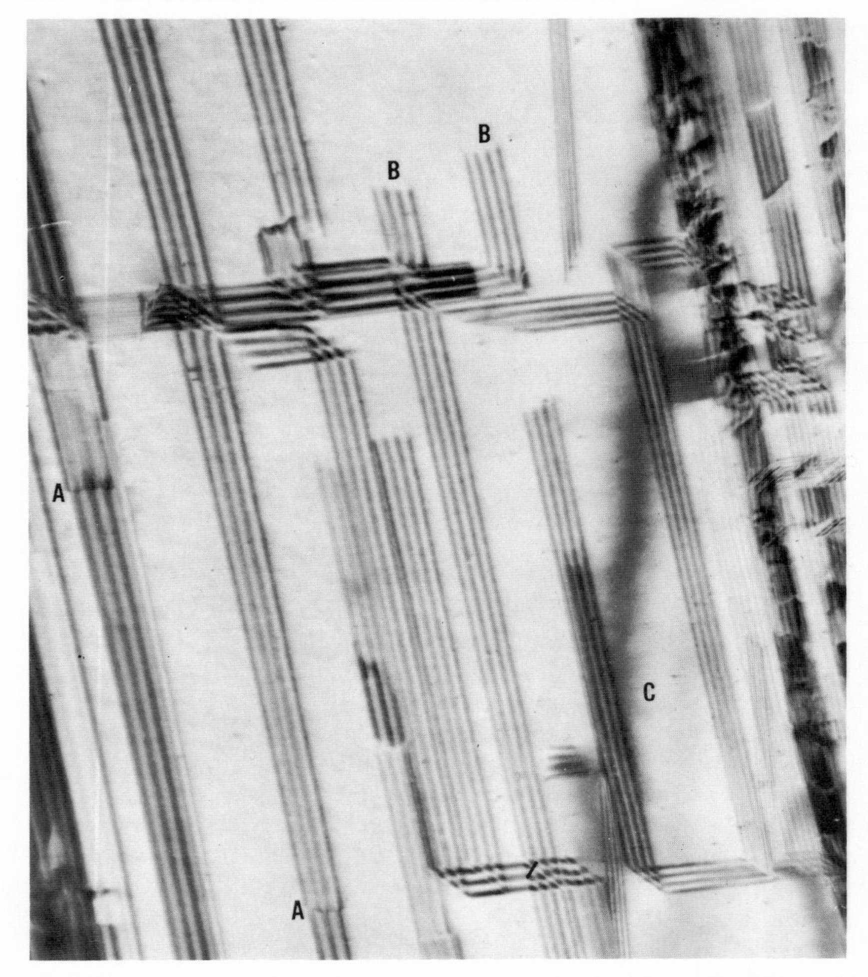

FIG. 104. Dissociated dislocations in Cu–10% Ge alloy after 5% tensile deformation. Notice the characteristic fringe contrast running parallel to the intersection of the stacking faults with the top and bottom surfaces of the foil ($\times$ 160,000). (See section 5.2.14.) Notice reversal of contrast at *A*, invisible partials at *B*, and high contrast at the extinction contour *C*. (After Swann and Nutting.[15])

stresses set up by deposition of carbonaceous matter onto the specimen from the impure vacuum may also be important in causing dislocations to move. In any case, the contrast at slip traces results from the interaction of the surface step, produced by slip, with the surface film (oxide layer or carbon). This can be seen from the illustration shown in Fig. 106. The

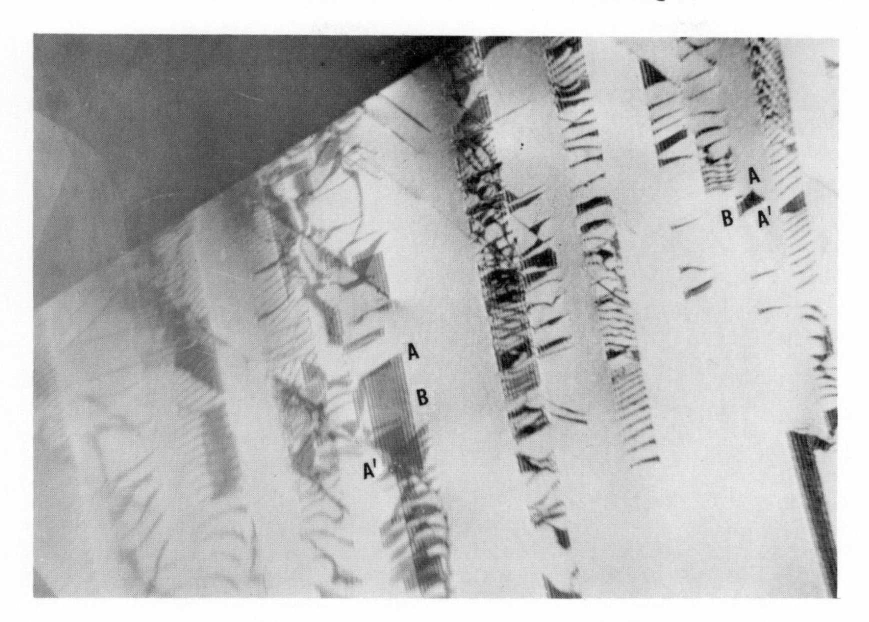

FIG. 105. Pile-ups of dislocations against a grain boundary in Cu–8% Al alloy after 3% tensile deformation. Dislocations lie in inclined planes. Note stacking faults *B* between partials *AA'*, and that the dislocations stay in the same slip plane (×15,000). (Courtesy P. R. Swann.)

intense contrast observed at the edges of slip traces is therefore interpreted as due to a dislocation line lying just below the surface. The dislocations shown in Figs. 105 and 107 are projections of such dislocations viewed in the direction of the electron beam, and Fig. 107 shows typical slip traces left behind moving dislocations. Recently, the movement of dislocations has been achieved by using straining devices in the specimen holder. In this way controlled deformation experiments are possible.[17,18] However, there are difficulties in mechanically straining specimens in the microscope because of the limited space available in the specimen chamber.

During movement of dislocations they are seen to be bowed out, e.g., Fig. 107 at *A* and *I*; this shows directly that the dislocations are pinned at the surfaces by some surface layer. From the radius of curvature *R* of the dislocation the local shear stress τ can be calculated from the relationship $\tau \cong G\mathbf{b}/2R$, where *G* is the shear modulus and **b** the Burgers vector.[13] In this way the local stresses in foils under the electron beam is about $G/1000$, which is often enough to cause the dislocations to move. In fact, it is often difficult to prevent dislocations from moving while the specimen

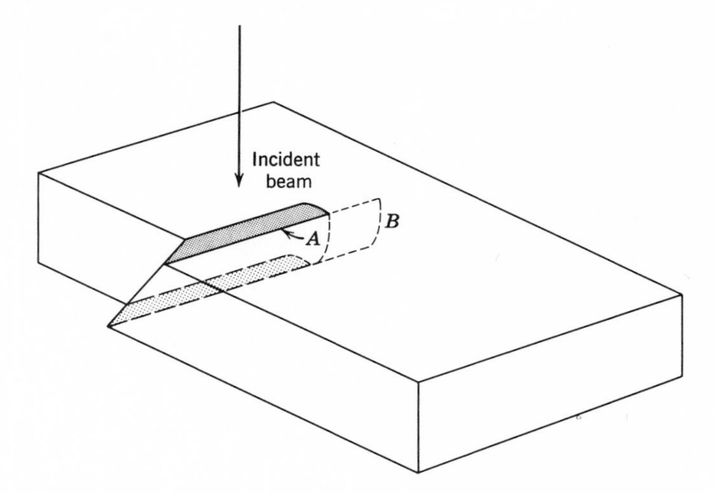

FIG. 106. A dislocation with a screw component terminates at a step on the surface (*A*). When this dislocation moves, (*B*), the step advances with it. However, if the surface film (e.g., oxide) is strongly bonded to the metal, the step may not penetrate the surface, and essentially two dislocation lines (shown dotted) are left, just below and parallel to the two surfaces. This illustrates the mechanism of contrast from slip traces at moving dislocations (Figs. 107, 109). (After Hirsch.[11])

is under observation. In Fig. 107 the dislocations at *A* and *I* are bowed out in opposite senses, showing that they are of opposite sign and therefore move in opposite directions for the same shear stress. Under continued deformation, the partials become further and further separated so that long ribbons of stacking fault are observed. As expected, cross-slip is not often seen in materials of low stacking fault energy.

In stainless steel and copper alloys the slip traces remain for long periods after the dislocations move, suggesting that the surface film is strongly bound to the metal, preventing the formation of a surface step. Stacking faults are always observed in solid solution alloys of copper containing zinc, aluminum, germanium, and silicon.[15] They are not often seen in copper.[19] This immediately points to the fact that alloying lowers the stacking fault energy of copper. Figure 104 shows extended stacking faults in a Cu–10% Ge alloy. In these alloys hardening can be (at least partly) attributed to the segregation of solute atoms to the region of the stacking fault, thereby producing Suzuki locking;[5] (in FCC alloys, atoms of hexagonal metals would be expected to go to the stacking fault). This has been demonstrated by Swann and Nutting[15] by allowing foils of copper alloys after deformation to dissolve in suitable etchants. The stacking

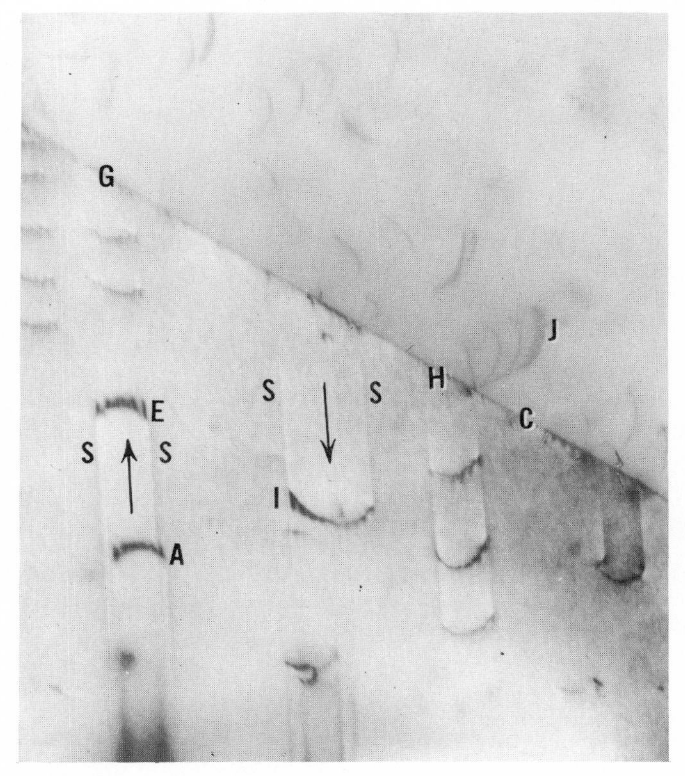

FIG. 107. Showing slip traces *S* in stainless steel. Grain boundary along *GC*. Dislocations *A* and *I* moving in opposite directions. Stacking fault at *E* (×40,000). (After Whelan et al.[13] Courtesy the Royal Society.)

faults or the regions between stacking faults dissolve preferentially, depending on the alloy system. An example is shown in Fig. 108, taken from a Cu–4% Si alloy, showing how the stacking faults *A* are left projected out of the foil, whereas the regions between them have been dissolved away. Observations such as these may also be helpful in analyzing problems of stress-corrosion in alloys.[15]

5.2.4 *Distribution of Dislocations and Work Hardening*

Theories of flow stress and work hardening depend on the particular dislocation distribution which is assumed to exist in the cold-worked metal.[20–26] In the long-range stress theories[20] dislocation pile-ups are

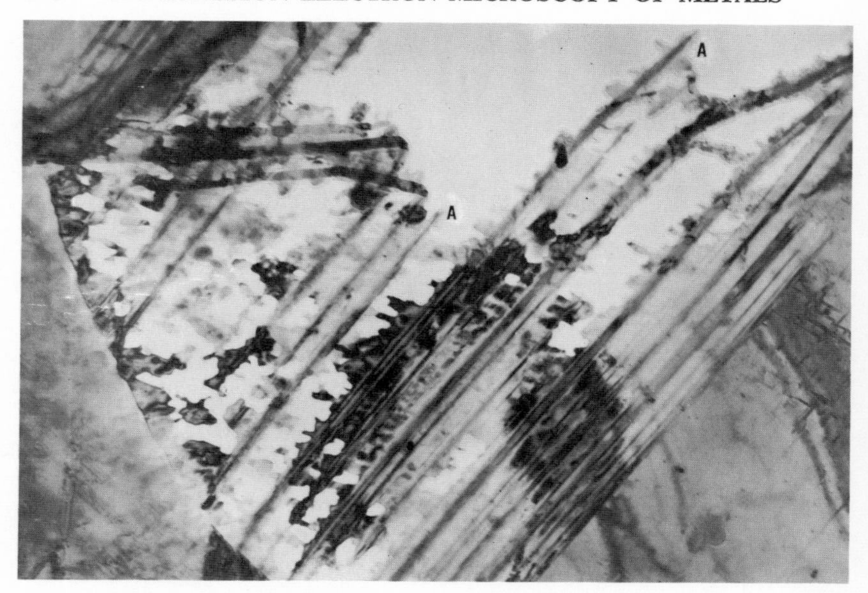

FIG. 108. Edge of thin foil of Cu–4% Si alloy after electroetching. Notice protruding undissolved, stacking faults at *A*. This shows that the stacking faults have a different chemical composition from unfaulted areas (×49,000). (After Swann and Nutting.[15])

assumed with the formation of immobile lengths of Lomer-Cottrell locks,[14] and these are the barriers to the glide dislocations. In stainless steel, pile-ups and dislocation interactions producing these locks have been observed and analyzed by Whelan,[14] who also showed that it is possible to make direct measurements of the stacking fault energy from the radii of curvature of the nodes (see section 5.2.6).

The dislocation distributions in metals other than cobalt, stainless steels and copper alloys, however, are quite different, which suggests that the type of dislocation distribution is determined by the stacking fault energy of the metal or alloy being considered. Thus, long-range stress theories of work-hardening can be applied only to metals of low stacking fault energy, in which pile-ups are definitely known to occur. The classification of metals into the three groups: (1) high, (2) intermediate, and (3) low stacking fault energies, suggested in section 5.2.2, may therefore be useful for considering work-hardening phenomena in metals. Transmission electron microscopy techniques can, of course, reveal directly the distribution of dislocations inside the specimens. This has led to the development of other work-hardening theories, postulating a considerable relaxation of internal stress, to account for plastic deformation in materials which do not show stacking faults or piled-up groups.[21–25] Earlier theories of this kind are based on considering the interactions between glide dislocations and those in the "forest," and are consequently known as "forest theories." More recently, as a result of electron microscope investigations, attention is being paid to the importance of jogs on dislocations,[24–26] the temperature dependence of the flow stress is then determined by the rate of climb or mobility of the jog.[24]

Experimental evidence in favor of the forest theory in cold-worked polycrystalline silver has been obtained by Bailey and Hirsch.[27] They showed that the distribution and densities of dislocations (determined by transmission electron microscopy), the flow stress, and stored energy measurements (by microcalorimetry) were correlated with each other. The electron micrographs showed that the dislocations were arranged in dense networks forming the boundaries of regions $\sim 1\ \mu$ diameter which are relatively dislocation-free (cell structure). The dislocation densities increased from 2.2×10^{10} to 6.8×10^{10} as the deformation was increased from 10 to 30%.† From these results, the flow stress was explained quantitatively in terms of the forest intersection mechanism at the boundaries. The considerable release of stored energy which was observed to occur during recovery was not accompanied by any rearrangement of the dislocations, showing that the long-range stresses must be largely relaxed. Dislocation arrangements similar to those in silver have also been observed in cold-worked copper,[19] nickel and gold;[28] an example for nickel is given in Fig. 123*a*.

The formation of jogs on dislocations in thin crystals of Zn and Cd has been demonstrated by Price.[29] During deformation inside the microscope, in the zinc crystals slip was observed by pyramidal glide on the $(11\bar{2}2)$ $[\bar{1}\bar{1}23]$ system. During glide many dislocation loops were formed as a result of the cross-slip of screw dislocations at jogs when they were held up at an obstacle. Initially, this produces an elongated narrow loop consisting of plus-minus pairs of edge dislocations which may later degenerate into rows of circular prismatic loops. It is clear that this process is strongly temperature dependent. Similar observations have been made in magnesium oxide,[30, 31] e.g., Fig. 128*a* and molybdenum.[40]

In aluminum the dislocations are never extended and cross-slip is frequently observed.[32, 33] Figure 109 shows an example of this process.

† These are calculated from measurements on the micrographs. If Rp is the mean projected length of a dislocation line through the foil, then the average true length R is given by $R = (4/\pi)Rp$.[27] The density is expressed in nR/cm^3 where n is the number of dislocation lines measured in obtaining the value of R, so that it is necessary to estimate the foil thickness. (See Appendix A.)

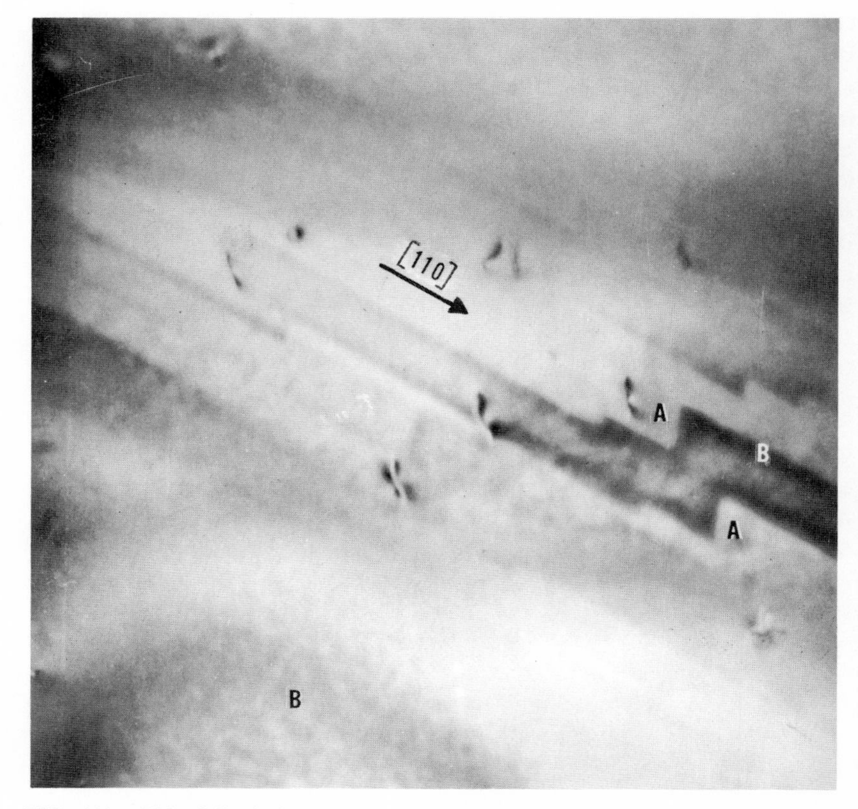

FIG. 109. Thin foil of aluminum, showing slip traces from dislocations moving in [110] and cross-slip at *A*. Note enhanced contrast at extinction contours *B* ($\times$ 40,000). (Courtesy *Phil. Mag.*)

Aluminum is a metal of high stacking fault energy and as such may be regarded as a "lone wolf" among the common metals. Alloying aluminum with other metals seems to have little effect† on the stacking fault energy,[34] whereas we have seen that additions of other metals to copper markedly affect its stacking fault energy. From the experimental results so far published it is now known that the distribution of dislocations is different in each group‡ and may be summarized as follows.[11] In group (1) pile-ups, stacking faults, complex networks, and deformation

† Except for Al–Ag (16 wt.%) alloys after quenching from near the melting point (see section 5.2.9).

‡ That is, the three groups: (1) low, (2) intermediate, and (3) high stacking fault energies.

twins are produced; in group (2) dislocations occur in irregular networks which build up into a cell structure with increasing deformation. This cell structure consists of boundaries containing a high density of dislocations separated by regions up to 1 μ diameter which are relatively dislocation-free[27] (e.g., see Fig. 123*a*). In metals of group (3), e.g., Al, after large deformations, dislocations form almost perfect subboundaries separating relatively dislocation-free subgrains.[32, 35] Thus the differences in dislocation distribution correlate with the differences in stacking fault energy. In metals of groups (2) and (3) the flow stress has been found to be approximately proportional to the square root of the dislocation

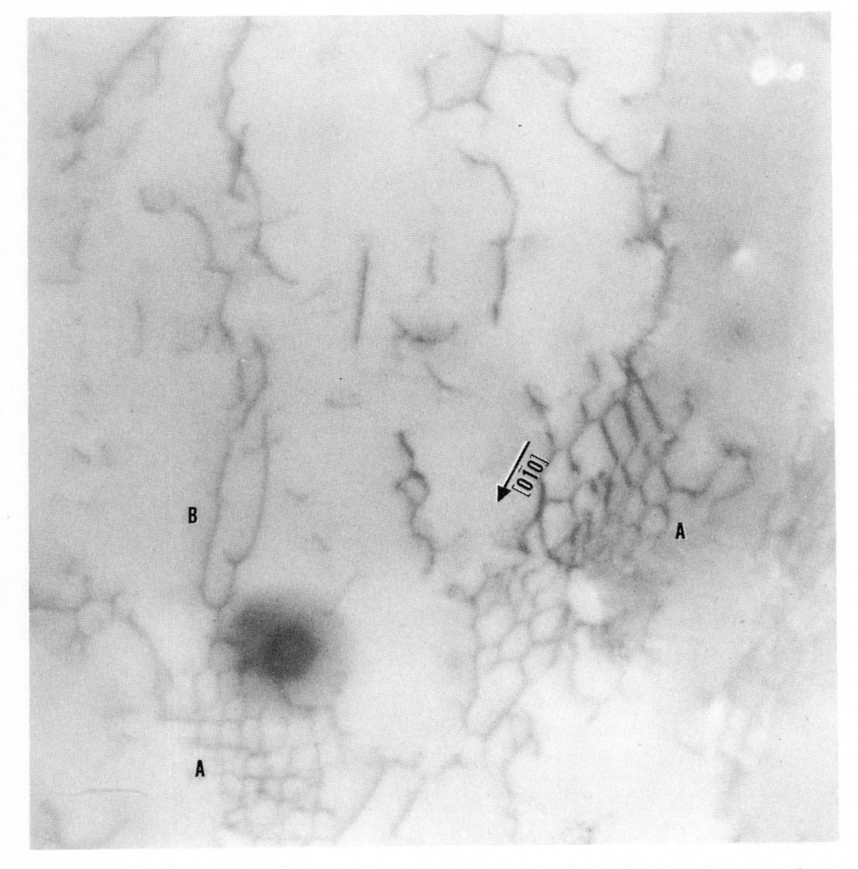

FIG. 110. Dislocations in Mo after 2% tensile deformation. Note hexagonal networks at *A* consisting of [111] and [100] dislocations and elongated loop *B* ($\times$23,500).

density,[11] which is consistent with "forest" and jog theories of the flow stress.[21, 22, 24, 27]

5.2.5 *Dislocations in Metals Other Than Face-Centered Cubic*

Apart from the face-centered cubic metals, iron seems to have been studied in more detail than any other.[36–39] In body-centered cubic metals many slip systems may operate during deformation, and by using the selected area diffraction technique it is possible to examine this in some

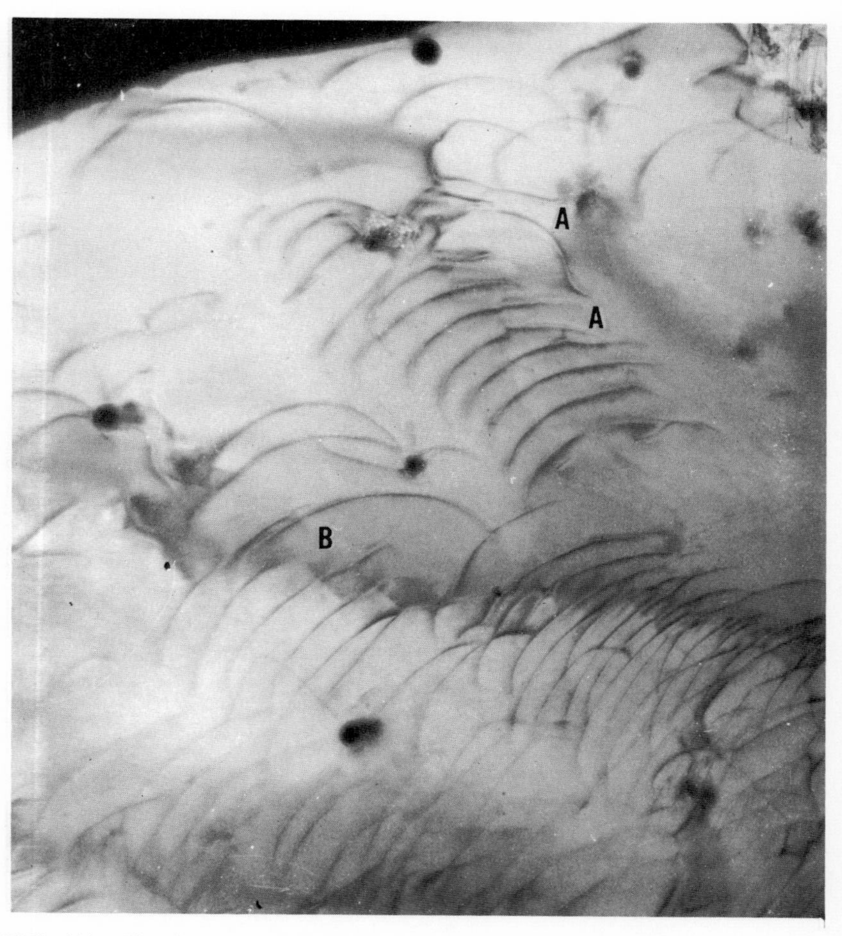

FIG. 111. Showing pyramidal slip in magnesium. The dislocations lie in $(10\bar{1}1)$ planes and glide along $[11\bar{2}0]$. Double images at *A*; image changes sides at *B* (×30,000).

detail. Generally speaking, dislocations in body-centered cubic metals are arranged either in regular or irregular networks, e.g., Fig. 110 shows dislocation networks and jogged dislocations in molybdenum deformed 2% in tension.[40] Recent work by Fourdeux and Berghezan[41] has demonstrated the existence of stacking faults in niobium. This is the first observation of dissociated dislocations in body-centered cubic metals.

Of other metals, isolated observations have been made on cobalt,[16] magnesium,[16, 40] and uranium.[42] An example of dislocations after glide in $(10\bar{1}1)$ planes in lightly deformed magnesium is shown in Fig. 111, and

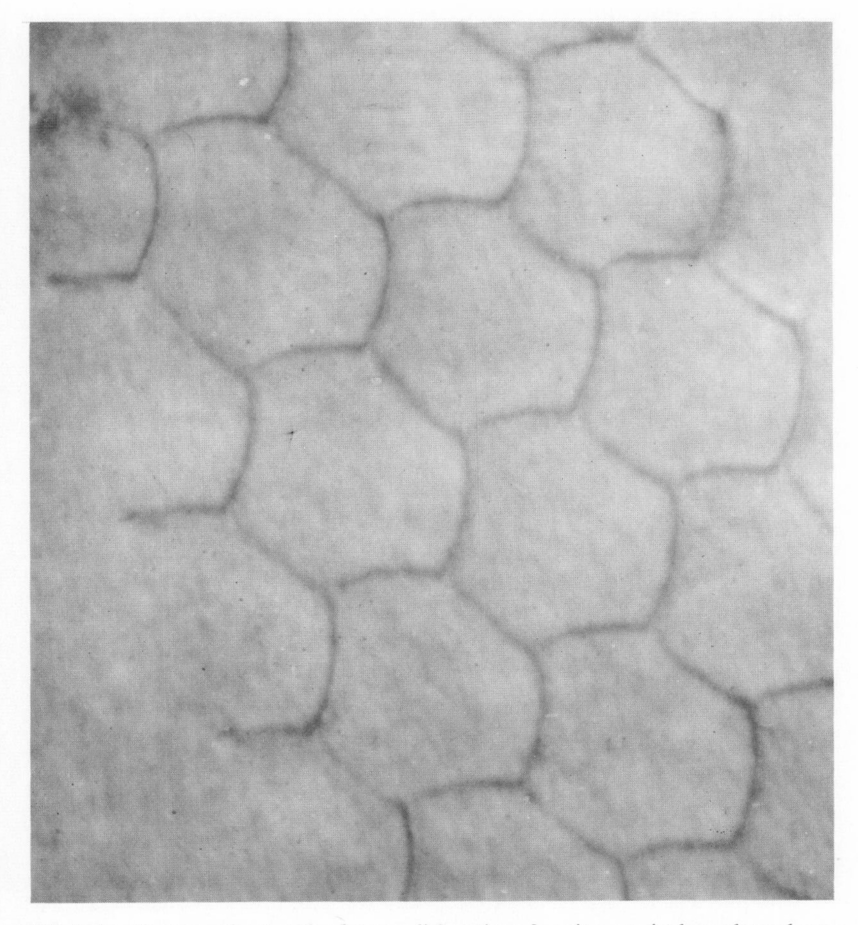

FIG. 112. Hexagonal network of screw dislocations forming a twist boundary along a (111) plane in aluminum (× 53,000).

it appears that this metal behaves in a similar manner to face-centered cubic metals of intermediate stacking fault energy. In zinc, however, Fourdeux, et al.[43] have shown the existence of stacking faults in the basal planes.

5.2.6 *Dislocation Networks*

Networks are formed by the interaction of dislocations.[5] In this way subboundaries and sessile dislocations can be formed.[14] A simple twist boundary in face-centered cubic metals is formed by the mutual interaction of screw dislocations forming a cross-grid in (100) or (111) planes. An example for aluminum is shown in Fig. 112. Examination of the nodes shows that the dislocations are not extended. In metals of low stacking fault energy, however, alternate nodes are extended and contracted respectively,[14] e.g., in stainless steels and copper alloys (Fig. 113). In Fig. 113 the stacking fault fringes can be clearly observed at the nodes. Provided these nodes are free from other sources of stress, they are in equilibrium, and the stacking fault energy can be deduced[14, 44] from the

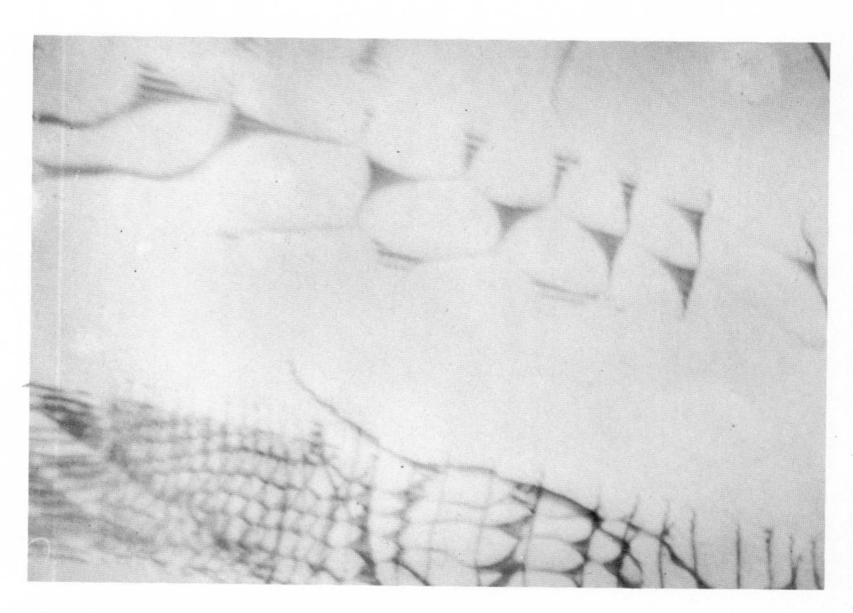

FIG. 113. Showing alternate extended and contracted nodes at hexagonal dislocation networks in Cu–8% Al alloy after 8% tensile deformation. Notice fringes at stacking faults. From the radius of curvature of isolated symmetrical nodes it is possible to determine the stacking fault energy (×29,000). (After Swann and Howie.[44])

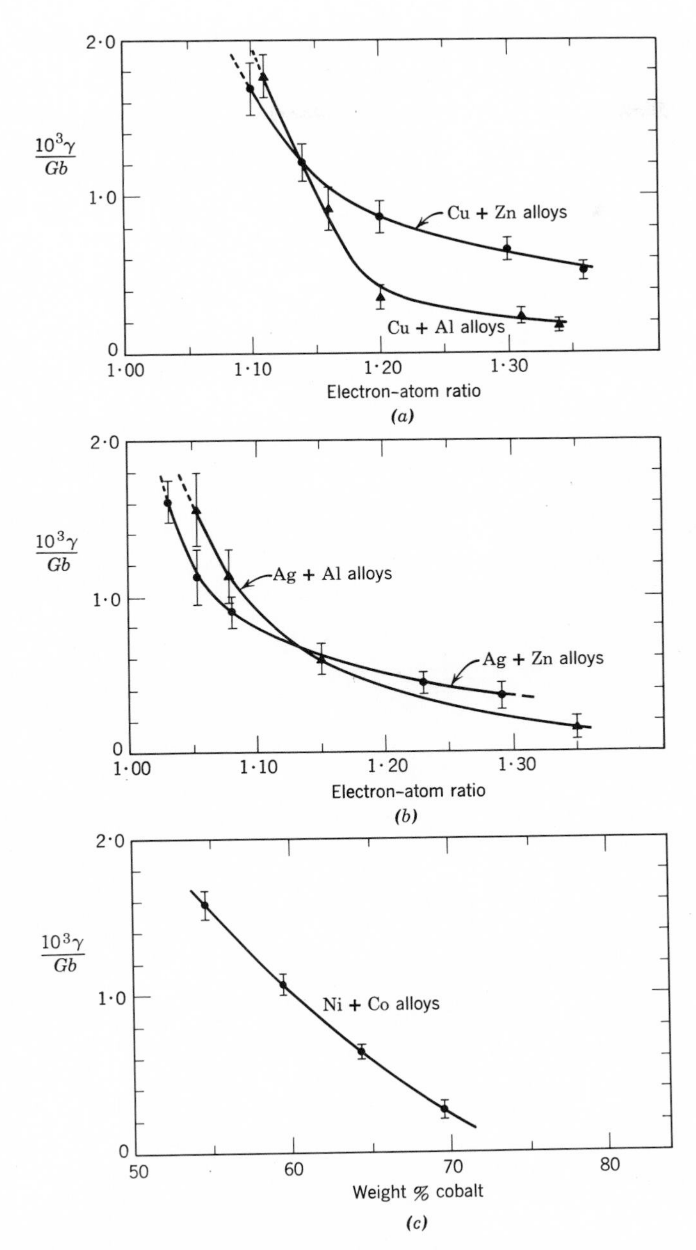

FIG. 114. Showing the variation of stacking fault energy with electron : atom ratio for (*a*) Cu–Zn and Cu–Al alloys; (*b*) Ag–Zn and Ag–Al alloys; and (*c*) Ni–Co alloys. Results obtained from measurements of radii of curvature of symmetrical dislocation nodes. γ is stacking fault energy, G, shear modulus and b the Burgers vector. (After Swann and Howie.[44])

relationship $\gamma = (Gb^2/4\pi K) \ln (R/b)$, where R is the radius of curvature at the node and K is a numerical constant[3] $= \frac{5}{6}$. For stainless steel γ has been estimated at ~ 13 ergs/cm^2,[11] and recent work by Swann and Howie[44] has shown that the stacking fault energy is a function of the electron atom ratio for copper, silver, and nickel-cobalt alloys. Their results are shown in Fig. 114.

In Fig. 104 the partials are widely separated as a result of the applied stress so that estimates of the stacking fault energy cannot be made from such micrographs. To do this therefore, one must search for areas similar to that shown in Fig. 113 where stress-free nodes are revealed.

FIG. 115. Showing parallel edge dislocations forming a tilt boundary in annealed iron ($\times$ 100,000). (After Brandon.[36])

Even so, errors in the estimates of γ from R arise because the exact position of the dislocation core may be unknown (due to contrast conditions, see section 5.2.14).

Perfect tilt boundaries can be formed by the alignment of edge dislocations on parallel slip planes, as shown in Fig. 115, which is taken from an annealed iron sample.[36] Other networks in body-centered cubic metals form from dislocations having [100] Burgers vectors in addition to those with $\frac{1}{2}$[111] Burgers vectors,[39, 40] e.g., Fig. 110.

Other types of interactions produce sessile dislocations, e.g., Lomer-Cottrell dislocations in face-centered cubic metals lying along [110] directions in a nonglide plane. Such interactions have been analyzed in detail by Whelan.[14]

5.2.7 *Dislocation Sources*

The operation of a classical Frank-Read source[45] has been observed by Wilsdorf[46] in thin foils of stainless steel. In this case, a length of dislocation anchored at a node was observed to throw off many loops, which then piled up in the slip plane. (Only parts of the loops are visible on inclined slip planes since the foil thickness is much smaller than the loop diameter.) In the same paper, Wilsdorf has described nucleation of dislocations at (1) grain boundaries, (2) twin boundaries, (3) polygonization walls, (4) precipitates, and (5) during glide; an example of the latter is shown in Fig. 116*a*.

Before Wilsdorf's results, dislocations had been frequently observed to nucleate at the edges of foils,[13, 32] at subboundaries[32] as well as at grain boundaries.[48] Experiments using straining devices inside the electron microscope have shown the nucleation of dislocations on the same slip plane[33, 46] and on many slip planes.[46] The Frank-Read mechanism has also been observed in quenched aluminum alloys.[49, 50] Figure 116*b* shows an example of a source having emitted at least six dislocation loops; the details at the center of the source are obscure. Thus it appears that alloys are the most suitable materials for the study of dislocation sources, because the dislocations appear to stick very easily in them so that the source configuration is preserved in the thin foils.

5.2.8 *Dislocations and Slip Lines*

As mentioned in section 4.1, most of the original research work done on plastic deformation of metals was carried out by using replicas to reveal the surface offsets caused by slip. Therefore it is important to investigate the relationship between the internal distribution of dislocations and the slip lines. In metals such as Cu, Ag, Au, Ni, and Al, the complicated networks which form after plastic deformation seem to bear little or no relationship to the slip line patterns revealed by replica studies. Such an investigation is possible by preparing foils in order that one surface is preserved (one-side technique, section 4.5). In this way it has been possible to reveal simultaneously slip lines and dislocations in the interior.[51, 52]

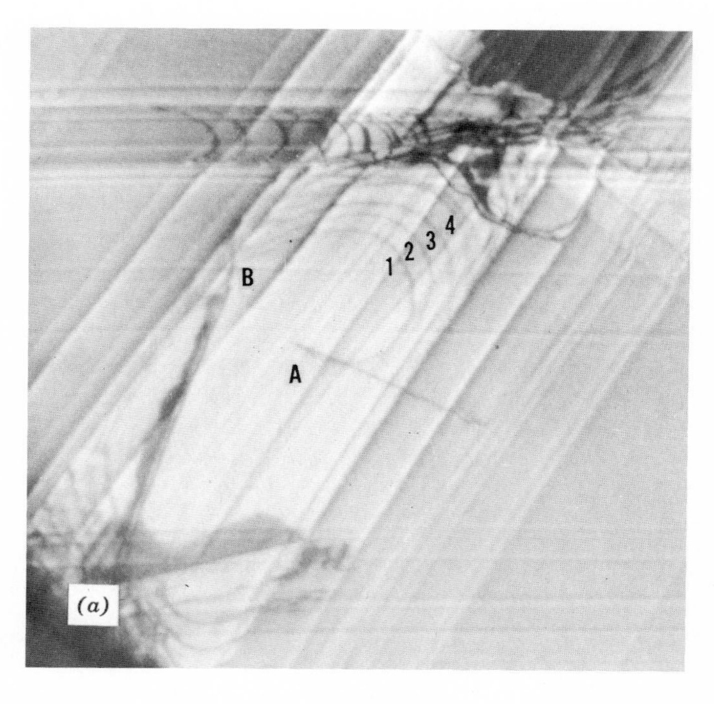

FIG. 116(*a*). A dislocation pinned at *AB* operates as a source by rotation about the anchor point *A*, throwing off successive loops 1, 2, 3, etc. This represents a spiral Frank Read source, since *B* is close to the surface. Thin foil of stainless steel (×40,000). (After Wilsdorf.[46])

An example of this in stainless steel is shown in Fig. 117, where at *A* the dislocations meet the surface in visible lines corresponding to the slip steps. This micrograph also shows the alternation of extended and contracted nodes in the network.

Dislocations have also been identified with the slip bands formed after fatigue deformation. Again polishing from one side has been carried out, and intrusions and extrusions have been revealed by this method.[52] In aluminum there is evidence that a large number of point defects are created during fatigue,[35, 53] since the specimens contain many dislocation loops.

5.2.9 *Defects in Quenched Metals*

If metals are quenched very rapidly from high temperatures, the concentration of vacancies present at the high temperature may be retained

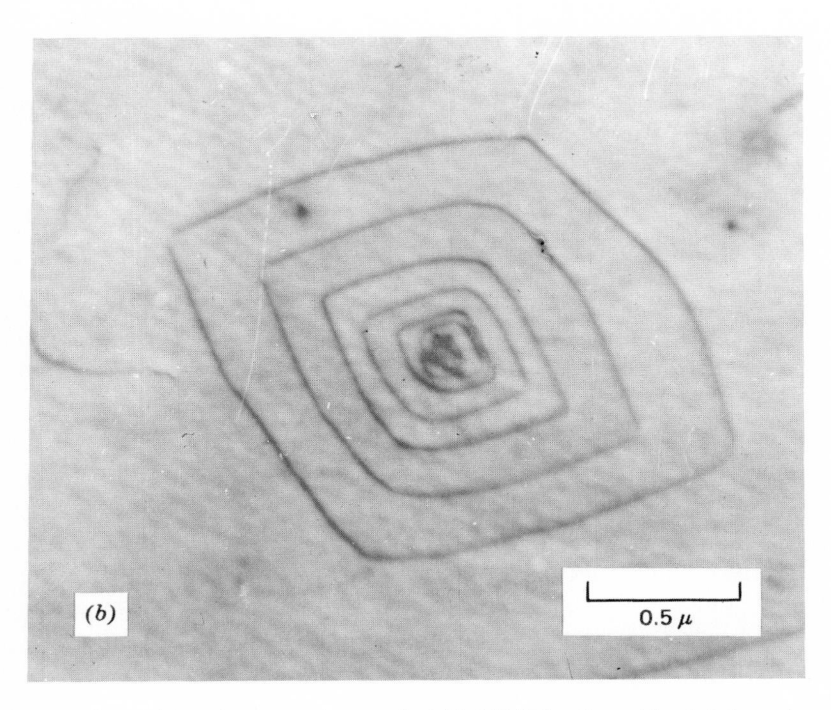

FIG. 116(*b*). Dislocation source in quenched Al–3.5% Mg alloy; at least six loops have been emitted (× 40,000). (Courtesy K. H. Westmacott, D. Hull, R. S. Barnes, and R. E. Smallman.)

in the sample.[54, 55] The quenched-in vacancies are now in supersaturation and will tend to anneal out. The concentration C of vacancies at temperature T° K is given by $C = A \exp -Ef/kT$, where Ef is the formation energy (~ 1ev for most metals), A the entropy factor, and k is Boltzmann's constant. Since vacancies occupy less space than atoms, they have strain energy and will tend to migrate in order to lower their energy. Suitable

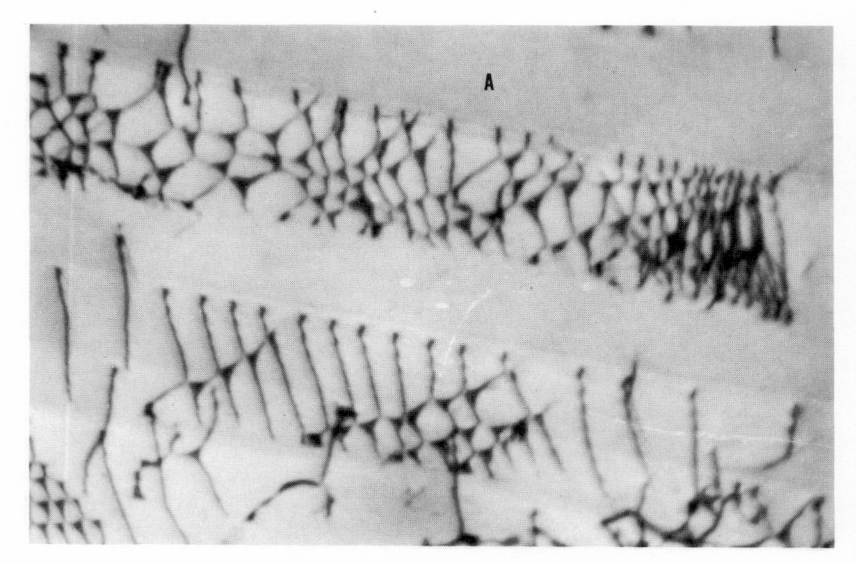

FIG. 117. Dislocation network in stainless steel associated with a slip line on the surface (after fatigue deformation). At *A*, the dislocations meet the original surface in visible lines corresponding to the slip steps. Specimen thinned from one side only. Note alternation of extended and contracted nodes (×46,000). (After Hirsch et al.[52] Courtesy *Phil. Mag.*)

sinks for this process are surfaces, grain boundaries, dislocations, other vacancies, and impurity atoms. Vacancies attracted to each other are observed to collapse as discs to form dislocation rings, as indicated for face-centered cubic metals in Fig. 118*a–c*. This process has been discussed in detail by Kuhlmann-Wilsdorf.[56] The situation represented by Fig. 118*c* is initially stable if the metal has a low stacking fault energy, i.e., there are dislocation rings (Frank-sessiles) with Burgers vector $\frac{1}{3}[111]$ normal to the plane of the loop surrounding a region of intrinsic stacking fault. Frank-sessile loops have recently been observed in Al–16% Ag alloys[101] where silver in solid solution reduces the stacking fault energy of aluminum so that Frank-sessile dislocations are stabilized by the migration of silver atoms to the loop. Figure 119*a* shows an example.

Since the Frank-sessile dislocation has a relatively large Burgers vector, it will tend to dissociate into (1) a low-energy stair-rod dislocation[57–59, 14] and (2) a Shockley partial dislocation on an intersecting slip plane, by the reaction

$$\underset{\text{Frank sessile}}{\tfrac{1}{3}[111]} = \underset{\text{Stair rod}}{\tfrac{1}{6}[101]} + \underset{\text{Shockley partial}}{\tfrac{1}{6}[121]} \qquad (79)$$

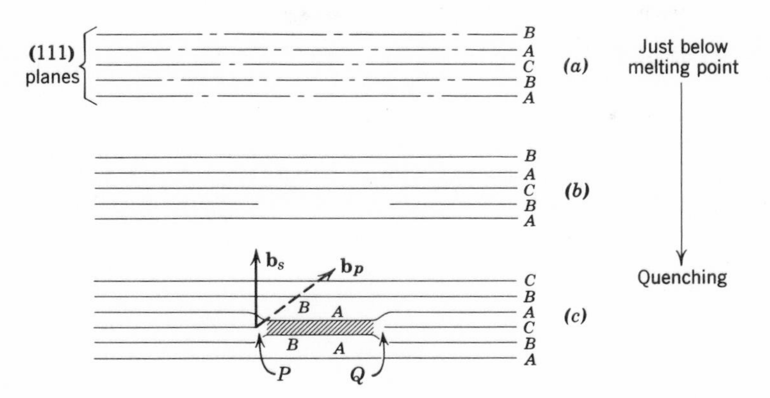

FIG. 118. Schematic illustration of the formation of dislocation loops in FCC metals after coalescence of vacancies by first forming small clusters (unspecified) (*b*), which later collapse as discs (*c*). State (*a*) represents vacancies existing near melting point. Note stacking fault between *P* and *Q*. $\mathbf{b}_s$ is Burgers vector of Frank sessile ($=\frac{1}{3}a[111]$); $\mathbf{b}_p$ is Burgers vector of glissile prismatic loop ($=\frac{1}{2}a[110]$).

The initial dislocation ring is expected to be in the form of triangles with edges parallel to [110] directions to facilitate dissociation, so that each side of the triangle can dissociate into stair rods and partials. Silcox and Hirsch have demonstrated that this happens in quenched gold[60] and that the partials attract each other in pairs to form stair rods along the [110] directions by reactions of the type

$$\underset{\text{Shockley partial}}{\tfrac{1}{6}[121]} + \tfrac{1}{6}[\bar{1}\bar{1}\bar{2}] = \underset{\text{Stair rod}}{\tfrac{1}{6}[01\bar{1}]} \tag{80}$$

Thus the final defect (originally a triangle of stacking fault) will be a tetrahedron of stacking fault having an inverse tetrahedron representing the Burgers vectors of the stair rods.[60] An example of tetrahedra in gold can be seen in Fig. 119*b*, in which the characteristic fringe contrast associated with the stacking faults is clearly resolved.

In metals of high stacking fault energy the situation shown in Fig. 118*c* is unstable. In this case, it is energetically favorable to nucleate a Shockley partial so as to remove the stacking fault by the reaction

$$\underset{\text{Frank sessile}}{\tfrac{1}{3}[111]} + \underset{\text{Shockley partial}}{\tfrac{1}{6}[11\bar{2}]} = \underset{\text{Prismatic dislocation}}{\tfrac{1}{2}[110]} \tag{81}$$

The resultant dislocation is a prismatic loop which can glide on the cylinder containing the loop and its Burgers vector. This was first demonstrated in quenched aluminum by Hirsch et al.[61] who also showed that as expected for loops on (111) planes they often have hexagonal shapes. Dislocation loops have also been observed in planes other than {111}.[62] Figure 120 shows an example of loops in quenched 99.995% aluminum; there is no stacking fault contrast, hence the loops must be prismatic. Very recently, however, Cotterill and Segall (private communication) have observed Frank-sessile dislocation loops in very high purity, zone-refined aluminum; the same effect is obtained if 99.995% pure aluminum is re-

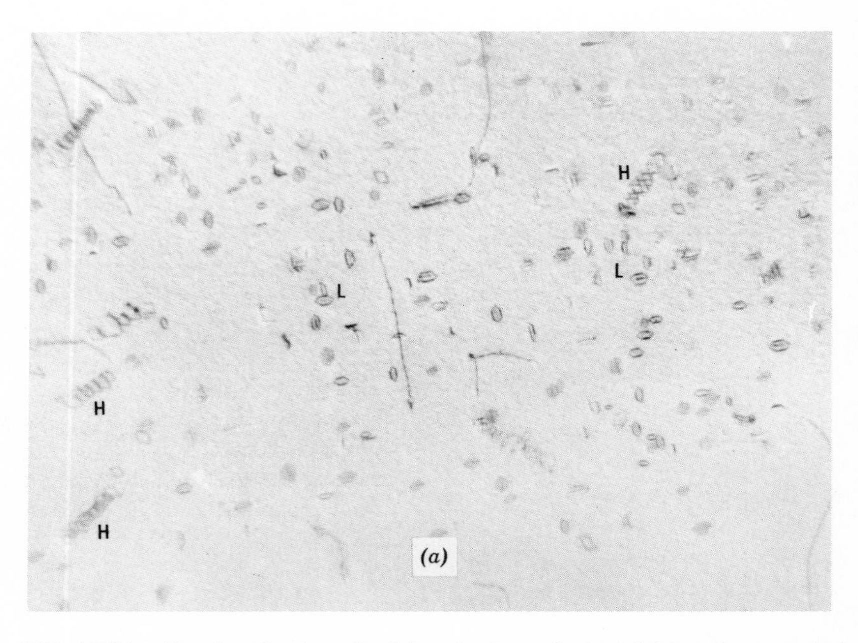

FIG. 119(*a*). Showing Frank-sessile dislocation loops in Al–16% Ag alloy quenched and aged 8 hr at 150° C. Note stacking fault contrast (fringes) inside loops *L* and helices *H* (×20,000).

peatedly quenched from 600° C. In both instances the loops are larger and their density is much lower than those shown in Fig. 120. This suggests that the type of loop produced by quenching is not governed solely by the stacking fault energy of the material, but also seems to be sensitive to nucleating conditions. For example, the results for aluminum can be understood qualitatively if nucleation occurs at impurities; a high density of nucleation sites seems to favor prismatic loops, whereas a lower density favors sessile loops. Figure 120 shows that there are loop-free regions adjacent to grain boundaries, and demonstrates clearly that boundaries are sinks for vacancies.

Dislocations may move nonconservatively by climb, either by emitting or absorbing point defects at jogs.[1] If point defects are absorbed by screw dislocations, these are converted into helices[50, 63] while edges become heavily jogged.[61] Figure 121 shows an example of helices in a quenched Al–4% Cu alloy where the quenched-in vacancies condense out on screw

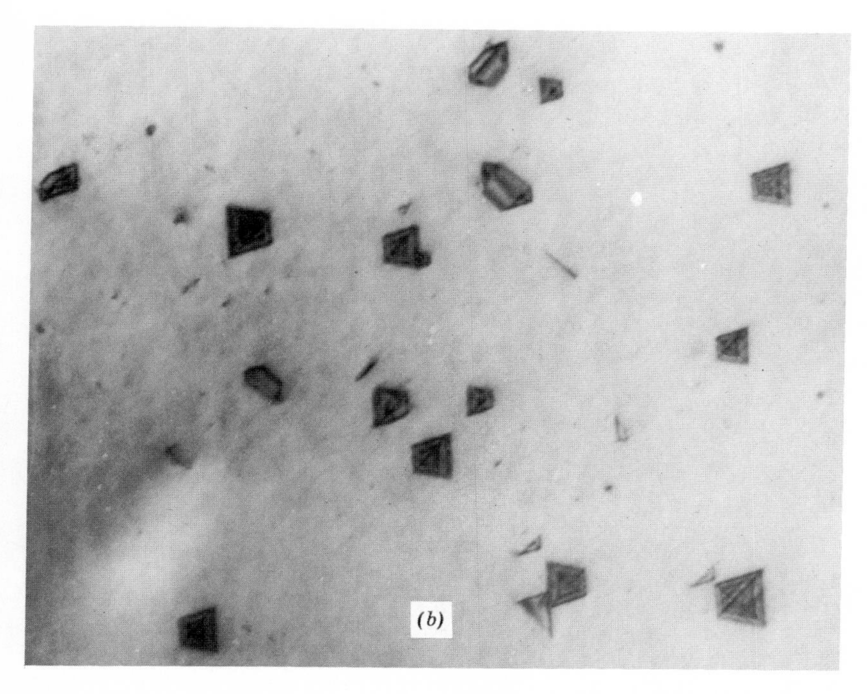

FIG. 119(*b*). Showing tetrahedra of stacking faults in quenched gold. Notice fringe contrast at faults (× 165,000). (After Silcox and Hirsch.[60] Courtesy *Phil. Mag.*)

dislocations. This also demonstrates that dislocations can be sinks (or sources) for vacancies.

At temperatures where self-diffusion occurs rapidly dislocations are expected to take up lower energy configurations by climb. Dynamic experiments using a hot stage have shown this process taking place in quenched aluminum heated to ~190° C.[64] The dislocation loops were observed to shrink probably by the evaporation of vacancies, and this work (recorded by ciné photography) has provided a direct demonstration of

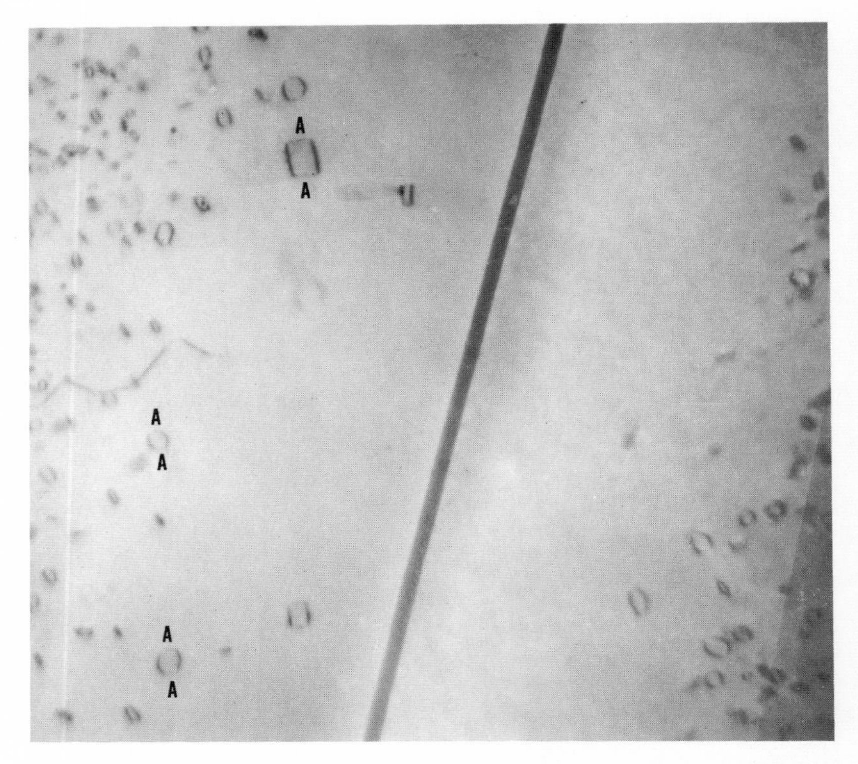

FIG. 120. Showing prismatic dislocation loops in quenched aluminum. Note absence of loops adjacent to grain boundary. Variation in visibility around loops is a contrast effect (e.g., at *A* segments steeply inclined to foil surface are practically invisible) (× 35,500). (After Hirsch et al.[61] Courtesy *Phil. Mag.*)

the climb process. Measurements made during these experiments enable the activation energy for self-diffusion to be calculated. Thus hot stage techniques seem to have great possibilities for future research with other metals.

Quenching defects have also been observed in copper,[65, 66] nickel,[67] and silver.[66, 67] In the former two metals the loops are similar to those in aluminum, whereas in quenched silver both loops and tetrahedra have been observed.[66, 67] This suggests that the energy of the two defects in silver is the same when collapse takes place, although local variations in composition, e.g., oxygen content, may be important in affecting the stacking fault energy. Quench-hardening in metals is probably due to the interaction between glide dislocations and these defects and accounts

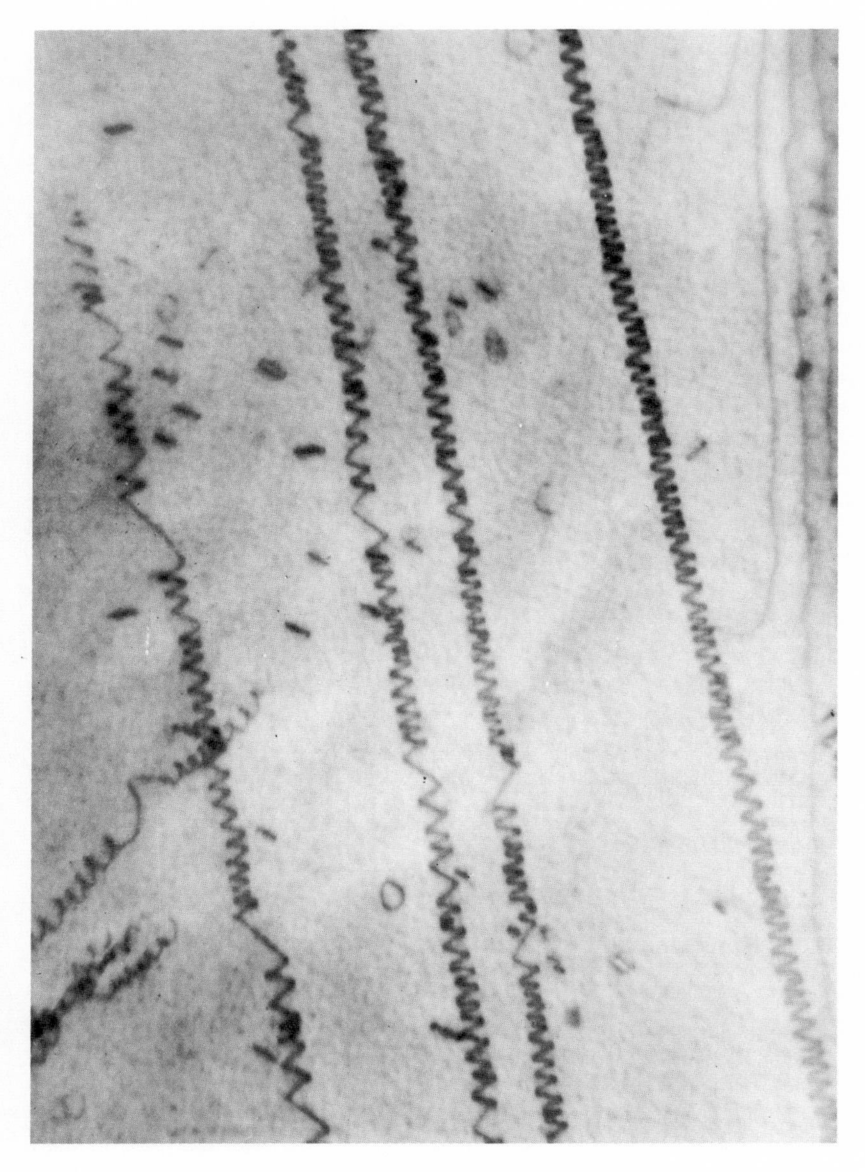

FIG. 121. Showing helical dislocations in quenched Al–4% Cu alloy. Screw dislocations assume partly edge character after climb. Axes of helices ⟨110⟩ (×58,500). (Courtesy *Phil. Mag.*)

for the yield points observed after quenching.[61] Dislocation loops may also be sources of slip and multipliers. They occur in densities of about $10^{10}/cm^2$, corresponding to those found in heavily deformed metals.[61]

Quenching defects have been investigated using transmission electron microscopy for alloys of aluminum containing various amounts of Cu, Zn, Ag, and Mg in solid solution.[34, 49, 50, 68] The vacancy concentration corresponding to the density of loops decreases with increasing solute content, suggesting that in concentrated alloys the vacancies are held, quenched-in, by association with solute atoms. Helical dislocations tend to be the preferred defects in these cases,[34] e.g., Fig. 121. This work suggests that alloying lowers the formation energy of a vacancy. The technique shows the possibilities of extending this work to a wide range of alloys. Since excess vacancies are retained by quenching, it is not surprising that age-hardening kinetics are sensitive to quenching conditions.

The concentration C of vacancies in quenched metals and alloys may be calculated from the expression $C \sim \pi R^2 \mathbf{b} N$, where R is the average loop (or helix) radius, **b** the Burgers vector, and N the density of loops (or number of turns of the helix per unit volume). Since R and N can be measured from the micrographs, C works out to be $\sim 10^{-4}$ for Al,[61] 10^{-3} for dilute alloys, and 10^{-5} for concentrated alloys.[34] An example of the high density of loops found in quenched dilute Al–Cu alloys can be seen in Fig. 122, which is to be compared with Fig. 120 for pure aluminum.

Dislocations formed as a result of quenching are expected to be favorable sites for capture of solute atoms since this would relieve elastic strains. If the binding energy between solute atoms and vacancies is high, then it is likely that capture occurs by the migration of solute atom-vacancy groups.[34, 68] In this way loops and helices become preferential sites for precipitation[69] (see section 5.3.2).

5.2.10 *Observations of Radiation Damage*

Studies of damage by α particles, fission fragments or neutron irradiation have been made on a number of metals.[66, 70, 71] As a result of collisions in the lattice, many vacancies and interstitials can be created. These are revealed as loops, black spots or regions of strain when they condense in sufficiently large numbers. Results on copper[70] indicate that the concentration of defects and the density of dislocations after heavy irradiation ($\sim 10^{20}$ n.v.t.) are similar to that found in quenched metals. The regions of strain are thought to be produced by aggregation of vacancy-interstitial pairs.

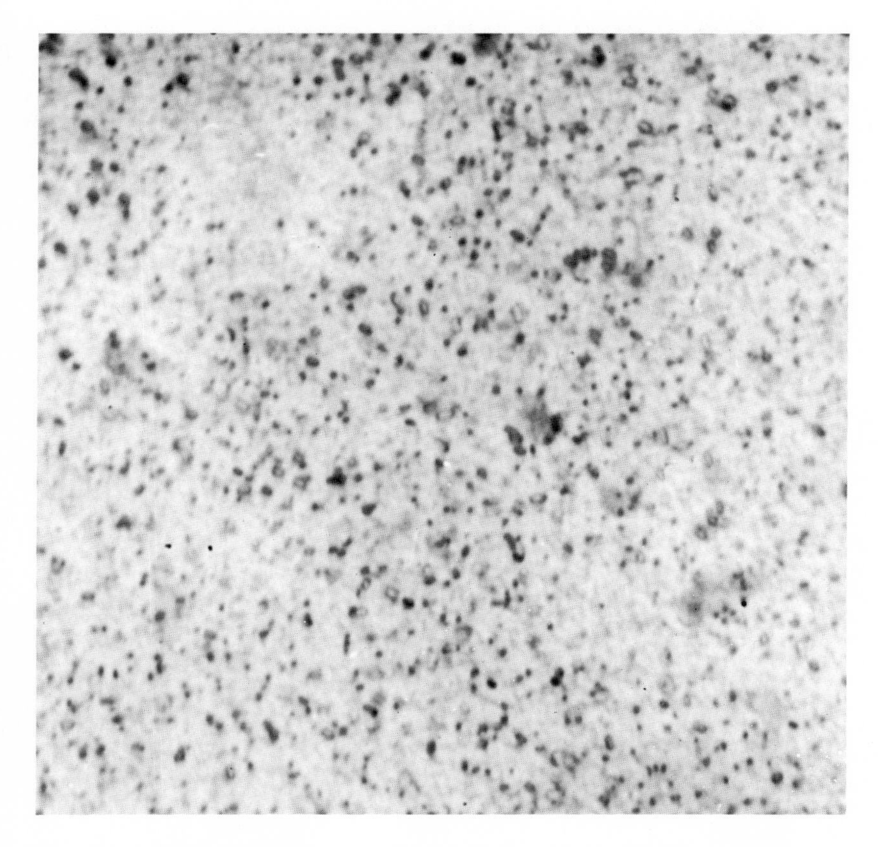

FIG. 122. Dislocation loops in Al–1% Cu alloy after quenching from 600° C (compare with Fig. 120). (× 80,000.)

Electron-microscopic observations of radiation damage in graphite† have been made by Bollmann.[72] In this work, Bollmann has shown that the dark field technique is especially useful for interpreting contrast effects observed in the images. The results indicate that the observed defects (black and white "dots") might be extrinsic and intrinsic stacking faults, or groups of these corresponding to interstitial and vacancy clusters (see also section 5.2.15).

† Reviews of work on radiation damage in graphite may be found in *Peaceful Uses of Atomic Energy*, Vol. 7, 1956 (United Nations, New York).

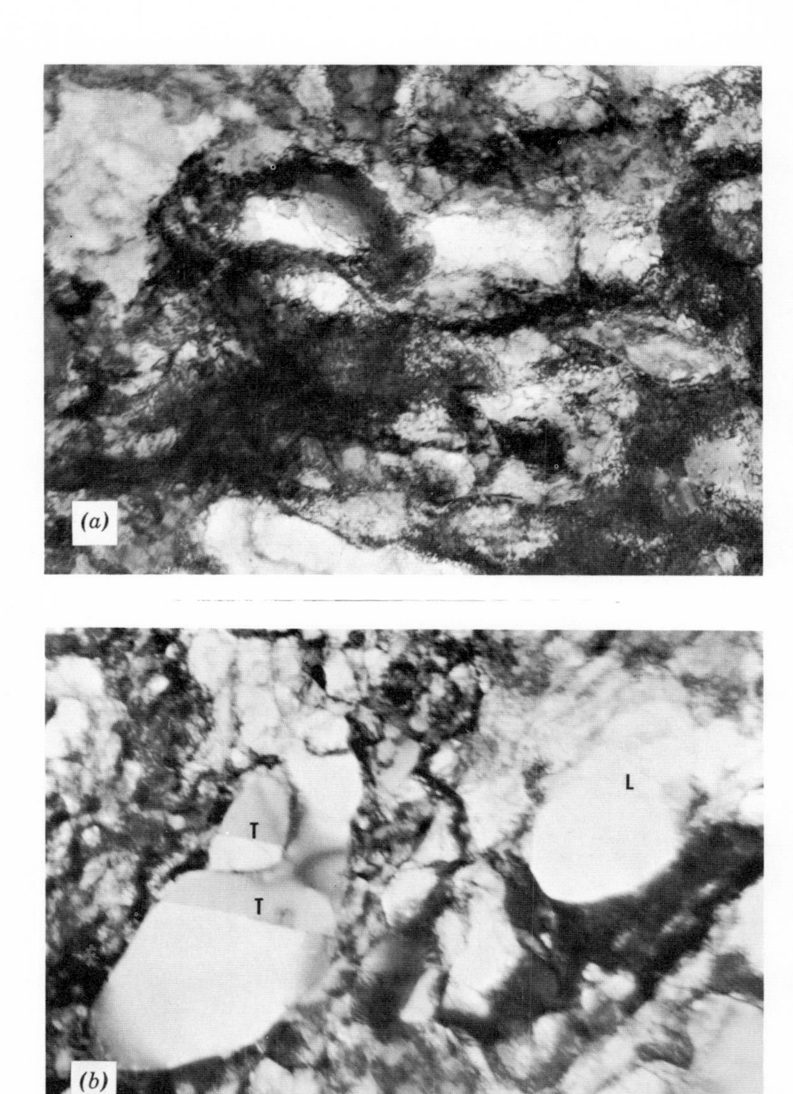

FIG. 123. Showing sequences during recrystallization of cold rolled nickel. (After Bollman.[74])

(*a*) Eighty per cent cold rolled showing dense networks of dislocations forming cell structure (× 37,500). (*b*) Annealed at 250° C. Note twin boundaries *T* and low angle boundary *L* (× 11,500). (*c*) Annealed at 300° C. Note twins *T* growing in incompletely polygonized matrix (× 14,500). (*d*) Fully recrystallized (after grain growth). Boundaries have probably migrated in regions *AA*. Notice fringes along the grain boundary *B* inclined to the surface of the foil (× 8500). (See section 2.3.)

T
(c)

A
B
A
(d)

5.2.11 *Observations of Recrystallization*†

The technique of transmission electron microscopy has been used by Bollmann[74] to study recrystallization in nickel and by Bailey and Hirsch for silver,[27] (section 5.2.4). In Bollmann's work, measurements were made on bulk samples after different treatments of cold work and annealing, and numerous foils prepared from these samples by electropolishing. The initial part of the recovery process was correlated with the process of polygonization starting from areas containing dense networks of dislocations, e.g., Fig. 123*a*. At the same time recrystallization occurred and proceeded until the networks were consumed. An early stage of the process is shown in Fig. 123*b*. As the new grains grew, twins were observed to form (Fig. 123*c*), and eventually secondary recrystallization and grain growth took place (Fig. 123*d*). In this way, Bollmann has been able to follow the recovery, recrystallization, and grain growth processes. Bailey has shown that recrystallization occurs by migration of the old grain boundaries.[27]

A high temperature stage may be used to follow these effects directly in the microscope, but because of the thinness of the section, the recrystallization rates are much slower than in bulk specimens.[27]

5.2.12 *Observations of Dislocations in As-Prepared Thin Foils*

As discussed earlier, the properties of thin foils prepared initially as thin sections (e.g., by evaporation) are different from those of bulk metals. However, bearing this in mind, much valuable information may be obtained from such investigations, particularly in cases where it may be difficult to prepare foils from bulk samples by conventional methods (see Chapter 4). Dislocations in thin foils of gold prepared by evaporation have been observed by Pashley.[75] It is interesting that in this work Pashley detected both stacking faults and cross-slip in his foils.

Indirect methods of resolving metal lattices have allowed dislocations to be revealed in the moiré patterns formed from overlapping crystals.[76] Basset et al.[77] have observed dislocations in overlapping foils of copper and palladium on gold and were able to determine the types of dislocation, their Burgers vectors, and the dislocation density. In this work it is possible to observe the extra half planes associated with edge dislocations. As we saw in Chapter 2, if the individual lattices have a spacing d and are rotated with respect to one another by a small angle ϵ about a normal to the specimen plane, the spacing of the moiré pattern is given by d/ϵ. Alternatively the parallel moiré patterns have a spacing $d_1d_2/(d_2 - d_1)$,

† For a review of recovery and recrystallization see reference 73.

where d_1 and d_2 are the individual spacings of two lattices. In this case the moiré magnification is $d_1/d_2 - d_1$ of the lattice spacing d_2 (Fig. 31). For small ϵ or $d_2 - d_1 \ll d_2$ the moiré magnification is much greater than unity.

Although the individual lattices may not be resolved directly in the microscope, it is possible by suitable choice of the moiré magnification to resolve the lattice indirectly via the moiré pattern. Thus if a dislocation is present in one of the crystals, it will be resolved and reproduced in an enlarged form in the moiré pattern. The optical analogue to illustrate this effect is shown in Fig. 124. Unfortunately, it is not possible to decide in which of the two crystals the dislocation lies. Figure 125 is an example of dislocations revealed in a parallel moiré pattern formed from a palladium layer deposited on a (111) gold film. Partial dislocations and stacking faults have been revealed using this technique.[77]

The resolved image of a single crystal lattice is formed in the electron microscope by interference in the image plane between the directly transmitted (zero-order) beam and the first order diffracted beam arising from the set of lattice planes resolved.[78] In effect, therefore, the crystal presents a plane grating to the electron wavefront. Recombination of these beams in the final image is only possible for crystals of large lattice spacings (>10 Å) and low Bragg angles, so that it is not possible to resolve planes in metal lattices (section 3.3). However, Menter[78] was able to

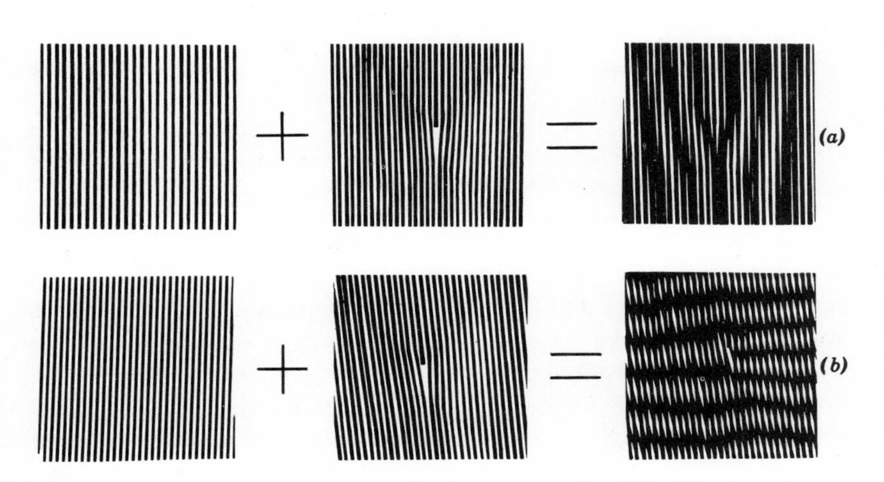

FIG. 124. Optical analogue showing formation of terminating half-planes in moiré pattern associated with a dislocation in one of the gratings. (*a*) Parallel gratings. (*b*) Rotated gratings. (After Bassett et al.[76] Courtesy the Royal Society.)

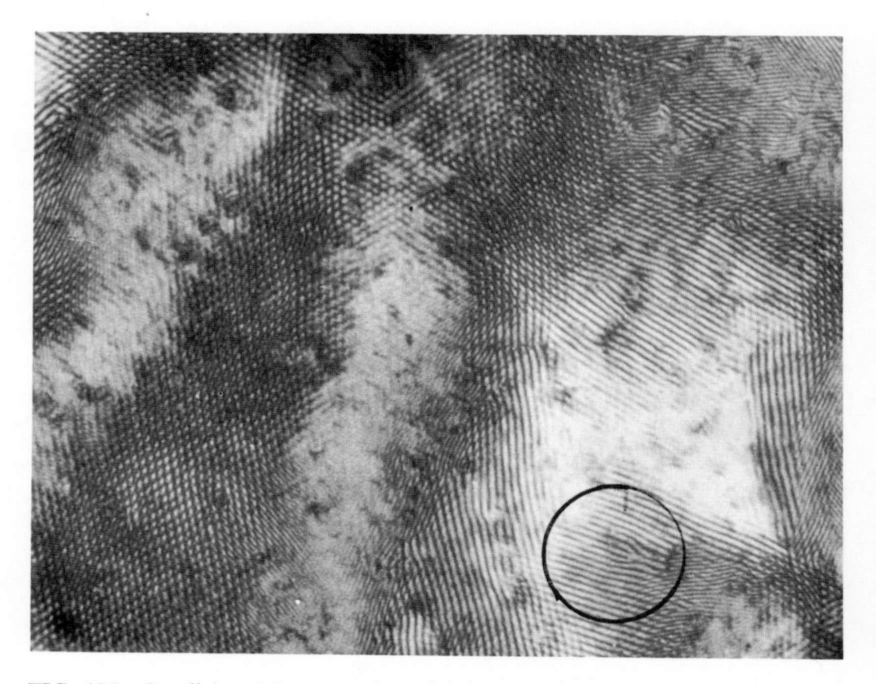

FIG. 125. Parallel moiré pattern from Pd deposited by evaporation onto (111) Au film. Circled area shows dislocation with two terminating half-planes. Variation in contrast is probably due to buckling and interference effects between films. Planes resolved inside circle are (220) of moiré spacing 29 Å (×270,000). (After Bassett et al.[76] Courtesy the Royal Society.)

resolve lattice planes in some organic crystals, e.g., platinum phthalocyanine, in which dislocations were often observed.

In the formation of moiré patterns from metal foils, the mechanism of imaging lattice planes is similar to that for a single layer crystal, except that the first order diffracted beam from the single lattice is replaced by a beam which has been diffracted in opposite senses by each of the superimposed lattices in succession.[79] Figure 126 shows a diffraction pattern obtained from overlapping gold-gold foils in [111] orientation in which the satellite spots due to double diffraction are clearly visible. A moiré image such as in Fig. 125 is obtained from this by allowing only the direct beam and its satellites to contribute to the image. (See section 2.2.) The moiré technique is thus a very useful method of indirectly resolving metal lattices, and the results obtained have provided beautiful examples of the very high resolutions that can now be attained in the electron

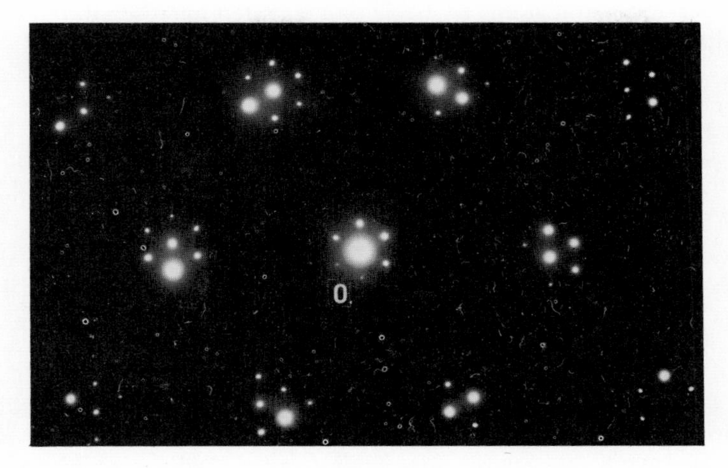

FIG. 126. Diffraction pattern from superposed (111) films of gold rotated by approximately 10°. Moiré pattern (bright field) is formed from the direct beam *O* and its satellites. Dark field image is formed from a diffracted beam and its satellites. Satellites are produced by double diffraction. (After Bassett et al.[77] Courtesy the Royal Society.)

microscope. Furthermore, details are made available of the structures and widths of dislocations, and future studies may provide enough experimental evidence to enable calculations to be made of their core energies.

Deformation experiments on thin foils prepared by evaporation and strained inside the microscope have shown that they have an abnormally high tensile strength, and catastrophic fracture occurs after $\sim 1\%$ strain. This has been explained in terms of thinning in localized regions due to slip[80] and is quite unlike the behavior of bulk specimens. Dynamic straining experiments with thin foils must therefore be treated as a special case of deformation.†

Price[29] has observed slip processes in thin crystals of zinc prepared by condensation from the vapor. The crystals were strained inside the microscope, and some deformed by twinning and others by pyramidal glide on the $(11\bar{2}2)$ $[\bar{1}\bar{1}23]$ system. In the twinning deformation Price observed that no glide was necessary to initiate the twins and neither did the twins grow in a jerky, high speed manner.

In the field emission microscope, Dreschler[81] has directly observed the surface steps associated with the emergence of a screw dislocation at a

† See also Neugebauer et al., *Structure and Properties of Thin Films*, John Wiley and Sons, New York, 1959.

metal surface, and it seems probable that emission microscopy will play a prominent part in future experiments with metals.

5.2.13 *Dislocations in Nonmetals*

Recently the electron microscope has been used for the examination of materials having hexagonal symmetry or layer structures (section 4.3.5). These materials include graphite,[82, 83] Bi_2Te_3,[84, 85] Sb_2Te_3,[85] MoS_2[86, 87, 88] mica,[89] and talc.[90] In each of these materials, except for the tellurides, single dislocations have been observed to split according to the reaction

$$a[100] \rightarrow a/3[210] + a/3[1\bar{1}0]\dagger \qquad (82)$$

and in talc, Amelinckx and Delavignette[90] have observed single dislocations split into four partials. From calculations based on the radii of curvature of dislocation nodes it has been shown that most of these materials have a very low stacking fault energy. Williamson,[83] e.g., has estimated γ to be only 0.1 erg/cm^2 in graphite. An example of one of his micrographs is shown in Fig. 127. In molybdenite (MoS_2) alternate basal planes in the layer contain extended and unextended dislocations, respectively.[88] This is expected since the MoS_2 structure consists of hexagonal sheets stacked in the order Mo–S–S–Mo–S–S–Mo, etc., and a stacking fault between S and Mo planes will have a higher energy than one between two S layers; thus dislocations between Mo and S planes are not expected to dissociate.[88] Many of the micrographs obtained from these investigations are complicated by rotational moiré patterns associated with twist boundaries consisting of dislocation networks similar to Fig. 127, but on a finer scale.

Ionic crystals have been widely studied for the last few years, using either etching or decoration techniques and light microscopy, and the results have provided information regarding their mechanical properties as well as on dislocation behavior. Recently, however, direct observations of dislocations in MgO have been made by Washburn et al.[30, 31] but so far this has been the only ceramic material successfully studied by transmission electron microscopy. One of the reasons for this is that other crystals (particularly the alkali halides) are severely damaged by electron bombardment during examination in the microscope. Figure 128*a* is an example of dislocations in (110) planes in MgO, and shows the formation of elongated loops of plus-minus pairs of edge dislocations and small prismatic loops. The Burgers vectors can usually be recognized by inspection; the slip planes are {110} and the slip directions ⟨110⟩, so that in specimens orientated with [100] parallel to the beam the slip planes are

† Vectors in cubic rather than hexagonal notation.

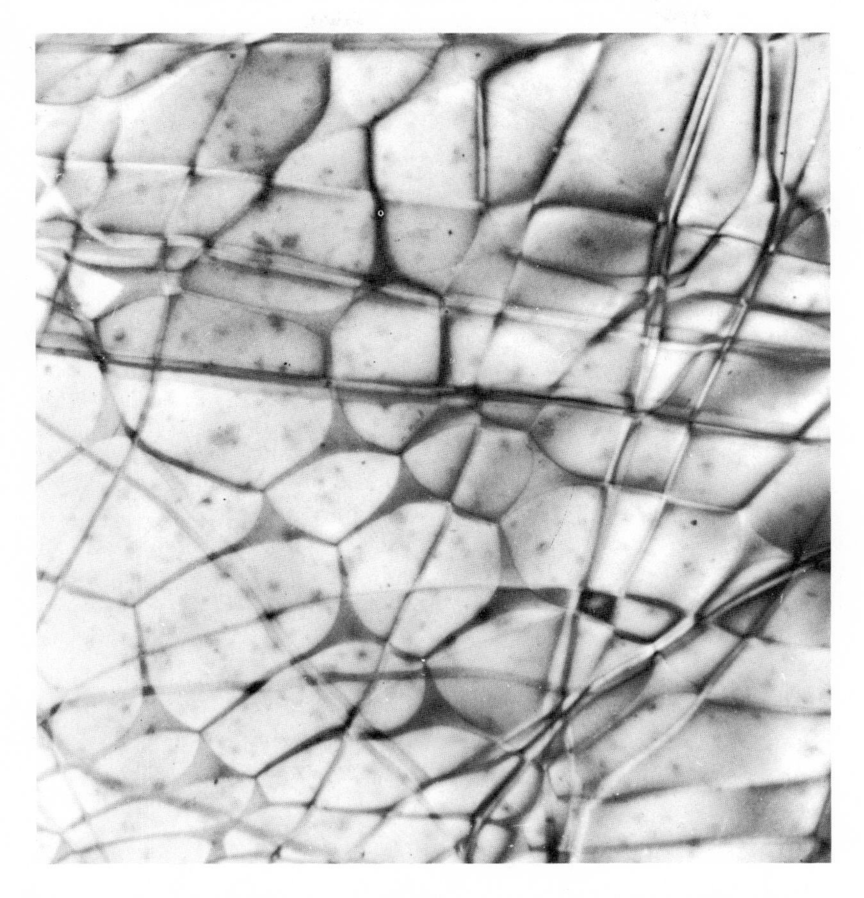

FIG. 127. Hexagonal dislocation networks in basal plane of graphite. Notice alternation of extended and contracted nodes ($\times$ 24,000). (After Williamson.[83] Courtesy the Royal Society.)

either parallel (with Burgers vectors perpendicular) to the beam or at 45° to the beam projected along $\langle 100 \rangle$ (Burgers vectors perpendicular to the projection of the slip plane). The dislocations are not extended and after passage along the slip plane they leave behind a high density of elongated plus-minus pairs of edge dislocations (Fig. 128*a*), so that interaction between slip planes occurs at relatively low strains. These elongated dislocations can form by cross-slip of part of a screw dislocation, leaving a segment of edge which acts as a pinning point so that after continued movement it leaves behind a plus-minus pair of edge dislocations

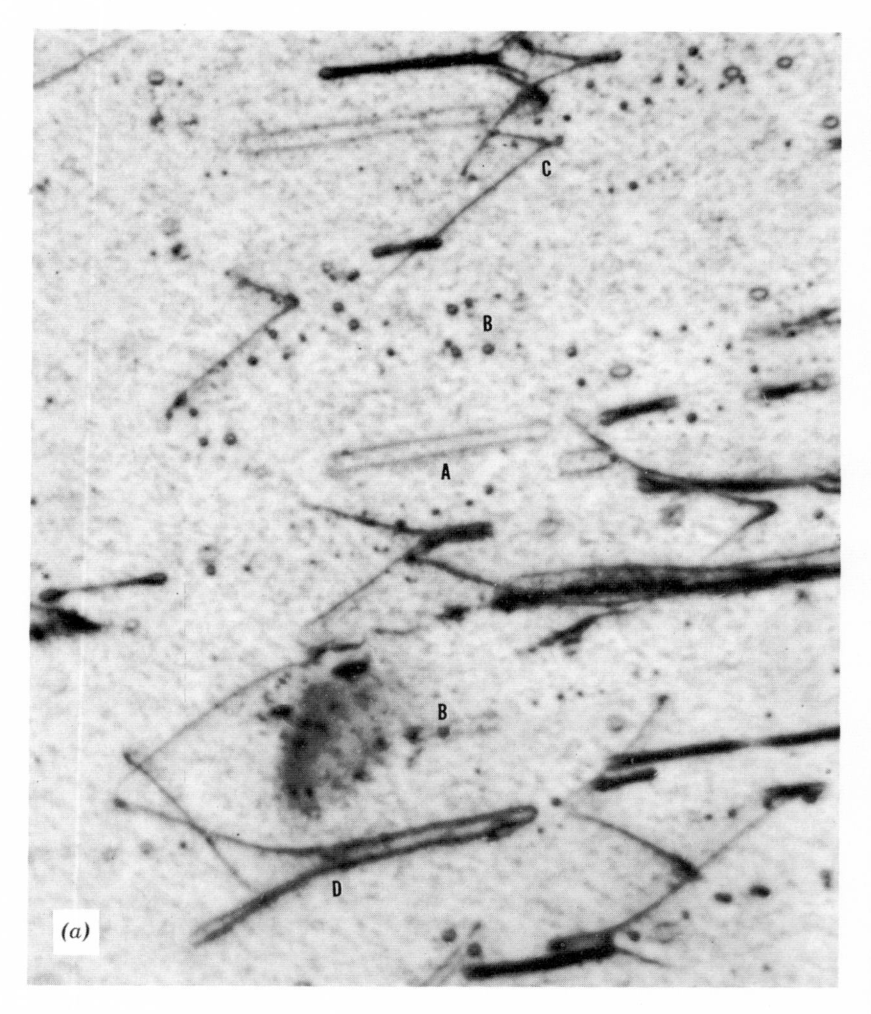

FIG. 128(*a*). Dislocations in MgO after bending at room temperature followed by annealing at 1000° C. Showing stages of formation of elongated loops of plus-minus pairs of edge dislocations *A*, at *C* and *D*. Pairs later collapse into small loops *B*. Orientation [001] (× 65,000). (After Washburn et al.[30] Courtesy *Phil. Mag.*)

connecting it to the jog (e.g., at *D*, Fig. 128*a*). Small precipitates may also act as pinning points[31] as shown in Fig. 128*b*. Future experiments will undoubtedly provide much needed information regarding strain hardening and fracture characteristics of MgO and other ceramic materials.

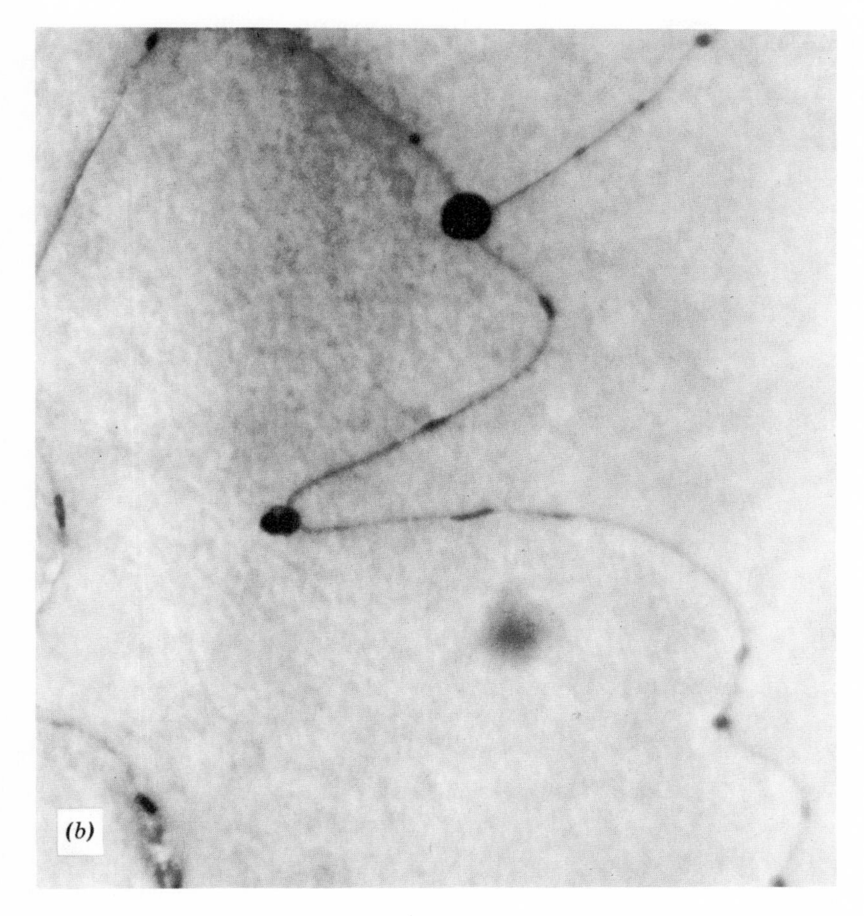

FIG. 128(*b*). Showing dislocations in MgO pinned by small precipitates (probably magnesium nitride) (×70,000).

5.2.14 *Kinematical Theory of Contrast from Defects in Thin Metal Foils*

In Chapter 2 we discussed the importance of diffraction theory in explaining some of the contrast effects met with in relatively perfect crystals, e.g., extinction contours in buckled foils, and fringes in wedge crystals, e.g., at foil edges, grain boundaries, etc., due to a discontinuity of structure on an inclined plane. Typical examples of these effects can be seen in Figs. 33 and 123*d*. Because of the phase contrast arising out of displacements of atoms from their ideal positions, defects in crystals are also revealed by Bragg diffraction contrast so that dislocations appear as dark lines, (bright field image) and stacking faults give rise to interference fringes. Examples of these have been discussed in the preceding paragraphs. This method of contrast formation has the advantage that the atomic array is not resolved so that resolution is not a limiting factor. Thus a projection of the three-dimensional arrangement of dislocation lines can be obtained in crystals that would be too thick for direct lattice resolution.[91]

An understanding of the nature of diffraction contrast is very important for the correct interpretation of electron microscope images of thin crystals.[91–95] In this section we shall describe electron diffraction from defects in terms of the simpler kinematical theory which provides a good qualitative understanding of the observed effects.[93] We shall use the method discussed in Chapter 1 (section 1.13).

From Fig. 20 we have seen that the amplitude scattered by a column of perfect crystal is given by expression (39), viz.,

$$A \sim \sum \exp\,[2\pi i(\mathbf{g} + \mathbf{s})\cdot\mathbf{r}_n]$$

However, if the atoms are displaced by a vector $\mathbf{R}$ from their ideal positions, the amplitude scattered is now

$$A \sim \sum \exp\,[2\pi i(\mathbf{g} + \mathbf{s})\cdot(\mathbf{r}_n + \mathbf{R})] \tag{83}$$

We can expand and write (83) as

$$A \sim \sum e^{2\pi i\mathbf{g}\cdot\mathbf{R}}\, e^{2\pi i\mathbf{s}\cdot\mathbf{r}_n} \tag{84}$$

where $e^{2\pi i\mathbf{g}\cdot\mathbf{r}_n}$ reduces to unity and neglecting the factor $\mathbf{s}\cdot\mathbf{R}$. Thus the displacement in the lattice produces a phase angle $\phi = 2\pi\mathbf{g}\cdot\mathbf{R}$ in the scattered wave, so that the resultant amplitude differs from that for a perfect crystal. Contrast thus arises through a phase contrast mechanism, the phase difference being produced by any atomic displacements $\mathbf{R}$. Rewriting equation (84) as an integral and neglecting the factor f/V (section 1.13.2) we get

$$A = \int_{\text{column}} e^{(i\phi)} e^{(2\pi i s z)}\, dz \tag{85}$$

where ϕ is a function of z (distance along the column), depending on both the reflection $\mathbf{g}$ and the displacement $\mathbf{R}$.

For a stacking fault (also see section 2.3.3), the displacement vector $\mathbf{R}$ corresponds to $\mathbf{b}_2$ and $\mathbf{b}_3$ of Fig. 103, which for face-centered cubic metals is $\frac{1}{6}\langle 1\bar{2}1\rangle$. A stacking fault thus produces a phase difference $\phi = \pi/3[h - 2k + 1]$ in waves diffracted from opposite sides of the fault equal to $\pm 120°$ or $0°$, depending on the indices of the reflection.[92] This phase difference can be represented by an amplitude-phase diagram as shown in Fig. 129. Here the point O corresponds to the middle of the column of crystal being considered (see Figs. 20, 21, Chapter 1), and the circle POP' is the amplitude-phase diagram for the perfect crystal [radius $(2\pi s)^{-1}$]. At the point Q, corresponding to the intersection of the column of crystal with the fault, an abrupt change of phase occurs (equal to $-120°$ in the case of Fig. 129). The resultant amplitude from the column with the fault is PP''; hence as Q varies, PP'' will oscillate, producing fringes as shown in Figs. 104, 113. In a perfect crystal (circle POP'), as s or t varies, PP' will oscillate also giving rise to fringes, and in both cases the fringes are explained by Fig. 34, where in the bright field image the first dark fringe occurs at a distance $t_0'/2$ from the top of the foil. The spacing of fringes corresponds to a depth periodicity at the fault equal to $t_0' = s^{-1}$. Where t_0' is greater than the foil thickness no fringe contrast will occur.† The factors affecting the contrast and the fringe spacings have already been discussed in Chapter 2.

For overlapping faults, if a single fault produces a phase angle $\phi = -120°$, two overlapping faults produce a phase angle $\phi = -240°$, which is equivalent to $+120°$. Under certain conditions reversing the sign of ϕ leads to reversal of intensity of the fringes. At each reversal the number of dark or light fringes changes by one.[92] This effect can be seen at A, Fig. 104. In relatively thick foils, reversals take place frequently. Clearly three overlapping faults will produce a total angle of $-360°$, which is equivalent to 0°, and so no fringes will result in this case. These effects are well known experimentally.[93]

In the electron diffraction patterns, streaks parallel to $\langle 111 \rangle$ can be

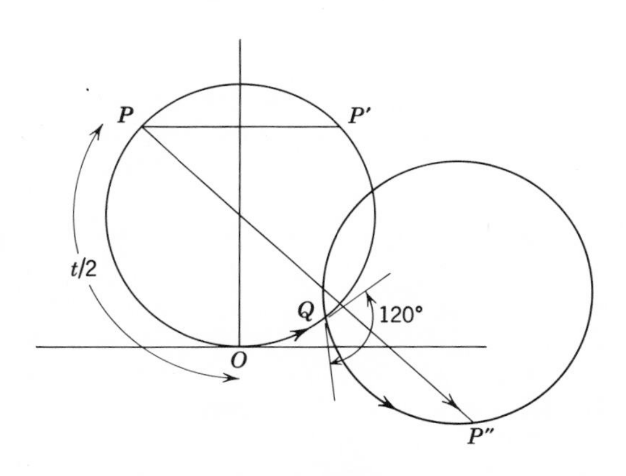

FIG. 129. Amplitude-phase diagram for a face-centered cubic crystal containing a stacking fault (see also Fig. 21). As PP'' oscillates (e.g., due to a change in thickness t), fringes appear in the image. (After Whelan and Hirsch.[92] Courtesy the Royal Society.)

† This only arises in very thin foils of metals of low atomic number (e.g., Al and Mg) and large Bragg angles (see Table II, Chapter 2).

observed (section 1.14.1), and the patterns should be checked with the corresponding microstructure to avoid false interpretation of the contrast effects.[92] As discussed in Chapter 2, reversal of contrast will occur upon going from the bright to the dark field image (Fig. 39) and, by tilting, it is possible to change the spacing of the extinction fringes, since this depends both upon t and the deviation from the Bragg condition. Since the contrast and intensity vary as s^{-2}, the contrast is particularly strong at the extinction contour, e.g., at C, Fig. 104.

Similar arguments apply to the case of antiphase domain boundaries in superlattices (see section 5.3.5). As an example, consider the ordered structure $AuCu_3$ which is simple face-centered cubic (Ll_2) with the corner sites occupied by gold atoms. The boundaries between the ordered domains are known as antiphase domain boundaries since the atoms change in phase or step at these points. The displacement vector **R** corresponding to a domain boundary in $AuCu_3$ is $a/2\langle 101\rangle$, so that the phase shift ϕ in waves diffracted from opposite sides of the boundary is equal to $\pi\langle 101\rangle \cdot (hkl)$. This is also true for AuCu superlattices. The possible values of ϕ are 0, or $\pm n\pi$, depending on the particular combinations of (hkl) and **R**, and where $n = 1, 2, 3$, etc. For mixed indices, i.e., superlattice reflections in $AuCu_3$ or AuCu, $\phi = 0$ or $\pm\pi$, whereas for unmixed indices, i.e., principal reflections, $\phi = 0$ or $\pm 2\pi$. This shows that domain boundaries will be observed only if the specimen is oriented for a strong superlattice reflection (e.g., at superlattice extinction contours) such that ϕ is not zero. In the case of the gold-copper series of superlattices, for $\mathbf{R} = a/2[101]$ such reflections should be (001) or (110) since then $\phi = \pi$ and contrast is possible, e.g., see Figs. 150 and 151. For reflections where $\phi = 0$, e.g., (010) for $\mathbf{R} = a/2[101]$, the antiphase domain boundaries will be invisible. Since the extinction distances corresponding to superlattice reflections are much greater than those corresponding to normal reflections† (these can be calculated following the procedures given in section 2.3.3), the appearance of fringe contrast typical of stacking faults is unlikely. As long as the displacement vector **R** is known, the contrast conditions for any lattice defect can easily be predicted by using the simple kinematical criterion for phase contrast.

In the problem of contrast at a dislocation, the strain field varies continuously near the dislocation (Figs. 100, 101). With an edge dislocation, consideration of Fig. 100 shows that the sense of rotation of lattice planes is opposite on opposite sides of the dislocation. In kinematical theory the crystal will now be rotated locally toward the reflecting position on

† For example, in the perfectly ordered $AuCu_3$ superlattice, t_0 for the (100) superlattice reflection is 860 Å, while t_0 for the normal (200) reflection is 263 Å {J. Marcinkowski and R. M. Fisher (to be published)}.

one side of the dislocation and away from it on the other side, e.g., if the electron beam is incident in the direction *GH* reflection will occur at *PQ* but not at *RS*, so that the contrast appears to the left of the dislocation. Since the *image is to one side of the dislocation core*, it is difficult to determine the exact position of the dislocation in the crystal.

The case of the screw dislocation (Fig. 101) is simpler to treat than that of the edge since in the former the strain field is radially symmetrical about the dislocation core and can be considered to be the same as in an infinite medium. Consider Fig. 130 which represents the nature of the displacements[93]: a column of crystal *CD* is deformed to a shape *EF* after introducing a screw dislocation *AB*. The displacement vector **R** is

$$\mathbf{R} = \mathbf{b}\alpha/2\pi = (\mathbf{b}/2\pi)\tan^{-1}(z/x) \tag{86}$$

where **b** is the Burgers vector of the dislocation. Hence

$$\phi = 2\pi\mathbf{g}\cdot\mathbf{R} = \mathbf{g}\cdot\mathbf{b}\tan^{-1}(z/x) = n\tan^{-1}(z/x) \tag{87}$$

Since **g** is a reciprocal lattice vector and **b** is an interatomic vector, *n* is an integer which can be positive, negative, or zero. Also since the scattering factor decreases rapidly with increasing scattering angle (section 1.8), the diffracted beams corresponding to small values of **g** are most important in producing contrast; hence only small values of *n* need be considered.

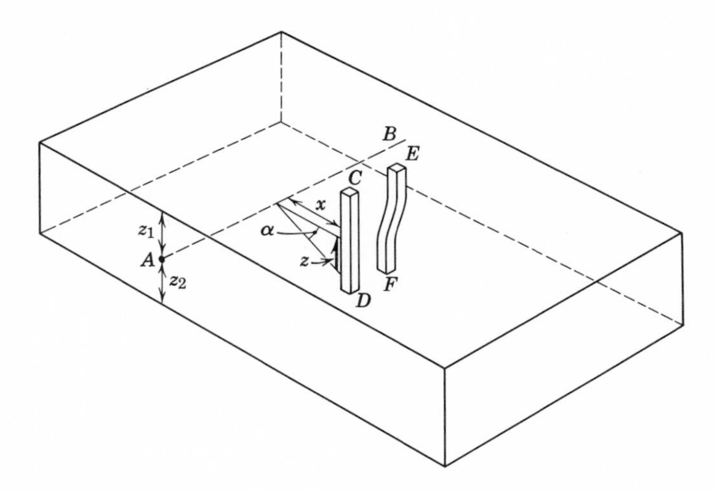

FIG. 130. Crystal containing a screw dislocation *AB* parallel to the surface at distances z_1 and z_2 from the top and bottom. The column of perfect crystal *CD* is deformed to shape *EF* after introducing the dislocation. (After Hirsch et al.[93] Courtesy the Royal Society.)

When $n = 0$, $\phi = 0$, hence $\mathbf{g} \cdot \mathbf{b} = 0$, i.e., $\mathbf{g}$ and $\mathbf{b}$ are mutually perpendicular, and the dislocation line is invisible (i.e., when the Burgers vector is parallel to the reflecting plane). *Thus, for maximum contrast the displacements must be normal to the reflecting plane.* This condition must be obtained to detect very thin precipitates in alloys. Since the Bragg angles are very small, displacements approximately normal to the foil surface produce no contrast.

To represent the amplitude-phase diagrams for columns of crystal close to the dislocation line we need to plot the amplitude-phase angle change along the column. Substituting equation (87) in equation (85) we get for the case shown in Fig. 130

$$A = \int_{-z_1}^{+z_2} \exp\left[(in \tan^{-1}(z/x) + 2\pi isz)\right] dz \tag{88}$$

Since this integral is an expression for the amplitude-phase diagram [z being the distance measured along the curve and $(n \tan^{-1}(z/x) + 2\pi sz)$ the slope] it is possible to calculate and plot the diagrams for various situations for n by the method given in section 1.13.2. Figures 131*a* and *b* are examples for a screw and an edge dislocation. It can be seen that the phase changes due to the lattice distortion will be of opposite sign on opposite sides of the dislocation; hence the diagram is an unwound spiral 1 or a wound spiral 2.[93] This means, for the case shown, that the contrast will be to the *left* of the dislocation. It follows that contrast effects are well explained qualitatively in terms of the amplitude-phase diagram, but it is well to remember that this analysis refers only to one diffracted beam, i.e., to one value of $\mathbf{g}$ and neglects interactions from other reflections. A complete analysis requires a dynamical treatment of the problem which takes into account absorption effects due to inelastic scattering. In this way the fine details of fringe patterns from stacking faults and other effects may be explained.[95] (See also section 2.3.5.) The calculations are extremely complicated, but details of the theory are now being made available. For example, for partial dislocations in the FCC lattice, Howie and Whelan (to be published) have shown that the condition for invisibility is given by $\mathbf{g} \cdot \mathbf{b} = \pm\frac{1}{3}$. Thus, if a whole dislocation, Burgers vector $\mathbf{b}_1$, splits into two partials of Burgers vectors $\mathbf{b}_2$, $\mathbf{b}_3$ separated by a ribbon of stacking fault, when one partial ($\mathbf{b}_2$) is invisible while the other ($\mathbf{b}_3$) is visible $\mathbf{g} \cdot \mathbf{b}_2 = \pm\frac{1}{3}$ and $\mathbf{g} \cdot \mathbf{b}_3 = \mp\frac{2}{3}$, i.e., for the unextended dislocation $\mathbf{g} \cdot \mathbf{b}_1 = +1$ and it is visible. If both partials are invisible (see Fig. 104 at *B*) then $\mathbf{g} \cdot \mathbf{b}_2 = \pm\frac{1}{3}$ and $\mathbf{g} \cdot \mathbf{b}_3 = \mp\frac{1}{3}$, i.e., for the unextended dislocation $\mathbf{g} \cdot \mathbf{b}_1 = 0$ and it is invisible. In all cases $\mathbf{b}_1 + \mathbf{b}_2 + \mathbf{b}_3 = 0$ (Fig. 103).

The dynamical theory also predicts that when the stacking fault extends through the section of the foil, if both the bright field image (symmetrical)

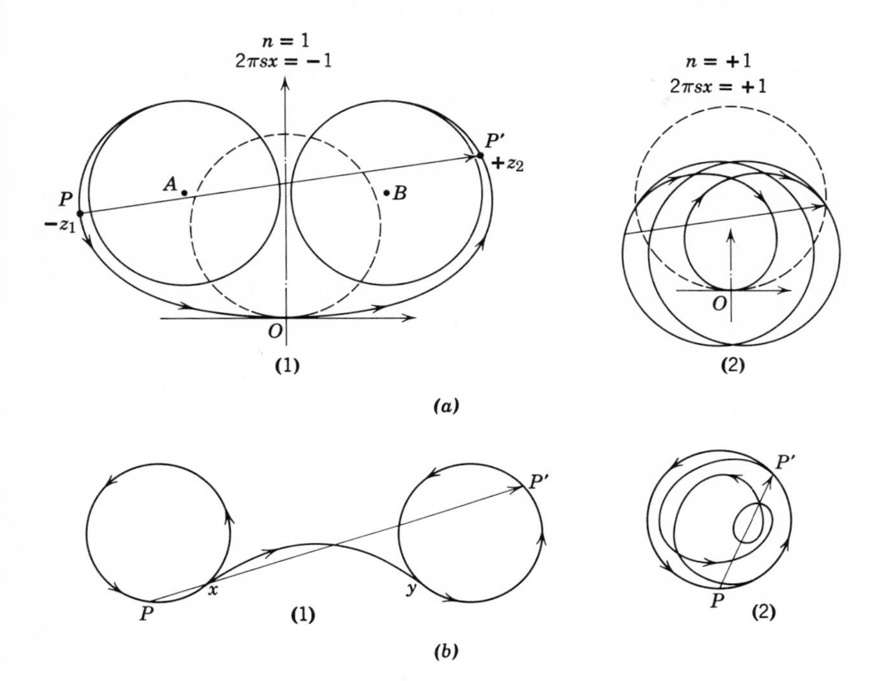

FIG. 131(*a*). Amplitude-phase diagram for a column of crystal near a screw dislocation (Fig. 130). The broken circle represents the perfect crystal. The diagram is an unwound spiral (1) or a wound-up spiral (2). There is a different diagram for each value of $2\pi sx$. The center of the dislocation corresponds to $2\pi sx = 0$. The amplitude diffracted by the column *EF* in Fig. 130 is given by the line *PP′* corresponding to the top and bottom of the column. Note that the amplitude scattered from one side of the dislocation (1) is greater than that from the other side (2). (*b*) Amplitude-phase diagram for the atom rows on opposite sides of the edge dislocation shown in Fig. 100, (1) to the left of *AB*, (2) to the right of *AB*. The length *xy* in (1) corresponds to the distorted segment of the atom row. The resultant amplitude is *PP′*. Again the amplitude in 1 is >2. (After Hirsch et al.[93] Courtesy the Royal Society.)

and the dark field image (asymmetrical, due to absorption) show the *same* intensity variation, i.e., alternating bright fringe, dark fringe (noncomplimentary), this corresponds to the top of the foil. If the bright and dark field images are complimentary at the opposite end of the fringe pattern, this corresponds to the bottom of the foil. At the present time this is the only way to determine which are the top and bottom surfaces of the crystal.

Returning again to kinematical considerations, it can be seen from

Fig. 131 that the amplitude scattered from one side of the dislocation (1) is greater than that from the other side (2). In other words, the dislocations scatter more intensity into the Bragg reflections than elsewhere and thus appear as black lines in bright field images and white lines in dark field images. Furthermore, because the intensity peak occurs for negative values of $2\pi sx$, the image of the dislocation lies to one side of the dislocation core. The displacement of the peak is of the order of the width of the image, which is represented in Fig. 132 for different values of n. This shows that for $n = 3$ and $n = 4$ there are two peaks on the same side of the dislocation, i.e., the image would show double lines. Usually, double images are observed when two sets of reflecting planes which produce contrast on opposite sides of the center of the dislocation are operating simultaneously (e.g., see Fig. 111 at A). Another contrast effect arises

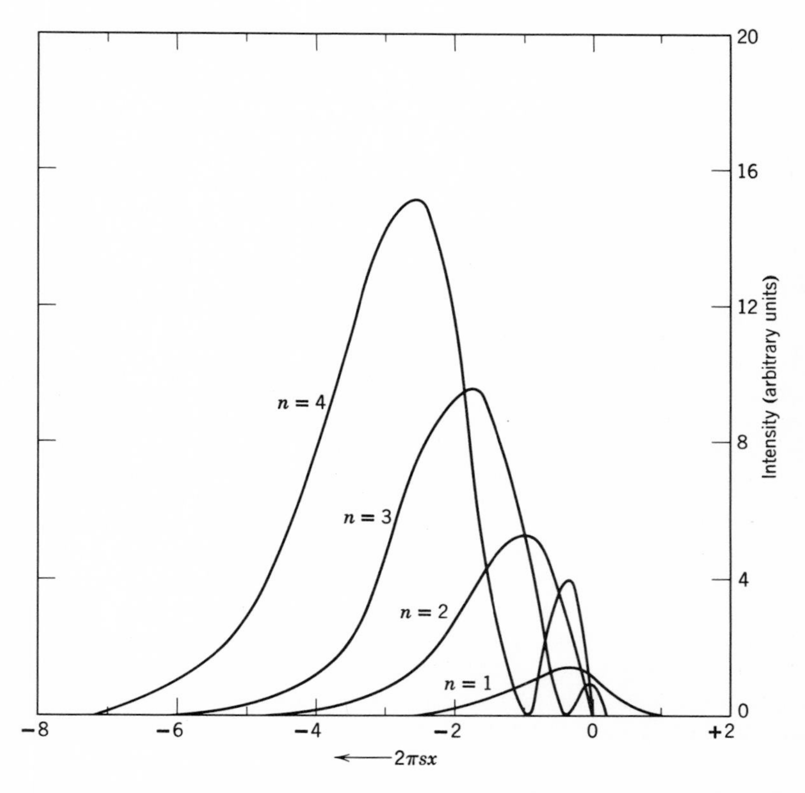

FIG. 132. Intensity profiles for images of screw dislocations for various values of n (i.e., $\mathbf{g} \cdot \mathbf{b}$). The center of the dislocation is at $2\pi sx = 0$. Note that the contrast lies to one side of the dislocation core. (After Hirsch et al.[93] Courtesy the Royal Society.)

when the dislocation is inclined to the plane of the foil. In this case, by analogy with Fig. 34 since the intensity of the scattered beam oscillates along the length of the dislocation line, dotted contrast can occur in the image, provided the extinction distance is less than the foil thickness. These effects are also well known experimentally.[93]

For an edge dislocation, the displacements parallel to the line are zero and have different values parallel to the Burgers vector and normal to the slip plane,[2] so that in this case even when $\mathbf{g} \cdot \mathbf{b} = 0$, contrast may still be obtained, e.g., with dislocation loops, contrast will be observed at edge components as a result of atomic displacements in the plane of, but normal to, the loop, as found in zinc.[29]

From the foregoing and the more detailed treatments given by the original authors[93] we may summarize contrast effects from imperfections predicted by kinematical diffraction theory as follows: (1) contrast due to mixed dislocations is similar to that from screws. The image of an edge dislocation may be slightly wider than for a screw whereas dislocations should become narrower and less visible the more steeply inclined they are to the plane of the foil (e.g., see Fig. 120 at *A*). (2) The contrast peak lies to one side of the center of the dislocations, the displacement of the peak being of the order of the width of the dislocation. (3) For certain reflections $\mathbf{g} \cdot \mathbf{b}$ may vanish, i.e., when $\mathbf{g} \cdot \mathbf{b} = 0$ the Burgers vector of the dislocation lies in the reflecting plane, and the dislocation is invisible. (4) Dislocations inclined to the plane of the foil will, in certain cases, give rise to a variation in visibility along its length. In this case, the dislocation has a dotted appearance. In other cases double images may be observed. Thus it is very important to be able to tilt the specimen and to record the diffraction patterns in order to decide whether double images are produced from two closely spaced dislocations or from a single dislocation. *In all cases it is advisable to tilt the specimen to see how the contrast varies before attempting to interpret the results, and the corresponding diffraction patterns should be photographed for each micrograph that is taken.* (5) The width and intensity of the image (Fig. 132) should increase as s decreases, i.e., when the crystal is tilted towards the reflecting position (e.g., at extinction contours). (See Figs. 104, 111.) The maximum width, for small s, is given by t_0/π (see Chapter 2) and is $\sim$100 to 200 Å. (6) Coherency strains due to thin precipitates should also give rise to contrast effects similar to those from dislocations.[91] (See also section 5.3.2.) (7) In all cases of Bragg diffraction contrast, the contrast may be reversed by changing from bright to dark field illumination using the operating reflection. (See Fig. 39.)

Since the contrast is a maximum at the Bragg reflecting position, it will be particularly strong at extinction contours getting weaker with increasing distance from them, i.e., as the deviation from the Bragg angles increases. This effect is shown at *C* in Fig. 104, and in Fig. 111, where the dislocation contrast is obviously strongest at the contours. The theory also predicts that the contrast due to a dislocation should lie to one side of the dislocation core. The side on which the contrast occurs depends on the relative signs of the phase angles ϕ and $2\pi sz$ in equation (85), i.e., on **g**, **b** and *s*. The contrast will change positions if any of these parameters change. For example, as a dislocation line crosses an extinction contour, **g** and **b** are fixed while *s* changes sign at the contour. Thus, the dislocation contrast will change sides as the dislocation crosses the contour. This effect can be seen at *B* in Fig. 111. During tilting of the foil the extinction contours may be observed to sweep across the field of view, revealing dislocations or other contrast effects which may not have been visible prior to tilting.

5.2.15 *Determination of Burgers Vectors*

Case (3) outlined above may be used to determine Burgers vectors of dislocations. This has been demonstrated by using foils of quenched Al–4% Cu[93] in which screw dislocations become helical as a result of vacancy climb.[50] Figures 133*a* and *b* show such helices running in [110] and $[\bar{1}10]$ directions.[50] The plane of the foil is (001). In Fig. 133*a* the diffraction pattern (inset) shows that the (020) reflection is producing the contrast. Since the dislocations *A* and *B* were originally screws, their Burgers vectors lie along their lengths; so $\mathbf{g}\cdot\mathbf{b} = 1$ for both dislocations. After tilting the specimen the (020) reflection has been removed, and the $(2\bar{2}0)$ reflection tilted in (Fig. 133*b*). This reflection has $\mathbf{g}\cdot\mathbf{b} = 0$ for dislocation $B(\frac{1}{2}[110]$ Burgers vector) and $\mathbf{g}\cdot\mathbf{b} = 2$ for dislocation $A(\frac{1}{2}[\bar{1}10]$ Burgers vector). Dislocation *B* thus vanished in agreement with theory, so that by careful tilting experiments it is possible to determine Burgers vectors. These results emphasize the necessity for having a tilting stage available in the specimen chamber of the microscope. Not all the dislocations (or other contrast effects) may be immediately visible on first examination of the specimen; however, tilting of the specimen may reveal them.

The same principle of Burgers vector determination may be followed using the dark field technique. By observing the diffraction pattern a 25 μ objective aperture can be placed over an intense diffraction spot. If the image due to this spot is then magnified (without tilting the specimen) so as to observe the dislocation structure, those dislocations having their Burgers vectors lying in this reflecting plane are invisible. The same area may then be compared to the bright field image. This procedure has been used by Delavignette and Amelinckx[85] in their analyses of dislocation

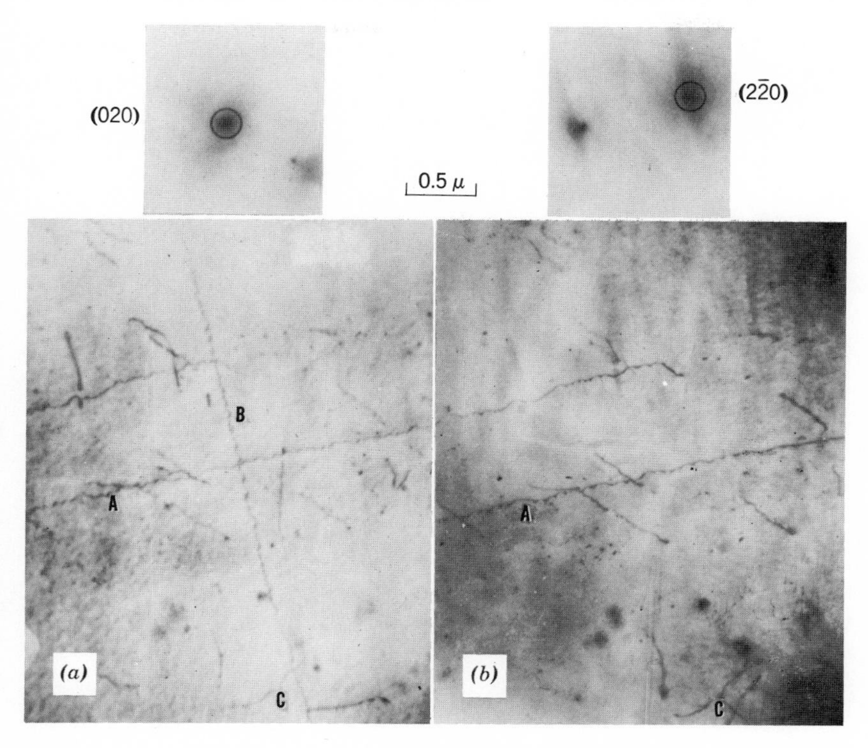

FIG. 133. Sequence of micrographs of quenched Al–4% Cu alloy showing two helical dislocations *A* and *B*. After tilting, dislocation *B* has vanished, whereas another has appeared at *C*. The diffraction patterns show the operating reflection in each case (× 13,500). (After Hirsch et al.[93])

networks in layer structures. These workers also point out the possibility of distinguishing between interstitial and vacancy dislocation loops in graphite because the vacancy loop has a component of its Burgers vector in the basal plane, whereas the interstitial loop has a component of its Burgers vector parallel to the *c*-axis. By dark field observation using a suitable reflection, it is possible to make a dislocation loop in the basal plane go out of contrast and hence to determine the character of the loop.

5.2.16 *Summary of Work on Dislocations*

It is clear from the preceding sections that much work is being done in the experimental field of dislocations. The results have confirmed the predictions made from theory, and a new mechanism, viz., formation of tetrahedra of stacking faults, has been discovered in quenched gold.[60] Thus the transmission electron microscope technique provides a powerful tool for studying in detail the fundamental properties of defects in crystals, thus providing much information regarding the mechanical properties of bulk materials. Direct observations can be made of changes taking place in metals under the influence of stress and temperature by using straining devices and hot stages. In many cases analysis of the results obtained by these techniques is complicated by the mass of detail revealed in the specimens. It should be possible to determine Burgers vectors of dislocations, and when this is done more details of the complex arrangements and interactions can be interpreted. The difficulties here lie in establishing the crystallographic features of the dislocations and, as we have seen, in applying diffraction theory to understand the observed contrast effects.

5.3 Studies of Transformations in Alloys

5.3.1 *Introduction*

The preparation of thin foils of alloys containing more than one phase becomes difficult when second phases are noncoherent with the matrix, since any chemical or electrochemical thinning process tends to preferentially attack the precipitates. However, many of the more interesting features of phase transformations occur on such a fine scale that this is not a very serious problem, particularly in the case of age-hardening alloys where the initially formed new phases remain coherent or partially coherent with the matrix. Extensive studies have been made of precipitation in light alloys,[48, 68, 96–101] titanium alloys,[102] copper alloys,[103] and steels,[104] using foils made from bulk specimens. Phase transformations in thin foils have been carried out on light alloys,[105–107] and steels,[108, 109] and studies of ordering in Cu–Au alloys have also been made.[110–113] In these last cases, the transformations may not be typical of those in bulk samples because of the limited thickness of the specimens. More recently, however, thin specimens of ordered alloys have been prepared from bulk material.[114–115] At the present time, except for martensitic structures, little has been reported for ferrous materials.

5.3.2 *Precipitation in nonferrous Alloys*

The decomposition of supersaturated aluminum alloys has been a topic widely investigated for the past fifty years. The changes in structure and properties that are induced by heat treatment are of great intrinsic interest as well as of practical importance, since they provide a method of obtaining high-strength alloys. The technique is to quench the alloy from a single-phase field to obtain supersaturation and then to age the alloy at various temperatures below about 300° C. Foils may then be prepared from the heat-treated bulk specimen.

During quenching, excess vacancies may be retained in solution, and when these anneal out (e.g., in forming dislocation loops or helices), their movement allows easy diffusion of solute atoms which tend to cluster to form zones. (For review, see references 97, 116.) The shapes of zones depend on the relative sizes of solute and solvent atoms; e.g., zinc and silver atoms have similar diameters to aluminum, and the zones are spherical†[97, 101] (see Fig. 134), whereas copper atoms have a smaller diameter than aluminum atoms and the zones (called G.P.1 zones) are disc-shaped[96] and form on {100} planes in densities of $\sim 10^{17}/cm^3$. An example, shown in Fig. 135, illustrates the necessity of operating the microscope for high resolution, since the discs of copper atoms forming the zones are only 1 to 2 atoms thick. As we saw in Chapter 1, these can only be resolved when the foil is oriented so that these discs lie in a direction parallel to the incident beam. Thus zones lying parallel to the surface cannot be detected. For example, when foils are in [001] orientation, zones along (100) and (010) are observed, whereas in Fig. 135, the orientation is [110] so that only (001) zones are observed. Spherical-shaped zones also occur in more complex alloys,[98] e.g., Al–Mg–Zn, Al–Mg–Zn–Cu, etc., whereas needle-shaped zones along ⟨100⟩ have been observed in Al–Mg–Si alloys.[99]

For thin plates or needle-shaped zones when they are favorably oriented they give one-dimensional diffraction effects as discussed earlier in Chapter 1 (see Fig. 23). Thus diffraction patterns from plates or needle-shaped zones show streaks (see Fig. 24), and as the zones thicken (with prolonged aging), the length of the streak decreases until eventually a two-dimensional pattern of a precipitate is obtained. Thus it may be possible to detect changes in structure from diffraction patterns before

† Friese et al. (*Acta Met.* 1961, **9**, p. 250) carried out X-ray small angle scattering measurements and transmission electron microscopy on the same specimen of Al–4.4 at. % Ag alloy. They showed that there are 2×10^{17} clusters/cm^3 with diameters of 20 to 60 Å and that the average composition of the clusters is greater than 90% Ag. From the X-ray intensity measurements they were also able to show that only ~10% of the alloy undergoes this segregation. This work shows the value of utilizing two techniques for obtaining the maximum information from a particular specimen, and is the only way of obtaining the composition of the zones with any accuracy. However, it is to be remembered that the Al–Ag alloy is particularly amenable to X-ray small angle scattering analysis because of the small lattice distortion associated with Ag in Al, and the large differences in scattering factors between Ag and Al atoms.

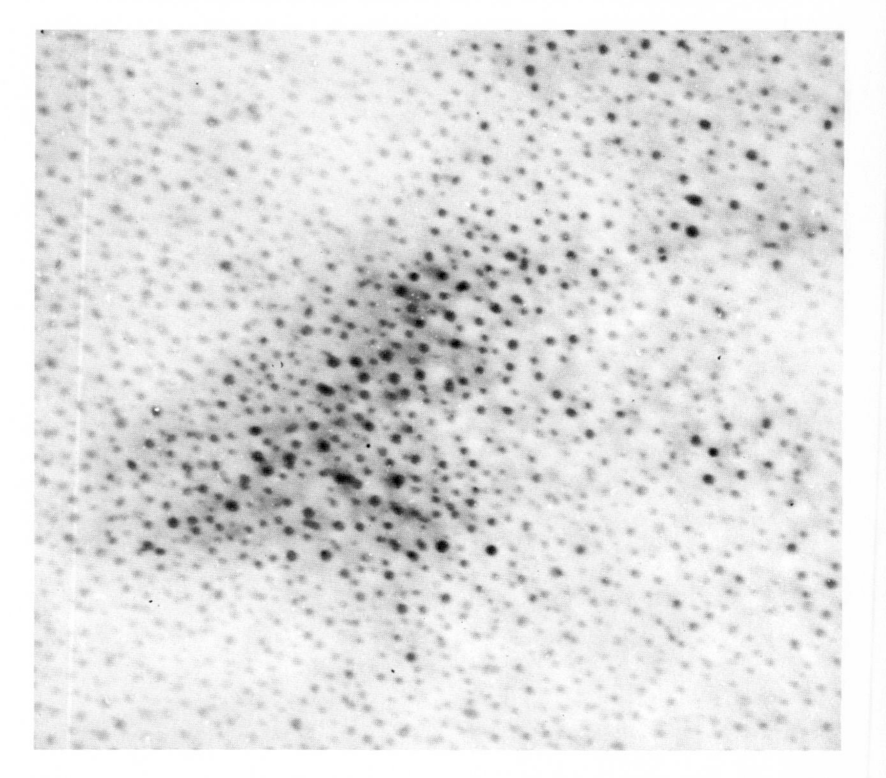

FIG. 134. Al–16% Ag; aged 5 days at 160° C, showing spherical-shaped Guinier-Preston zones (×160,000). The dark contrast from the silver-rich zones can be attributed mainly to the larger scattering factor of silver compared to aluminum (mass thickness contrast). However, it can be seen that enhancement of contrast occurs at extinction contours (diffraction contrast). (After Nicholson et al.[97, 101])

they can be resolved in the micrograph, so that the use of selected area diffraction is of great advantage in studies of very small structures.

The formation of zones corresponds to the first increase of hardness which occurs during the aging treatment (age-hardening). The hardening due to zone formation is attributed to one or more of three mechanisms: (1) dispersion hardening, (2) internal strain-hardening, and (3) chemical hardening. In dispersion hardening the zones or precipitates must be nondeformable so that they provide obstacles to the movement of dislocations.[117–119] Alloys of the Al–Mg–Zn type are thought to harden in this way.[98] In internal strain-hardening the elastic coherency strains resulting from any misfit between the zones and the matrix will tend to

FIG. 135. Al–4% Cu; aged 16 hr at 130° C, showing Guinier-Preston zones ~4 Å wide. Notice contrast effects around zones indicating presence of coherency strain-fields. Foil orientation [110], zones parallel [001]. Zones along [100] and [010] at 45° to foil surface not detected. Figure 24 is a diffraction pattern from an area similar to this, but in [001] orientation (× 575,000). (After Nicholson and Nutting.[96] Courtesy *Phil. Mag.*)

oppose movement of dislocations,[97] so that the flow stress of the alloy will increase. Results on Al–4% Cu alloys provide some evidence that hardening occurs by this mechanism; e.g., in Fig. 135 the coherency strains around the G.P.1 zones are revealed by the presence of light or dark contrast effects similar to those found for dislocations.[96] The ability to detect zones is therefore strongly dependent on orientation, and it is usually necessary to tilt the foil in order to see them, e.g., in the case of G.P.1 zones the foils must be exactly in ⟨100⟩ or ⟨110⟩ orientation. In chemical hardening, the increase in strength is thought to be due to the gain in chemical energy (and entropy) arising from increasing the number of solute-solvent bonds as dislocations cut through the zones. However, differences in elastic moduli between zones (or precipitates) and the matrix will also contribute to the hardening. The chemical hardening theory can only apply to coherent zones or precipitates, i.e., to second phases which deform homogeneously with the matrix.[120, 121] Direct observations of dislocations cutting through zones have now been made.[48, 97]

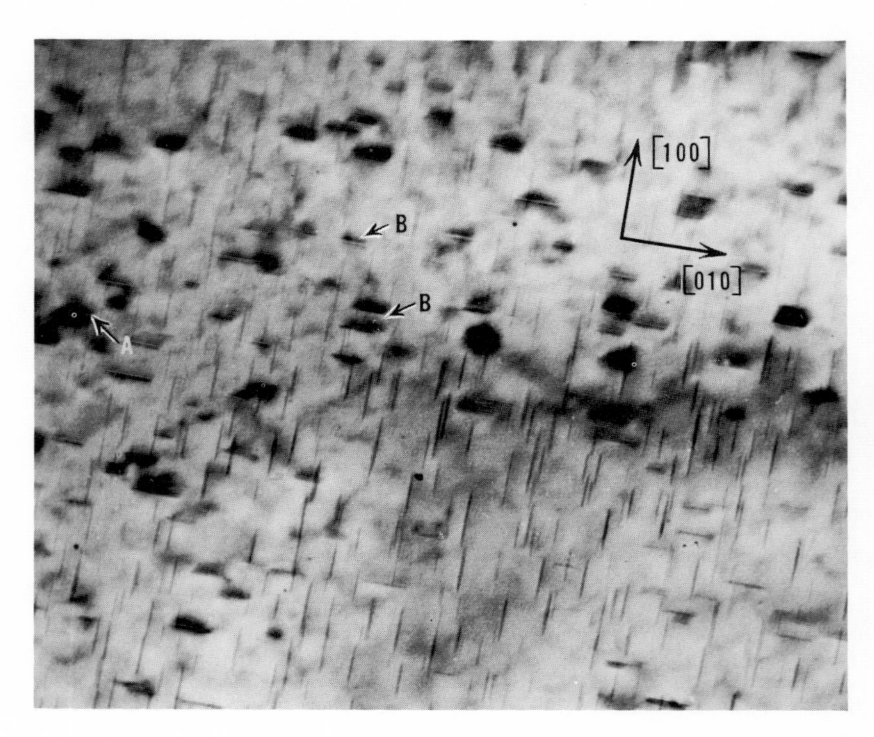

FIG. 136. Al–4% Cu; aged for 1 day at 130° C, showing θ'' plates lying parallel to electron beam along [100] and [010]. Foil in [001]. Note contrast effects around precipitates. For θ'' in center of foil there are equal amounts of dark contrast on either side (e.g., at *A*). If precipitates are at the top or bottom surfaces, contrast occurs on one side only (e.g., *B*). The contrast arises from displacement of atoms normal to the plane of the precipitate (coherency strains) and is similar to contrast from dislocations (×240,000). (See Fig. 137.) (After Nicholson and Nutting.[96] Courtesy *Phil. Mag.*)

When Al–Cu alloys are aged a little longer, e.g., 1 day at 130° C, second zones or a coherent intermediate precipitate called θ'' are formed, around which contrast effects are easily detected when thin foils in [100] orientation are examined in a microscope. An example is given in Fig. 136. As for G.P.1 zones, these contrast effects are associated with the coherency strain fields produced by the θ'' structure.

The explanation of the contrast around thin coherent zones or precipitates may also be obtained from the amplitude-phase diagram.[91] Figure 137*a* shows schematically how a precipitate formed with a decrease in volume, i.e., under tension, distorts columns of crystal *AB*, *CD* on either side of it (as in G.P.1 zones). For precipitates formed under compression the displacement of the columns would be reversed. In either case the elastic strain field is similar to that around a prismatic dislocation loop. Thus the amplitude-phase diagram for the length of the column *AO* may be an unwound spiral similar to Fig. 131*b* at (1), and for *OB* it will be a wound-up spiral similar to Fig. 131*b* at (2). The resultant amplitude-phase diagram for the whole column would then be as shown schematically in Fig. 137*b*. For the column *CD* on the other side of the precipitate the amplitude-phase diagram is shown at *c* and is the mirror image of Fig. 137*b*.

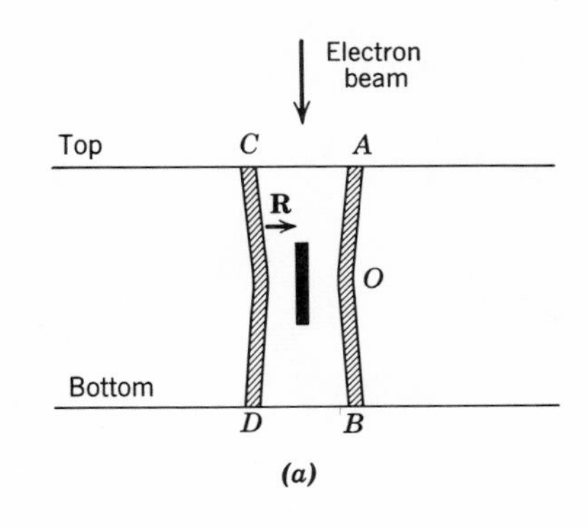

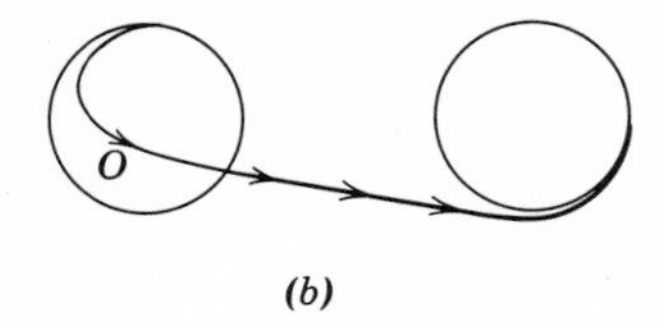

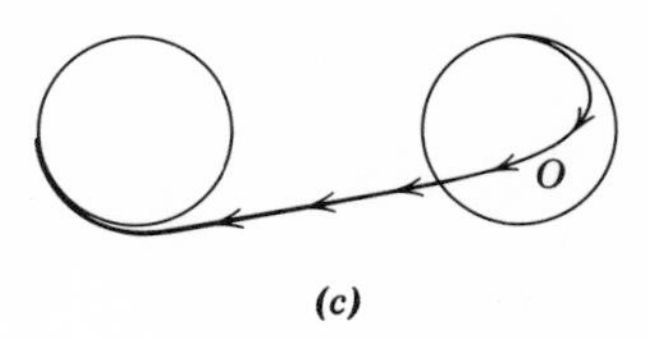

FIG. 137. (*a*) Thin foil containing a coherent zone or thin precipitate viewed edge-on. (*b*) and (*c*) Schematic amplitude-phase diagram for the two columns *AB*, *CD* in (*a*). (After Whelan.[91])

A precipitate in the center of the foil viewed edge-on would then give rise to equal amounts of dark contrast (bright field image) on either side. However, if the precipitate is at the top or bottom of the foil, contrast is obtained on one side only. Examples of these effects can be seen in Fig. 136. As for a dislocation, maximum contrast is obtained when the displacement vector **R** (expression 84) is normal to the incident electron beam, and when a strong reflection is operating, i.e., at the extinction contour. This effect is also clearly demonstrated by Fig. 136 where it can be seen that the contrast gets weaker with distance from the contour, i.e., with increasing deviation from the Bragg condition.

Aging light alloys beyond the zone stage results in the formation of intermediate precipitates, which are usually partially coherent with the matrix. In this case the misfit is too great to be accommodated by coherency strains, thus the interface is expected to consist of a dislocation network.[122] Confirmation of this has been obtained in Al–4% Cu alloys

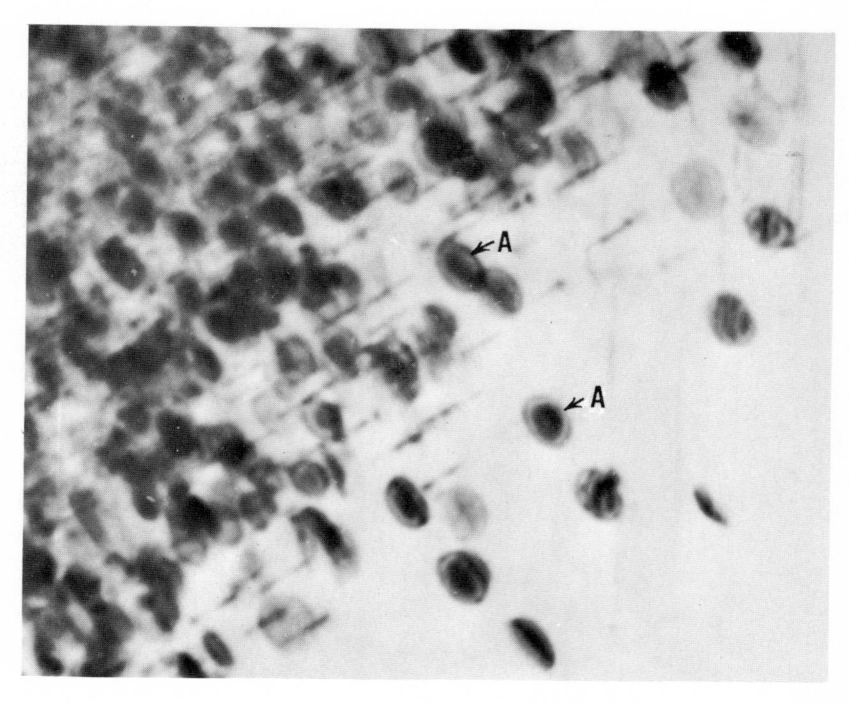

FIG. 138. Al–4% Cu; aged 12 hr at 200° C, showing θ'–$CuAl_2$ particles. Notice possible dislocation rings around precipitates parallel to surface (e.g., at *A*) (× 32,000). (After Nicholson et al.[97])

aged to produce the intermediate phase θ', which also forms on {100} planes. For example, in Fig. 138 the plates of θ' lying parallel to the foil surface are surrounded by dislocation rings. The dislocation networks are also revealed in the moiré patterns produced from the θ' and matrix lattices when the θ' plates lie parallel to the surface of the foil, e.g., at *A*, Fig. 139. Dislocations and cross-slip may also be observed in this micrograph. Thus the moiré technique may be utilized to study the interfaces between two phases, provided one phase exists as a thin plate or disc and the lattice planes to be resolved lie parallel to the beam. In Al–Ag alloys the intermediate phase γ' is hexagonal and forms in thin plates on (111) planes.[97, 101] Since this structure corresponds to a face-centered cubic lattice containing growth faults on every other (111) plane,

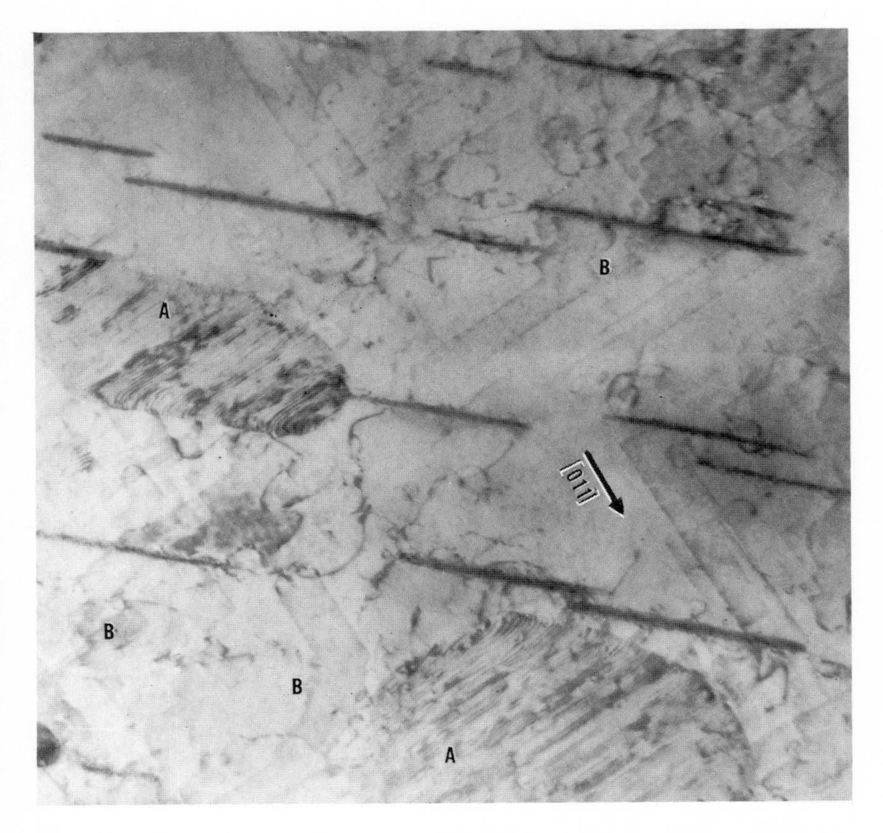

FIG. 139. Al–4% Cu; aged 1 hr at 300° C, showing θ'–$CuAl_2$ precipitates {orientation (001)}. Notice moiré pattern in precipitates parallel to surface *A*, and cross-slip of dislocations *B* (×38,500).

the contrast due to these precipitates, is exactly like that observed from stacking faults and as shown in Fig. 140*a* characteristic fringes are produced. These are only clearly observable when a strong Bragg reflection is operating, e.g., at extinction contours (section 2.3.3). The precipitate at *A* shows strong fringe contrast at either end of the plate but none in the center. This means that the total number of overlapping faults in the center is 3, 6, 9, etc., so that the total phase difference in the diffracted wave is zero, i.e., $\phi = \pm 360°$ (see section 5.2.14) and there is no contrast. Hence the shape of the plate *A* must be lenticular and thicker in the center than at the edges by one unit cell.[101]

The observations are that the γ' phase only nucleates heterogeneously at stacking faults, e.g., at Frank sessile dislocations (Fig. 119*a*) or at dissociated helical dislocations.[101] Figure 140*b* shows an early stage in the formation of the γ' precipitate at helices *H* and loops *F*. These become favorable nucleation sites since diffusion of silver atoms to the defects will locally lower the stacking fault energy; hence the helices can dissociate by the reverse reaction of expression (81), producing a Frank sessile dislocation and a stacking fault on (111). These faults then become the basal planes of the platelets. It should be noticed that the contrast at *F* in Fig. 140*b* is analagous to that in Fig. 119*b* for stacking fault tetrahedra in quenched gold.

The movement of dislocations in aged alloys is generally confusing because of the mass of detail that is revealed in the micrographs. However, dislocations have been observed to move through coherent and partially coherent precipitates, but not through noncoherent ones.[48] Noncoherent phases are usually produced during the final stages of aging (i.e., overaging), e.g., θ-$CuAl_2$ in Al–Cu, the γ phase in Al–Ag, etc. These are usually large enough to be resolved in the light microscope. Preparation of foils from alloys containing these phases is very difficult, but nevertheless results have been obtained which confirm that dislocations prefer to cross-slip around these obstacles rather than through them (e.g., Fig. 139). In this way, they can act as dislocation sources by pinning segments of dislocation line.[49, 119] One might expect that the hardest alloys would be those containing very small, closely spaced, noncoherent phases, e.g., as is found in some internally oxidized metals and alloys.[123]

One of the difficulties commonly encountered in using high-strength light alloys is that they are susceptible to intergranular failure. Electron microscope studies have shown that this susceptibility is associated with the formation of precipitate-free regions adjacent to grain boundaries and when the matrix and grain boundaries are densely populated with precipitates.[98] An example of this condition is shown in Fig. 141, taken from a high-strength Al–Mg–Zn alloy. The reason for intergranular failure is

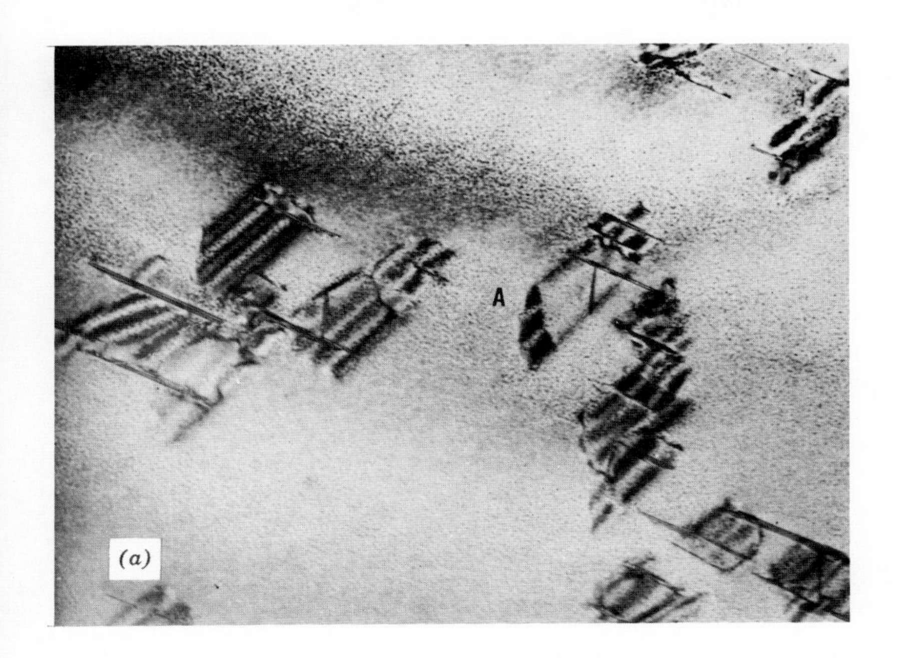

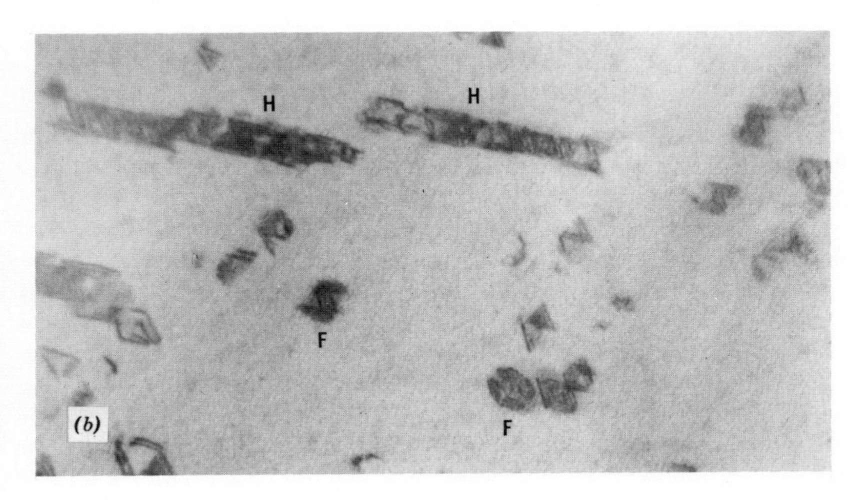

FIG. 140. (*a*) Al–16% Ag; quenched from 520° C, aged 5 days at 160° C, showing spherical-shaped zones and γ' precipitates (note fringe contrast) ($\times$ 30,000). (After Nicholson and Nutting.[101]) (*b*) Al–16% Ag; showing nucleation of γ' plates on Frank-sessile loops *F* and dissociated helices *H* ($\times$ 21,000).

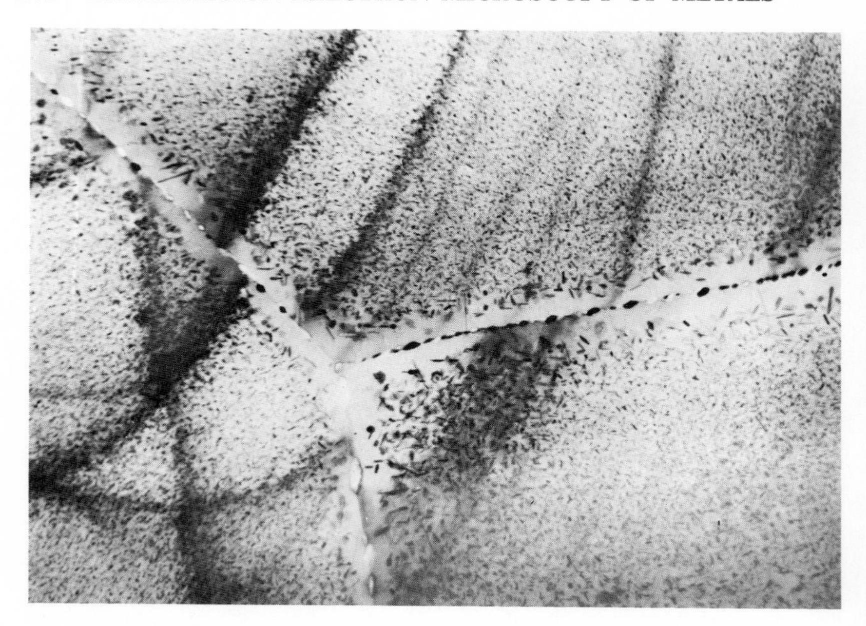

FIG. 141. Al–3% Mg–6% Zn alloy; aged 1 day at 160° C, showing three grains, preferential precipitation along grain boundaries, and precipitate-free regions adjacent to boundaries. Condition similar to quenched metals (see Fig. 120). Dark bands are bend extinction contours (×12,500). Notice that precipitates can still be clearly observed in regions away from the contours.

that plastic flow occurs preferentially in the precipitate-free, relatively soft regions. This is confirmed by replica studies after deformation, and also by the fact that a high density of dislocations is often observed in regions adjacent to grain boundaries. These regions have also been observed to act as sources of dislocations.[48]

One of the interesting features of the results reported for alloy studies is that most of the "grown-in" dislocations eventually become preferred sites for precipitation.[68, 97] Examples of this are shown in Fig. 142 taken from an Al–Mg–Zn alloy, and Fig. 143 taken from an Al–4% Cu alloy, in which θ' precipitates have formed on helical dislocations.[50, 105] It is possible to separate the effects due to dislocations and precipitates by performing tilting experiments: i.e., by tilting, the dislocation contrast is made to disappear, whereas the precipitate contrast is usually not affected appreciably, e.g., compare areas *A* and *B*, Fig. 143. These results show that most of the original dislocations become immobilized as a result of preferential precipitation and that for plastic flow to occur new disloca-

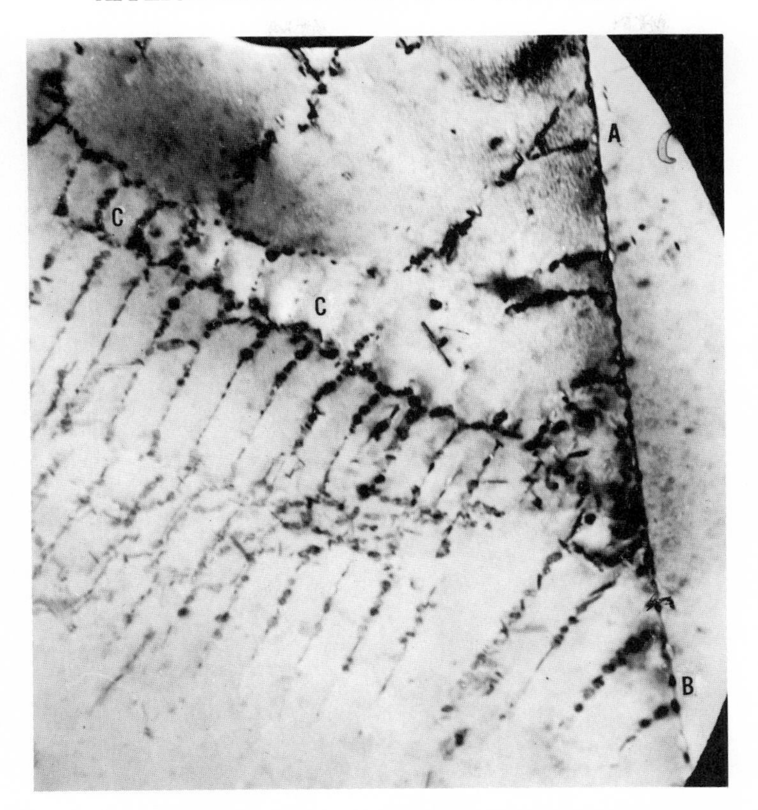

FIG. 142. As Fig. 141, after aging for 5 days. Notice preferential precipitation of M′– $MgZn_2$ phase along dislocations and extended hexagonal network *C*. *AB* is a grain boundary (×10,000). (Courtesy *Phil. Mag.*)

tions must be created. Thus the question, "what and where are the sources of glide dislocations," is still not fully answered and more work with foils is required.

Experiments involving direct observations of phase changes in Al–4% Cu alloys[105] using a high temperature stage have shown that because of the thinness of the foils (~1000 Å) precipitation occurs mainly at the surfaces. Correlation with bulk behavior in such experiments therefore is not always possible. However, diffusion data may be obtained by measuring the rates of dissolution and precipitation of second phases. This is done by recording these changes on ciné film and measuring the change in size of particles with time directly from the printed film. In this way the diffusion coefficient for Cu in Al has been obtained by

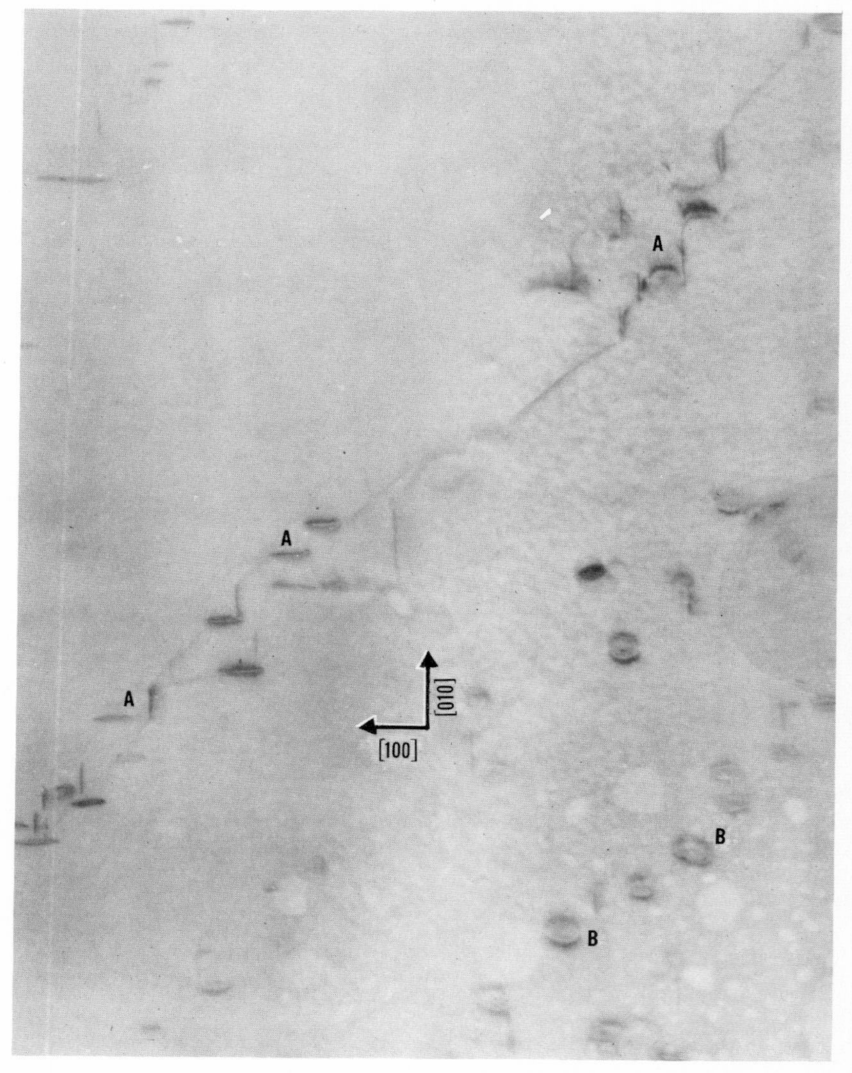

FIG. 143. Showing an early stage in precipitation of θ' phase in Al–4% Cu along helical dislocations *AA*. Note loops around particles *B* and that connecting dislocation is out of contrast (×42,000). (Courtesy *Phil. Mag.*)

measuring the rates of dissolution of $CuAl_2$ particles at $\sim 500°$ C. A typical sequence is shown in Fig. 144*a–e*. However, only in a few cases has the rate change $d(r^2)/dt$ been observed to be constant (r is the pre-

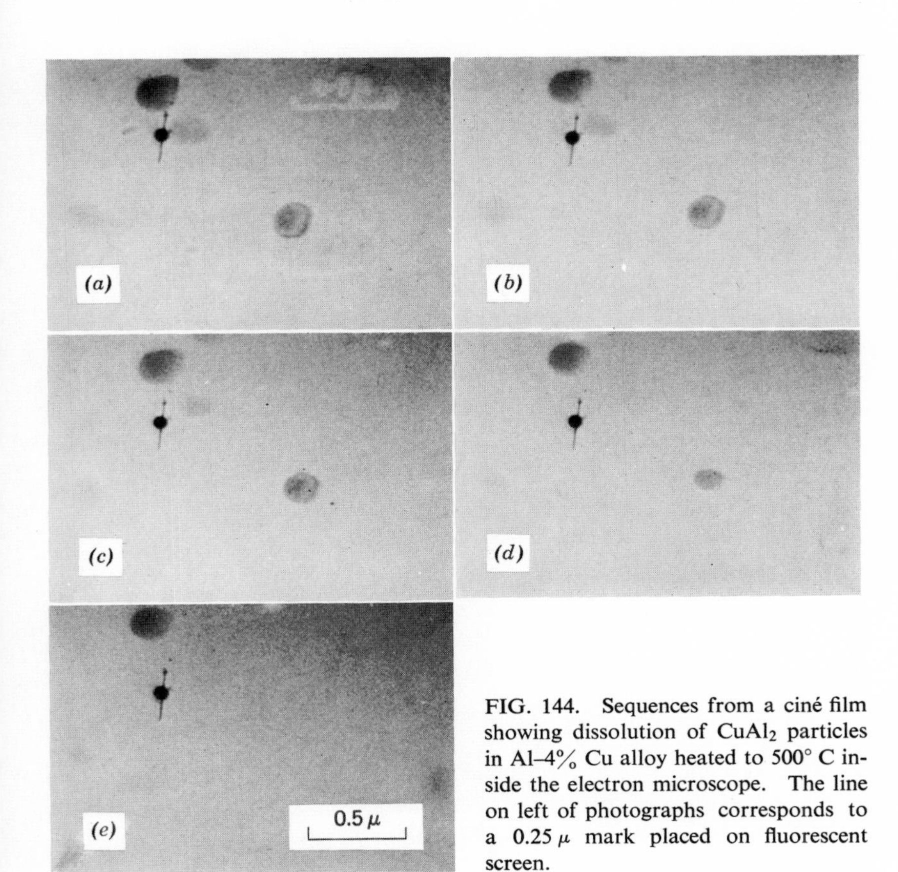

FIG. 144. Sequences from a ciné film showing dissolution of $CuAl_2$ particles in Al–4% Cu alloy heated to 500° C inside the electron microscope. The line on left of photographs corresponds to a 0.25 μ mark placed on fluorescent screen.

cipitate radius at time t), i.e., where dissolution can be considered to be the result of a diffusion-limited process. In this case, $r^2 = kDt$, where k is the rate constant and D is the coefficient for diffusion of material to or away from the precipitate.† The growth-rate constant k depends on the shape of the precipitate, and can be shown to be equal to $2(\rho_s - \rho_e)/(\rho_c - \rho_s)$ for spherical particles.[105] From this the diffusion coefficient is given by:

$$D = -\frac{1}{2}\frac{\rho_c - \rho_s}{\rho_s - \rho_e}\frac{d(r^2)}{dt} \tag{89}$$

† See, e.g., F. C. Frank, *Proc. Roy. Soc.* A, 1950, **201**, p. 586 and F. S. Ham, *J. Phys. Chem. Solids* 1958, **6**, p. 335.

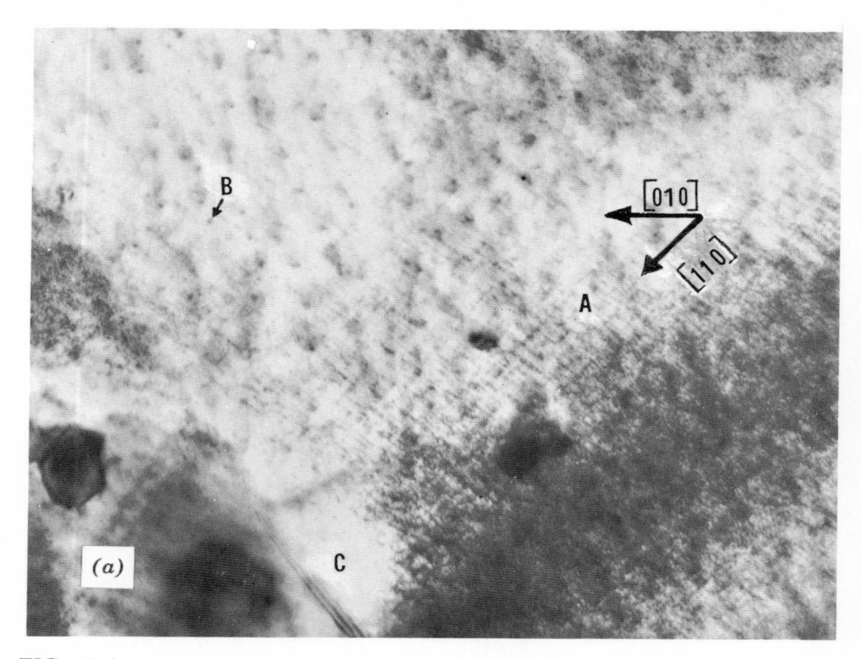

FIG. 145(*a*). Cu–2%Be–0.2% Co alloy; aged 1 hr at 100° C, showing zones parallel to ⟨100⟩ and aligned on traces of ⟨110⟩. Note (1) fringes at twin boundary *C*, (2) resolved zones lying parallel to surface *B*, (3) enhanced contrast *A* near bend extinction contours (×45,000).

where ρ_c, ρ_s, and ρ_e are the atomic concentrations of diffusing material inside, at the surface, and at large distances from the precipitate. Since $d(r^2)/dt$ can be determined from ciné-film sequences such as that illustrated in Fig. 144, and ρ_c, ρ_s, and ρ_e can be determined from the equilibrium phase diagram, it is possible to obtain values for D. This has been done for an Al–1.6 at % Cu alloy by Thomas and Whelan[105] who obtained D for Cu in Al (500° C) at about 2×10^{-9} cm^2 sec^{-1}, which is of the same order of magnitude as the values for D reported in the literature.†

With further refinement, the use of high temperature stages in electron microscopy might provide a suitable means of measuring diffusion coefficients. However, it appears to be limited at present by (1) uncertainties in the theoretical interpretation, (2) because effects at the surfaces (particularly in influencing the rates of dissolution) are neglected, (3) the difficulties of determining the true particle shape, and (4) the practical

† For example, see G. J. Smithells, *Metals Reference Book* (Butterworths, London), 1955, **2**, p. 552.

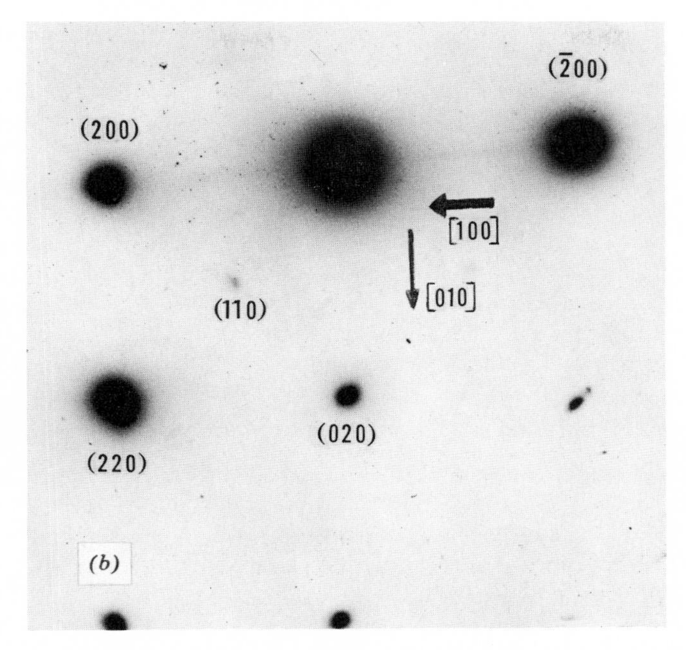

FIG. 145(*b*). Selected area diffraction pattern from region close to *A* (Fig. 145(*a*)). Foil surface approximately (001). Streaks running parallel to ⟨100⟩ similar to those found for G.P.1 zones in Al–4% Cu alloys (see Fig. 24). Note also the superlattice (110) reflection.

difficulties of estimating the temperature accurately. We must also remember that the analysis outlined above is based on considering the reverse of the precipitation case, i.e., for small values of the rate constant, and this is only approximate, since so far, no exact solution of the dissolution problem has been given.

One other fact arises out of these experiments, namely, that the temperature rise in the Al–Cu specimens due to the electron beam using double condensers (10 μ diameter spot, 30 μa beam current) is very small (probably less than 40° C).

Alloys based on titanium–aluminum are also heat-treatable alloys, and are of great interest because of their lightness coupled with high strength. Studies of aging phenomena in these alloys[102] have revealed interesting features of the β (BCC) to α (CPH) transformation; in particular, ordered substructures were discovered in the α phases of titanium containing up to 15% Al. In this work the use of selected area diffraction techniques proved to be invaluable in determining the crystallographic features of the transformation. In fact, this technique is a great advantage over standard diffraction techniques for studying phase changes in alloys.

The Cu–Be and Cu–Be–Co alloys are interesting since they are hardened by precipitation and also, depending on composition, the dislocations in the alloy are dissociated.[103] These materials are thus suitable for investigating the interaction between extended dislocations and precipitates. In the early stages of aging, the zones or precipitates exhibit contrast effects around them similar to those observed for G.P. zones in Al–Cu alloys,[96] and the diffraction patterns also exhibit streaks through {200} along ⟨200⟩. An example of zones in a Cu–2% Be–0.2% Co alloy is shown in Fig. 145*a* and the corresponding diffraction pattern in *b*. The appearance of {110} reflections shows that the zones are ordered.

5.3.3. *Electron Diffraction; Dark Field Techniques*

In alloys containing second phases, provided there are enough precipitates present (or that they are large enough) in the area being examined in the electron microscope, extra reflections from the precipitates will be observed in the selected area diffraction patterns. It is possible to analyze these patterns and index them by the methods given in Appendix A, and in this way the orientation relationships and other crystallographic features can be established. Figure 146 is an example of a selected area diffraction pattern obtained from an Al–4% Cu alloy aged to produce only θ–$CuAl_2$ precipitates.[105] The Bravais lattice of θ–$CuAl_2$ is body-centered tetragonal with $a = 6.054$ Å and $c = 4.864$ Å,† whereas the aluminum matrix is face-centered cubic with $a = 4.04$ Å. From these facts and knowing the structure factor, we know what reflecting planes are possible and what the interplanar or *d*-spacings are. Thus, in Fig. 146 the matrix reflections at A, A', C, $\bar{A}$, $\bar{A}'$, C' can be identified as (200), (020), ($\bar{2}$20),

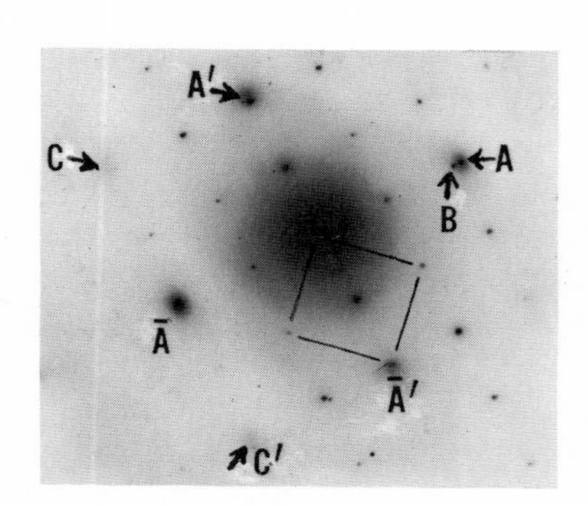

FIG. 146. Selected area diffraction pattern from a foil containing single crystals of θ–$CuAl_2$ precipitate in Al–4% Cu alloy, showing cross-grating pattern of θ reflections. The matrix reflections are at A, C, etc., whereas B is the (220)θ spot. The *a*-axes of θ are outlined. Electron beam parallel to [001] Al and θ (*c*-axis). (Pattern obtained from area similar to A, Fig. 147.)

† X-ray results of A. J. Bradley and P. Jones, *J. Inst. Metals* 1933, **51**, p. 131.

$(\bar{2}00)$, $(0\bar{2}0)$, $(\bar{2}\bar{2}0)$ respectively. This shows that the foil surface is approximately (001), i.e., the electron beam is parallel to [001] matrix. All the extra reflections forming the cross-grating pattern can now be identified and indexed, e.g., *B* is the (220) θ reflection. From the diffraction pattern it is seen that the *a* axes (outlined in Fig. 146) lie normal to the electron beam so that the *c* axes are also parallel to [001] matrix. From this and the other features which have been established, the orientation relationships are:

[100] θ parallels [110] matrix,

[001] θ parallels [001] matrix.

By taking many more diffraction patterns of differing orientations from the same specimen it is then possible to define completely the orientation relationships. Although this is comparatively easy to do for the case illustrated here, the situation is much more complicated when the structure of second phases is unknown, since structural analysis by electron diffraction is a difficult procedure (see Chapter 1). In these cases it is necessary to use X-ray diffraction techniques.

The use of dark field illumination is often valuable when carrying out observations on foils containing more than one phase. For example, if there are two different kinds of precipitates in the foil, e.g., as in Fig. 147, and each gives rise to diffraction spots other than those from the matrix, then by using one of the precipitate diffraction spots to form the dark field image it is possible to show up the location of many of the corresponding precipitates in a given area. In this way it has been possible to differentiate between θ' and θ $CuAl_2$ precipitates in Al–Cu alloys[105] in areas similar to that shown in Fig. 147 (bright field image).

In order to estimate the thickness of a foil from the projected widths of precipitate particles running through the section of the foil one must first ascertain that the precipitate does, in fact, cut the top and bottom surfaces. This can be done by using one of the matrix diffraction spots to form a dark field image. If the precipitate does extend completely through the foil, it will then behave like a hole in the film, i.e., it scatters a negligible amount of electrons into the matrix reflections. The foil thickness can then be estimated in a similar way to that for dislocation traces as shown in Appendix A.

5.3.4 *Martensitic Transformations*

Martensitic transformations have been widely studied using physical and X-ray diffraction methods, but only recently has it been possible to reveal directly the fine details associated with the martensitic structure

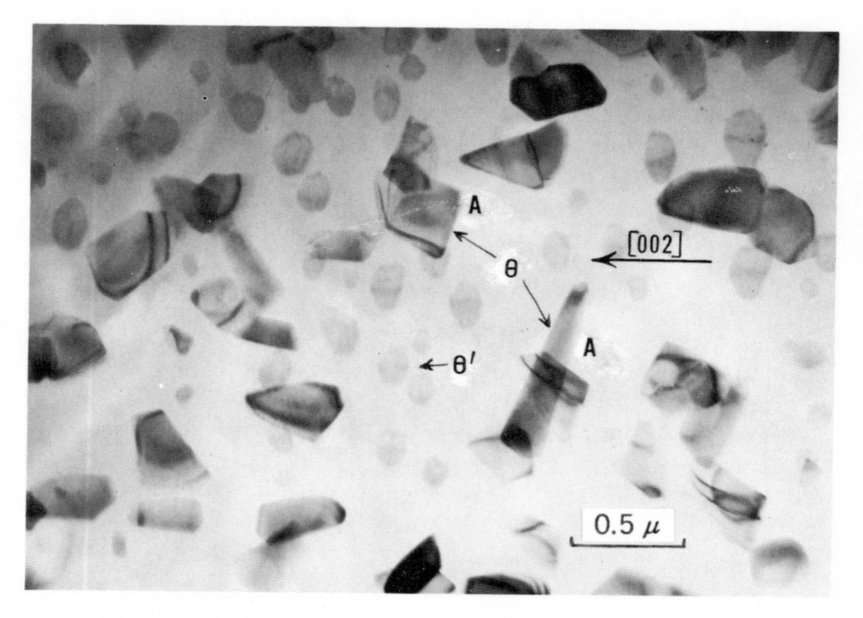

FIG. 147. Bright field image of foil of Al–4% Cu alloy, containing θ (at *A*) and θ'–$CuAl_2$ particles. By using one of the θ or θ' reflections we obtain a dark field image enabling us to make a distinction between θ and θ'. Note extinction contours in precipitates (×30,000).

because of the difficulties of preparing foils from bulk samples. The results of Pitsch[108, 109] on transformed thin foils of iron alloys containing carbon and nitrogen, showed that the shear transformation austenite-martensite during quenching from about 700° C obeyed orientation relationships of $[\bar{1}10]$ austenite parallel to $[11\bar{1}]$ martensite and (110) austenite parallel to (112) martensite. This is quite different from any orientation relationships found in massive material (e.g., see references 124 and 125). The difference is explained by the fact that in thin metal films the transformation mechanism is less hampered by the surrounding parent lattice than it is in massive material. This further goes to show how careful one must be in interpreting the behavior of thin films with that of bulk metal. However, Kelly and Nutting[104] prepared foils of a variety of carbon and nickel steels from bulk specimens by the jet-polishing method described in section 4.3.4. This work showed that in low carbon steels the martensite formed as long needles parallel to (111) α, whereas in high carbon steels the martensite occurred as plates which were internally twinned on a fine scale. The orientation relationships deduced from micrographs

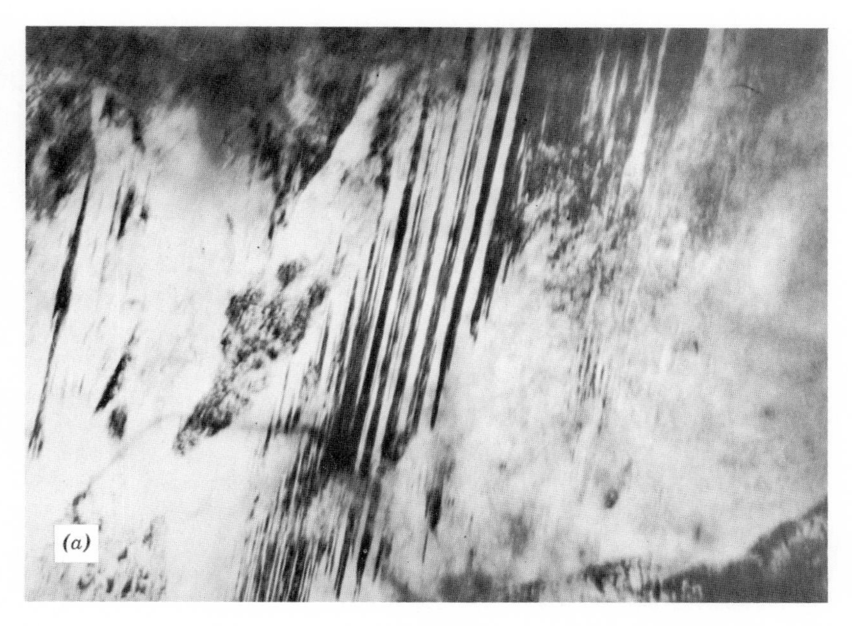

FIG. 148(*a*). 0.985% C steel quenched from 850° C showing twins parallel to {112} martensite (×61,000). (After Kelly and Nutting.[104] Courtesy the Royal Society.)

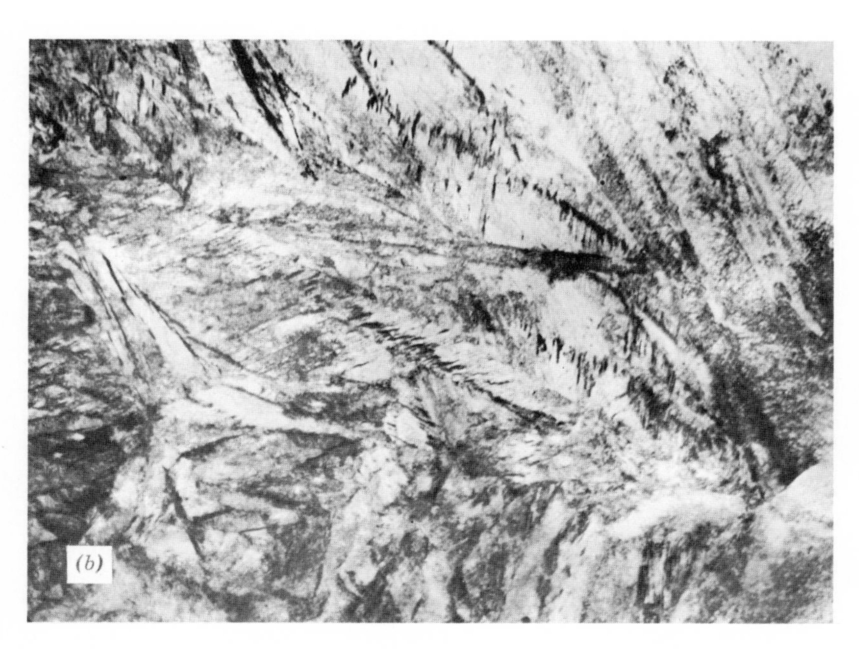

FIG. 148(*b*). Region similar to (*a*) after tempering for 1 hr at 300° C, showing precipitation of Fe_3C particles along twin boundaries of martensite (×15,000). (After Kelly and Nutting.[104] Courtesy the Royal Society.)

and selected area diffraction patterns agreed with the {225} austenite habit expected in high carbon steels.[124] An example of this is shown in Fig. 148*a*, in which twins are visible parallel to {112} planes in martensite. When this material is aged, carbide particles form along the twin boundaries (Fig. 148*b*).

In the theories of the martensite transformation[124, 125] the total transformation strain has been deduced to consist of two components: a homogeneous macroscopic strain which produces the characteristic change of shape, and a further shear, homogeneous in small regions only, which produces the correct lattice without further change of shape. This second shear has been thought to occur either by slip or by twinning in the martensite lattice. The electron microscope results have proved conclusively that in ferrous alloys[104, 108, 109] the second shear occurs by twinning. The fact that the structure shown in Fig. 148*a* is twinned can easily be shown from the selected area diffraction patterns where twin reflections are observed (see section 1.14.2). Also, since the contrast produced by the twin bands must be due entirely to the difference in orientation between the twinned and untwinned parts of the crystal, the twins may appear light or dark and the contrast may be reversed by changing from bright to dark field illumination. In copper–12% aluminum alloys,[126] however, the micrographs show stacking faults inside the martensite plates as shown in Fig. 149. The transformation can thus be considered to be BCC (ordered)→ faulted FCC (ordered). (Swann and Walimont, private communication.)

Kelly and Nutting[104] have considered the factors responsible for the hardness of martensitic steels and conclude from their observations that the numerous twin interfaces which are present when the martensite occurs in the form of plates (high carbon steels) are mainly responsible for the increase in hardness with carbon composition. Other factors which contribute to the hardening are the solid solution strengthening produced by supersaturated carbon in the martensite lattice, and the internal strain hardening resulting from the transformation.

Since there are many other systems which show martensitic transformations, there are great possibilities of extending this kind of work, thereby enabling metallurgists to understand more fully the complex nature of these transformations. Here again the use of a hot stage would be helpful for observing directly tempering phenomena and for obtaining kinetic data.

5.3.5 *Observations of Ordering in Thin Foils*

In many solid solution alloys both solvent and solute atoms may, below some critical temperature, take up a preferred arrangement in the lattice,

FIG. 149. Cu–12% Al alloy quenched from 800° C, showing stacking faults in plates of martensite (×40,000).

producing a superlattice by a process of ordering. The degree of order depends on temperature, time of annealing, and composition, and quite often the superlattices have complicated structures (antiphase superlattices). An example of an antiphase superlattice is the CuAu 11 structure in Cu–Au alloys. This forms a simpler CuAu 1 structure at temperatures between 380° C and the disordering temperature of 410° C. The ordered structures in Cu–Au alloys have been studied in detail using foils prepared by evaporation in order to obtain thin specimens of the required orientation and which are suitable for high resolution.[110–113] The ordering of the atoms produces domain structures from which the contrast in the electron microscope image has been attributed either to the interaction between the incident beam and electrons diffracted from the long period spacing in the ordered structure,[111] or to interference between singly and doubly diffracted electrons.[112] However, a simpler explanation of the various types of contrast observed can be found by using the kinematical theory as shown in section 5.2.14.

For CuAu 11, alternate (002) planes contain only gold and only copper atoms respectively, but half way along the a_1 axis of the unit cell the copper planes revert to gold planes and vice-versa. At these places antiphase domain boundaries are formed.[113] The CuAu 11 superlattice can therefore be described as a CuAu 1 superlattice with a periodic arrangement of antiphase domain boundaries every 20 Å lying perpendicular to [100]. Figure 150 shows the appearance of these boundaries in an electron micrograph taken in bright field. In order to reveal these structures the foils must be uniformly thin, in exactly the right orientation to produce the

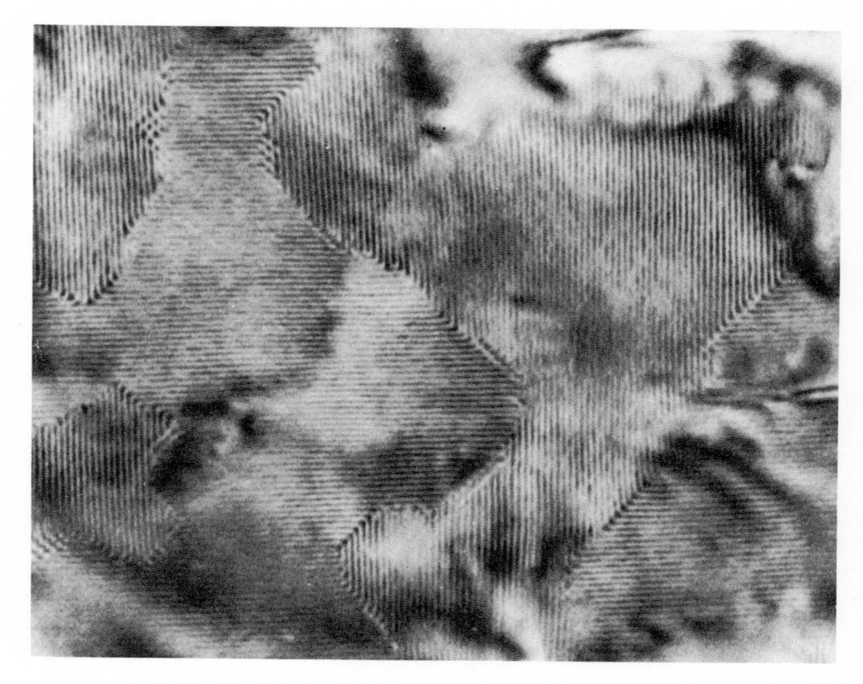

FIG. 150. Electron micrograph (bright field image) of CuAu 11 superlattice. White lines represent domain boundaries (×400,000). (After Pashley and Presland.[113]) The image is obtained using the spots inside the circle A of Fig. 26.

contrast and the microscope operated for resolutions better than 20 Å. In the case of CuAu 11 the orientation must be [100] and the diffraction pattern corresponding to this structure is seen in Fig. 26 in which the satellite spots are a direct consequence of the 20 Å periodicity of the domain boundaries. Hence for imaging these, only one diffraction spot and its satellites are utilized (see section 2.2).

As was shown in section 5.2.14, antiphase domain boundaries will be visible only when a strong superlattice reflection is operating and when the phase shift in waves diffracted from opposite sides of the boundary is not equal to zero. High contrast is obtained by using dark field techniques with tilted electron illumination to maintain high resolution.

Antiphase boundaries also occur in simpler superlattices, e.g., CuAu 1, but in this case they are not periodic as shown in Fig. 151. With the use of a hot stage in their microscope, Pashley and Presland[113] followed the

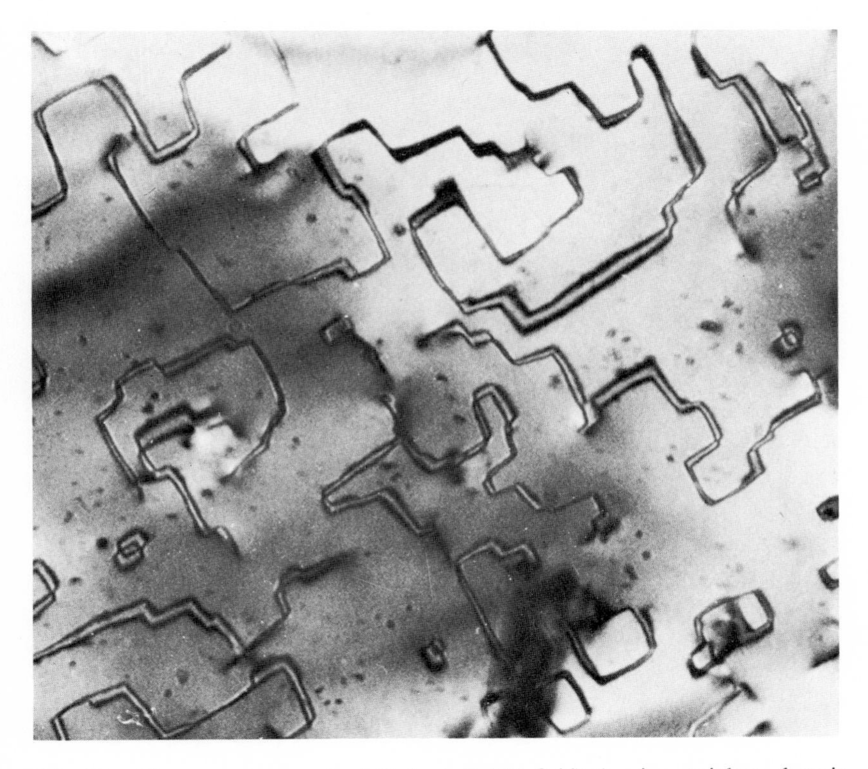

FIG. 151. Electron micrograph of CuAu 1 (bright field) showing antiphase domain boundaries. The tilted boundaries appear as pairs of black fringes (contrast effect similar to stacking faults) (×250,000). (After Pashley and Presland.[113])

transition CuAu 1–CuAu 11 and showed that the transformation does not involve any dislocation mechanism. Dislocations do, however, modify the growth of the CuAu 11 structures. More recently, Pashley and Presland[114] have examined the growth of the tetragonal CuAu 1 structure in foils prepared from the bulk and have observed twins in the lamellae of CuAu 1. Antiphase domain boundaries associated with dislocations were also observed similar to those found in evaporated foils. These boundaries have also been observed in $AuCu_3$ and Ni_3Mn.[115]

It is well known that since a single dislocation moving through an ordered lattice will leave a strip of antiphase boundary in its wake,[1] dislocations in ordered alloys will tend to travel in pairs (superdislocations). These have been observed for the first time by Marcinkowski and Fisher[115] in both $AuCu_3$ and Ni_3Mn, and an example for $AuCu_3$ is shown in Fig. 152.

Several means have been used for observing domains and domain walls in thin ferromagnetic films,[127, 128] but it is only recently that the transmission electron microscope has been employed for this purpose. Hale

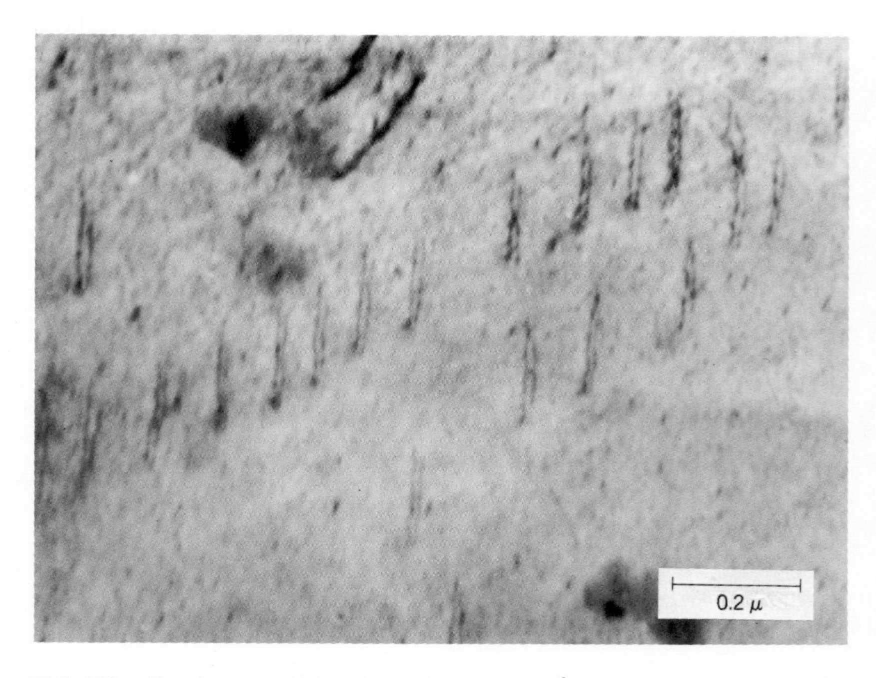

FIG. 152. Showing superdislocations of spacing 130 Å forming boundaries of antiphase domains in ordered $AuCu_3$. (After Marcinkowski et al.[115])

et al.[29] showed that when electrons pass through a magnetic film, the direct interaction of the electrons with the specimen magnetization is responsible for the image. As a result, domain walls have been observed in permalloy and iron-nickel films.[129, 130] It is interesting to note that these authors prefer to use an electrostatic transmission electron microscope so as to allow the use of full objective power without influencing the magnetization distribution of low coercive force films.[130]

5.3.6 *The Pearlite Transformation*

The transformation of austenite to pearlite in iron alloys has been the subject of extensive investigation for many years, partly because of its close association with the hardenability of low alloy steels and because of its intrinsic interest as an important type of solid state transformation. The decomposition of austenite to pearlite

$$\gamma \rightarrow \alpha + Fe_3C$$

gives a lamellar structure of ferrite (α) and cementite (Fe_3C) which is a very familiar one to metallurgists and has been the subject of many light optical and electron optical investigations (e.g., see reference 14, Chapter 4). However, only very recently has this transformation been studied by means of transmission electron microscopy using thin foils.†

Using the results of electron microscopy in conjunction with electron diffraction, Darken and Fisher showed that the orientation relationships between ferrite and cementite of pearlite and prior austenite are approximately:

$$(111)_\gamma \text{ parallel } (110)_\alpha \text{ parallel } \sim (001)_c$$

$$[110]_\gamma \text{ parallel } [111]_\alpha \text{ parallel } [010]_c$$

Ferrite is BCC and cementite is orthorhombic; the electron diffraction patterns from cementite show extensive streaking in $\langle 100 \rangle$ which these workers attribute to the presence of stacking faults within the cementite plates. This has been confirmed by their electron micrographs. Thus, the authors suggest that the nuclei for the transformation are stacking faults in the austenite, and not ferrite or cementite as has been thought previously. These faults could be generated by the deformation which accompanies the transformation (e.g., due to volume changes and differential thermal contraction). In many ways, therefore, the precipitation of cementite is analagous to that of γ' in Al–Ag alloys (section 5.3.2).

Other interesting features of Darken and Fisher's work are that subboundaries can arrest pearlite growth and that the carbide lamellae are

† L. S. Darken and R. M. Fisher (private communication).

often offset, probably as a result of interactions during growth with intersecting dislocations and stacking faults. An example of these latter effects is illustrated in Fig. 153. The typical fringe contrast from faults can be seen in this micrograph.

This work also shows that a thin zone of ferrite separates the cementite

FIG. 153. Thin foil of Fe–0.9C–0.3Mn steel austenitized at 940° C and slowly cooled to produce pearlite. Cementite shows as dark plates: note offsets and stacking faults inside plates (×41,500). (Courtesy of L. S. Darken and R. M. Fisher, U.S. Steel Corporation, Monroeville, Pennsylvania.)

from the austenite during growth of pearlite. Since this zone is about one-tenth of the pearlite spacing, it cannot be observed in the light microscope. As a result of this discovery, it has been possible to compute a growth rate which is in agreement with that observed.

5.3.7 *Summary*

In order that the reader should not become confused by the mass of experimental data that is continually becoming available, not all of the many results which have been obtained using thin foil microscopy have been discussed.† Even those described above are only outlined very briefly, since many of the explanations of the observed phenomena are detailed and complicated and require considerable background knowledge of physical metallurgy or solid state physics. Nevertheless it is hoped from this that most of the effects commonly observed in micrographs can now be recognized and understood. We have seen that the use of thin foil techniques is providing answers to many previously unsolved questions. Although the use of electron microscopy alone will not necessarily give the complete solution to a problem, it is of great advantage to be able to see directly the substructures in crystals, and, when combined with other physical techniques, it is indeed a powerful tool in research. One thing is clear, namely, there is a dearth of information regarding thin foil observations of ferrous alloys; here is a field well worth extensive future work in view of the importance of these materials commercially.

For metallographic work the electron microscope is as necessary as a light microscope, for there is always a stage in any investigation where we want to look beyond what we can see at present. We expect, therefore, that electron microscopes will soon be as widely used as light microscopes, even allowing for the fact that they are considerably more expensive.

5.4 Other Examples of Examining Metals

We have now seen how the direct examination of thin specimens by transmission electron microscopy has many applications in the examination of structure and properties of crystalline materials. However, there are other methods of examining metal surfaces directly, which, when coupled with other physical experiments and high-resolution microscopy, all add up to presenting a more complete picture of the fundamental properties of materials comprising the solid state. In conclusion, therefore, we

† Examples of the many diverse problems now being investigated can be found in the literature listed in Appendix C.

present a list of other methods of examining metal surfaces using electrons, with suitable references to enable the reader to pursue these topics further:†

(1) Field emission microscopy.[81, 131–134]
(2) Secondary emission microscopy.[135]
(3) Reflection electron microscopy.[136–142]
(4) Thermal emission microscopy.[143]
(5) Scanning electron probe microanalysis.[144–146]
(6) Scanning electron microscopy (ultimate resolution ~50 Å).[147]

References

1. A. H. Cottrell, *Dislocations and Plastic Flow in Crystals* (Clarendon Press, Oxford), 1953.
2. W. T. Read, *Dislocations in Crystals* (McGraw-Hill Book Co., New York), 1953.
3. J. Friedel, *Les Dislocations* (Gauthier-Villars, Paris), 1956.
4. A. Seeger, *Kristallplastizität—Handbuch der Physik*, vol. VII-2 (Springer-Verlag, Berlin), 1958.
5. *Dislocations and Mechanical Properties of Crystals*, Lake Placid Conf. 1956 (John Wiley and Sons, New York), 1957.
6. *Defects in Solids*, Report of Bristol Conference, 1954 (London Phys. Soc.), 1955.
7. A. J. Forty, *Advances in Physics* 1954, **3**, p. 1.
8. P. B. Hirsch, *Prog. Met. Physics* 1956, **6**, p. 236.
9. J. W. Menter, *Advances in Physics* 1958, **7**, p. 299.
10. P. B. Hirsch, *Met. Reviews* (Inst. of Metals, London), 1959, **4**, p. 101.
11. P. B. Hirsch, *J. Inst. Metals* 1959, **87**, p. 406.
12. W. Bollmann, *Phys. Rev.* 1956, **103**, p. 1588.
13. M. J. Whelan, P. B. Hirsch, R. W. Horne, and W. Bollmann, *Proc. Roy. Soc.* A 1957, **240**, p. 524.
14. M. J. Whelan, *ibid.* 1958, **249**, p. 114.
15. P. R. Swann and J. Nutting, *J. Inst. Metals* 1960, **88**, p. 478.
16. H. M. Tomlinson, *Phil. Mag.* 1958, **3**, p. 867.
17. H. Wilsdorf, L. Cinquina, and C. J. Varker, *Proc. 4th Int. Conf. Electron Microscopy* (Springer-Verlag, Berlin, 1960), p. 559. See also reference 19, Chapter 3.
18. D. W. Pashley, *ibid.*, p. 562; also with G. A. Bassett, *J. Inst. Metals* 1959, **87**, p. 449.
19. E. Votava and A. Fourdeux, *ibid.*, p. 556.
20. A. Seeger, see p. 243, reference 5.
21. P. R. Thornton and P. B. Hirsch, *Phil. Mag.* 1958, **3**, p. 738.
22. Z. S. Basinski, *ibid.* 1959, **4**, p. 393.
23. P. B. Hirsch, *Int. Conf. Stresses and Fatigue in Metals*, Amsterdam, 1959 (Elsevier, Amsterdam), p. 139.
24. P. B. Hirsch, *Int. Crystallography Congress*, Cambridge, England, 1960. Abstracts to be published by Univ. of Denmark, Copenhagen (ed. R. W. Asmussen).
25. D. Kuhlmann-Wilsdorf, *ibid.*, reported by R. Maddin.

† The publications of Conference Proceedings (see Appendix C) should also be consulted for further information.

26. A. Howie, *European Congress Electron Microscopy*, Delft, Holland, 1960, De Nederlandse Vereniging voor Electronenmicroscopie 1961, p. 383.
27. J. E. Bailey and P. B. Hirsch, *Phil. Mag.* 1960, **5**, p. 485, see also J. E. Bailey, *ibid.*, p. 833.
28. P. B. Hirsch, P. G. Partridge and H. M. Tomlinson, *Proc. 4th Int. Conf. Electron Microscopy* 1958 (Springer-Verlag, Berlin, 1960), p. 536.
29. P. B. Price, *European Congress Electron Microscopy*, Delft, Holland, 1960, De Nederlandse Vereniging voor Electronenmicroscopie 1961, p. 370. See also, *Phil. Mag.* 1960, **5**, p. 873, *J. Appl. Physics* 1961, **32**, p. 1746.
30. J. Washburn, G. K. Williamson, A. Kelly, and C. W. Groves, *Phil. Mag.* 1960, **5**, p. 991.
31. J. Washburn, W. Elkington, and G. Thomas, to be published.
32. P. B. Hirsch, M. J. Whelan, and R. W. Horne, *Phil. Mag.* 1956, **1**, p. 677.
33. H. G. F. Wilsdorf, *A.S.T.M. Spec. Publication*, No. 245, p. 43, 1958.
34. G. Thomas, *Phil. Mag.* 1959, **4**, p. 1213.
35. R. L. Segall and P. G. Partridge, *ibid.*, p. 912.
36. D. G. Brandon, *Ph.D. Thesis*, Cambridge University, 1960.
37. D. G. Brandon and J. Nutting, *Acta Met.* 1959, **7**, p. 101, *J. Iron Steel Inst.* 1960, **196**, p. 106.
38. A. S. Keh, *AIME Symposium on Dislocations in Metals* (St. Louis, 1961), Trans. AIME 1961, in press.
39. W. Carrington, K. F. Hale, and D. McLean, *Proc. Roy. Soc.* A 1960, **259**, p. 203.
40. G. Thomas, J. Nadeau, and R. Benson, *European Congress Electron Microscopy*, Delft, Holland, 1960, De Nederlandse Vereniging voor Electronenmicroscopie 1961, p. 447. R. Benson, G. Thomas, and J. Washburn, *Dislocation Substructures in Deformed and recovered Molybdenum* (Symposium on Dislocations in Metals, St. Louis, 1961). *Trans. AIME* 1961 (in press).
41. A. Fourdeux and A. Berghezan, *J. Inst. Metals* 1960, **89**, p. 31.
42. J. Silcox, *Proc. 4th Int. Conf. Electron Microscopy* (Springer-Verlag, Berlin, 1960), p. 552.
43. A. Fourdeux, A. Berghezan, and W. W. Webb, *J. Appl. Physics* 1960, **31**, p. 918.
44. P. R. Swann and A. Howie, *Phil. Mag.*, to be published.
45. F. C. Frank and W. T. Read, *Phys. Rev.* 1950, **79**, p. 722.
46. H. G. F. Wilsdorf, *Structure and Properties of Thin Films*, ed. Neugebauer et al. (John Wiley and Sons, New York), 1959, p. 151.
47. A. Berghezan and A. Fourdeux, *Comptes Rendus* 1959, **248**, p. 1333.
48. R. B. Nicholson, G. Thomas, and J. Nutting, *Acta Met.* 1960, **8**, p. 172.
49. K. H. Westmacott, D. Hull, and R. S. Barnes, *Phil. Mag.* 1959, **4**, p. 1089.
50. G. Thomas and M. J. Whelan, *ibid.*, p. 511.
51. K. Thomas and K. F. Hale, *ibid.*, p. 531.
52. P. B. Hirsch, R. Segall, and P. G. Partridge, *ibid.*, p. 721.
53. R. N. Wilson and P. J. E. Forsyth, *J. Inst. Metals* 1959, **87**, p. 336.
54. F. C. Frank, *Symposium on Plastic Deformation of Crystalline Solids* (Washington, D.C.: U.S. Government Printing Office), 1950, p. 89.
55. F. Seitz, *Phys. Rev.* 1950, **79**, p. 890.
56. D. Kuhlmann-Wilsdorf, *Phil. Mag.* 1958, **3**, p. 125.
57. N. Thompson, *Proc. Phys. Soc.* B 1953, **66**, p. 481.
58. N. Thompson, *Report of Conf. on Defects in Cryst. Solids.* (*Phys, Soc.*, London), 1955, p. 153.
59. J. Friedel, *Phil. Mag.* 1955, **46**, p. 1169.

60. J. Silcox and P. B. Hirsch, *ibid.* 1959, **4**, p. 72.
61. P. B. Hirsch, J. Silcox, R. E. Smallman, and K. H. Westmacott, *ibid.* 1958, **3**, p. 897.
62. D. Kuhlmann-Wilsdorf and H. G. F. Wilsdorf, *J. Appl. Physics* 1960, **31**, p. 516.
63. S. Amelinckx, W. Bontinck, W. Dekeyser, and F. Seitz, *Phil. Mag.* 1957, **2**, p. 355.
64. J. Silcox and M. J. Whelan, *ibid.* 1960, **5**, p. 1.
65. P. B. Hirsch and J. Silcox, *Rep. of Conf. on Growth and Perfection of Crystals* (John Wiley and Sons, New York 1958), p. 262.
66. R. E. Smallman and K. H. Westmacott, *J. Appl. Phys.* 1959, **30**, p. 603.
67. R. E. Smallman, K. H. Westmacott, and J. A. Coiley, *J. Inst. Metals* 1959, **88**, p. 127.
68. G. Thomas, *Phil. Mag.* 1959, **4**, p. 606.
69. G. Thomas and J. Nutting, *Acta Met.* 1959, **7**, p. 515.
70. J. Silcox and P. B. Hirsch, *Phil. Mag.* 1959, **4**, p. 1356.
71. H. G. F. Wilsdorf, *Phys. Rev.* 1959, **3**, p. 172.
72. W. Bollmann, *Phil. Mag.* 1960, **5**, p. 621.
73. P. A. Beck, *Adv. in Phys.* 1954, **3**, p. 245.
74. W. Bollmann, *J. Inst. Metals* 1959, **87**, p. 439.
75. D. W. Pashley, *Phil. Mag.* 1959, **4**, p. 324.
76. D. W. Pashley, J. W. Menter, and G. A. Bassett, *Nature* 1957, **179**, p. 752.
77. G. A. Bassett, J. W. Menter, and D. W. Pashley, *Proc. Roy. Soc.* A 1958, **246**, p. 345.
78. J. W. Menter, *ibid.* 1956, **236**, p. 119.
79. W. C. T. Dowell, J. L. Farrant, and A. L. G. Rees, *Proc. Int. Conf. Electron Microscopy* 1954 (Roy. Mic. Soc., London, 1956), p. 279.
80. G. A. Bassett and D. W. Pashley, *J. Inst. Metals* 1959, **87**, p. 449. See also D. W. Pashley, *Proc. Roy. Soc.* A 1960, **255**, p. 218. J. W. Menter and D. W. Pashley, *Structure and Properties of Thin Films*, ed. Neugebauer et al. (John Wiley and Sons, New York), 1959, p. 111.
81. M. Drechsler, *Z. Metallk.* 1956, **47**, p. 305.
82. S. Amelinckx and P. Delavignette, *Nature* 1960, **185**, p. 603.
83. G. K. Williamson, *Proc. Roy. Soc.* A 1960, **257**, p. 457.
84. G. A. Geach and R. Phillips, *Proc. 4th Int. Conf. Electron Microscopy* 1958 (Springer-Verlag, Berlin, 1960), p. 572.
85. P. Delavignette and S. Amelinckx, *Phil. Mag.* 1960, **5**, p. 729. *Phys. Rev.* (letters) 1960, **5**, p. 50.
86. P. Delavignette and S. Amelinckx, *European Congress Electron Microscopy*, Delft, Holland, 1960, De Nederlandse Vereniging voor Electronenmicroscopie 1961, p. 413.
87. F. C. Boswell, *ibid.*, p. 409.
88. D. W. Pashley and A. E. B. Presland, *ibid.*, p. 417.
89. S. Amelinckx and P. Delavignette, *J. Appl. Phys.* 1960, **31**, p. 2126.
90. S. Amelinckx and P. Delavignette, *Phil. Mag.* 1960, **5**, p. 533.
91. M. J. Whelan, *J. Inst. Metals* 1959, **87**, p. 392.
92. M. J. Whelan and P. B. Hirsch, *Phil. Mag.* 1957, **2**, pp. 1121, 1303.
93. P. B. Hirsch, A. Howie, and M. J. Whelan, *Phil. Trans. Roy. Soc.* A 1960, **252**, p. 499.
94. R. D. Heidenreich, *J. Appl. Phys.* 1949, **20**, p. 993.
95. A. Howie and M. J. Whelan, *Proc. European Conf. Electron Microscopy*, Delft, Holland, 1960, De Nederlandse Vereniging voor Electronenmicroscopie 1961

pp. 181, 194, 207. See also H. Hashimoto, A. Howie, and M. J. Whelan, *Phil. Mag.* 1960, **5**, p. 967.
96. R. B. Nicholson and J. Nutting, *Phil. Mag.* 1958, **3**, p. 531.
97. R. B. Nicholson, G. Thomas, and J. Nutting, *J. Inst. Metals* 1959, **87**, p. 429.
98. G. Thomas and J. Nutting, *ibid.* 1959, **88**, p. 81.
99. G. Thomas, *J. Inst. Metals* 1961, in press.
100. B. Genty, R. Graf, and G. Lenoir, *Comptes Rendus* 1958, **247**, pp. 1731, 2126.
101. R. B. Nicholson and J. Nutting, *Acta Met.* 1961, **9**, p. 332. See also G. R. Frank, D. L. Robinson, and G. Thomas, *J. Appl. Phys.* 1961, **32**, p.1763.
102. A. Saulnier and M. Croutzeilles, *Proc. 4th Int. Conf. Electron Microscopy* 1958 (Springer-Verlag, Berlin, 1960), p. 588.
103. G. Thomas and P. Roy (to be published).
104. P. M. Kelly and J. Nutting, *J. Iron and Steel Inst.* 1959, **192**, p. 246; *ibid.* 1961, **197**, p. 199. *Proc. Roy. Soc.* A 1960, **259**, p. 45.
105. G. Thomas and M. J. Whelan, *Phil. Mag.* 1961, in press.
106. N. Takahashi and K. Mihama, *Acta Met.* 1957, **5**, p. 159.
107. N. Takahashi and K. Ashinuma, *Proc. 4th Int. Conf. Electron Microscopy* 1958 (Springer-Verlag, Berlin, 1960), p. 610.
108. W. Pitsch, *Phil. Mag.* 1959, **4**, p. 577.
109. W. Pitsch, *J. Inst. Metals* 1959, **87**, p. 444.
110. D. Watanabe, *Acta Cryst.* 1957, **10**, p. 483.
111. S. Ogawa, D. Watanabe, H. Watanabe, and T. Komoda, *ibid.* 1958, **11**, p. 872.
112. A. B. Glossop and D. W. Pashley, *Proc. Roy. Soc.* A 1959, **250,** p. 132.
113. D. W. Pashley and A. E. B. Presland, *J. Inst. Metals* 1959, **87**, p. 419.
114. D. W. Pashley and A. E. B. Presland, (*a*) *European Congress Electron Microscopy*, Delft, Holland, 1960, De Nederlandse Vereniging voor Electronenmicroscopie 1961, p. 429; (*b*) *Structure and Properties of Thin Films*, ed. Neugebauer et al. (John Wiley and Sons, New York), 1959, p. 211.
115. M. J. Marcinkowski and R. M. Fisher, *ibid.* (*a*), p. 400, with N. Brown, *Acta Met.* 1961, **9**, p. 129.
116. A. Guinier, *Advances in Solid State Physics* 1959, **9**, p. 293 (Academic Press, New York).
117. E. Orowan, *Internal Stresses in Metals and Alloys* (Inst. of Metals, London) 1948, p. 447.
118. J. C. Fisher, E. W. Hart, and R. H. Pry, *Acta Met.* 1953, **1**, p. 336.
119. G. Thomas, J. Nutting, and P. B. Hirsch, *J. Inst. Metals* 1958, **86**, p. 13.
120. A. Kelly and M. E. Fine, *Acta Met.* 1957, **5**, p. 365.
121. A. Kelly, *Phil. Mag.* 1958, **3**, p. 1472.
122. J. H. Van der Merwe, *Proc. Phys. Soc.* A 1950, **63**, p. 616.
123. M. F. Ashby and G. C. Smith, *Phil. Mag.* 1960, **5**, p. 298.
124. J. S. Bowles and J. K. Mackenzie, *Acta Met.* 1954, **2**, pp. 129, 138; 1957, **5**, p. 137.
125. B. A. Bilby and J. W. Christian, *Mechanism of Phase Transformations in Metals* (Inst. of Metals, London), 1955, p. 121.
126. G. Thomas and M. C. Huffstutler (to be published): P. R. Swann and H. Warlimont (to be published).
127. C. A. Fowler and E. M. Fryer, *Phys. Rev.* 1955, **100**, p. 746.
128. L. Mayer, *J. Appl. Phys.* 1957, **28**, p. 975.
129. M. E. Hale, H. W. Fuller, and J. Rubenstein, *ibid.* 1959, **30**, p. 789.
130. H. W. Fuller and M. E. Hale, *ibid.* 1960, **31**, pp. 238, 1699.
131. E. W. Müller, *Ergeb. Exakt Naturwiss.* 1953, **27**, p. 290.

132. E. W. Müller, *J. Appl. Phys.* 1956, **27**, p. 474.
133. E. W. Müller, *Structure and Properties of Thin Films*, ed. Neugebauer et al. (John Wiley and Sons, New York), 1959, p. 476.
134. See *Proc. 4th Int. Conf. Electron Microscopy* (Springer-Verlag, Berlin, 1960), Section E, pp. 780–848.
135. G. Möllenstedt and H. Düker, *Optik* 1953, **10**, p. 192.
136. J. S. Halliday and R. C. Newman, *Brit. J. Appl. Physics* 1960, **11**, p. 158.
137. J. W. Menter, *J. Inst. Metals* 1952, **81**, p. 163.
138. M. E. Haine and W. Hirst, *Brit. J. Appl. Physics* 1953, **4**, p. 239.
139. C. Fert, *Proc. Stockholm Conf. Electron Microscopy* (Academic Press, New York), 1957, p. 8.
140. J. A. Chapman and J. W. Menter, *Proc. Roy. Soc.* A 1954, **226**, p. 400.
141. J. S. Courtney-Pratt, J. W. Menter, and M. Seal, *Proc. 3rd Int. Conf. Electron Microscopy* (Roy. Mic. Soc., London), 1956, p. 645.
142. G. Thomas, N. P. Sandler, and I. Cornet, *J. Inst. Metals*, 1961, **89**, p. 253.
143. G. W. Rathenau and C. Baas, *Acta Met.* 1954, **2**, p. 875.
144. R. Castaing and J. Descamps, *J. Phys.* 1955, **16**, p. 304.
145. V. E. Cosslett and P. Duncumb, *Proc. Stockholm Conference Electron Microscopy* (Academic Press, New York), 1957, p. 12.
146. D. A. Melford and P. Duncumb, *Metallurgia* 1960, **61**, p. 205.
147. Papers independently reported by R. F. M. Thornley and K. C. A. Smith at *European Congress Electron Microscopy*, Delft, Holland, 1960, De Nederlandse Vereniging voor Electronenmicroscopie 1961, pp. 173, 177, 547.

Appendix A

Analysis of Micrographs with Corresponding Selected Area Diffraction Patterns

For the purposes of illustrating the methods employed in this kind of analysis, consider Fig. A.1 which shows a micrograph of a thin foil of aluminum. In this, dislocations have moved producing the slip traces *A* and *B*. The corresponding selected area diffraction pattern is shown in Fig. A.2. A typical problem which may be encountered is to determine the slip planes corresponding to the traces, and when this is established, to calculate the thickness of the foil. To do this the following procedures must be adopted:

1. Determination of Foil Orientation

This is done by making measurements on the diffraction pattern, and in the first method the orientation will be obtained without using the camera constant. For aluminum (FCC, reflections allowed when h, k, l are all even or all odd) the first seven reflections are:

Reflection	$(h^2 + k^2 + l^2)$	d-spacings of planes (Å)	$(h^2 + k^2 + l^2)^{1/2}$
111	3	2.34	1.736
200	4	2.02	2.000
220	8	1.43	2.828
311	11	1.22	3.316
222	12	1.17	3.464
400	16	1.01	4.000
313	19	0.93	4.358

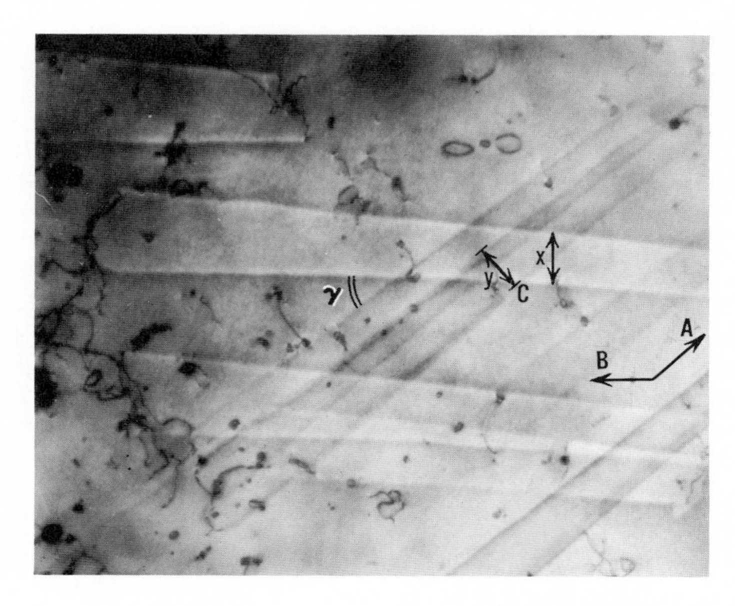

FIG. A.1. Showing slip traces *A*, *B* in a thin foil of pure aluminum (×29,000).

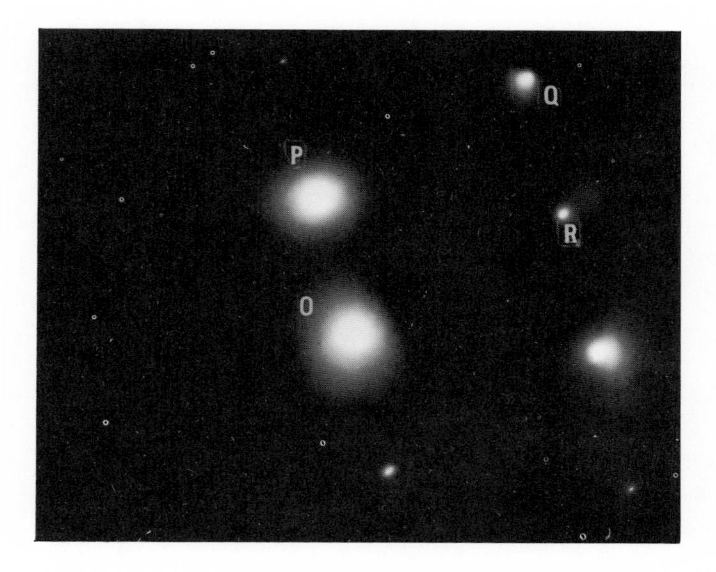

FIG. A.2. Selected area diffraction pattern corresponding to Fig. A.1.

If we consider the three reflections P, Q, R in Fig. A.2, then since we know that OP, OQ, OR each represents the reciprocal of the d-spacings of the reflecting planes giving rise to these spots, and since d is inversely proportional to $(h^2 + k^2 + l^2)^{1/2}$ for different planes, it is possible to assign tentatively the form of the Miller indices to the spots.† From the diffraction pattern $A2$ the average values are $OP = 1.925$ cm, $OQ = 4.25$ cm, $OR = 3.25$ cm, so that $OQ/OP = 2.208$, $OR/OP = 1.688$, $OQ/OR = 1.308$. By inspection of the ratios of $(h^2 + k^2 + l^2)^{1/2}$ for the following reflections, we get for (313):(200), (113):(200) and (313):(113), 2.179, 1.658, and 1.314 respectively. Thus the spots P, Q, R, are probably of the {200}, {313}, and {113} form. By measuring the distance from the origin of the other spots on the pattern, it is then possible to index the pattern completely, remembering that the values of the indices must be consistent along each row of spots, e.g., in Fig. A.3 along the rows OP, RQ, h decreases by -2, whereas k and l are unchanged, since each spot

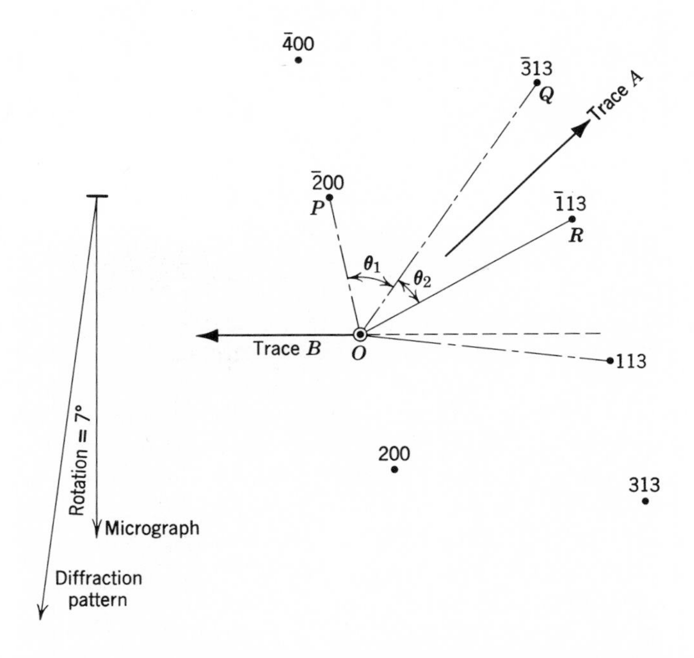

FIG. A.3. Analysis of Fig. A.2 showing directions of slip traces and angle of rotation. Orientation is approximately [0$\bar{3}$1].

† In cubic crystals, the d-spacing is given by:

$$d_{hkl} = \frac{a}{(h^2 + k^2 + l^2)^{1/2}},$$ where a is the length of the cube edge $(=d_{100})$.

in the row belongs to the same zone. To check that the indices have been correctly assigned, the angles between the planes should be measured and then calculated using the cosine θ relationship. In the case considered here, if $h_1k_1l_1$ and $h_2k_2l_2$ are the Miller indices of two intersecting planes, then the angle between them is given by

$$\cos\theta = \frac{h_1h_2 + k_1k_2 + l_1l_2}{\{(h_1^2 + k_1^2 + l_1^2)(h_2^2 + k_2^2 + l_2^2)\}^{1/2}}$$

These angles can be obtained directly from the diffraction pattern, e.g., consider the angle between the spots P and Q (Fig. A.3) $= \theta_1 = 47°$. The calculated value for the angle between $(\bar{2}00)$ and $(\bar{3}13)$ is 46° 30′, and this shows that the correct indices have been assigned to the spots P and Q. Similarly for Q and R, $\theta_2 = 26°$ so that the spot R must be $(\bar{1}13)$. Figure A.3 shows the complete indexing of the diffraction pattern in Fig. A.2.

The foil orientation is then obtained by using the zone law†: the zone axis $[uvw]$ contains the two planes $(h_1k_1l_1)$ and $(h_2k_2l_2)$ if $u{:}v{:}w = (k_1l_2 - k_2l_1){:}(l_1h_2 - l_2h_1){:}(h_1k_2 - h_2k_1)$. In other words, the zone axis is given by the cross product of two different lattice vectors; thus by taking the reflections P and Q we get $[uvw] = [\bar{2}00] \times [\bar{3}13] = [0\bar{3}1]$. This means that the direction of the beam and the normal to the foil is $[0\bar{3}1]$, i.e., the foil surface is $(0\bar{3}1)$.

This orientation is only approximate to within about $\pm 5°$. If the foil was exactly in $[0\bar{3}1]$, a symmetrical diffraction pattern about vertical and horizontal lines drawn through the center should be observed (e.g., as in Fig. 16) and the reflections $(3\bar{1}\bar{3})$, $(1\bar{1}\bar{3})$ corresponding to the points opposite to P and Q would also be present. The fact that these reflections are not present in Fig. A.2 shows that the electron beam is not exactly parallel to $[0\bar{3}1]$. This example has been purposely chosen to illustrate how approximate the orientation determination can be; of course, the foil can be tilted into a more exact $[0\bar{3}1]$ orientation.

λL for the pattern can be obtained from the relation $\lambda L = rd$; hence for the (200) reflection $\lambda L = 2.024 \times 1.925$, i.e., 3.90 cm Å; the average value is 3.93 cm. Å. Once λL is known for the microscope settings, the diffraction patterns may be quickly indexed simply by dividing λL with the measured distances r of the spots from the origin. In this way the d-spacings are obtained directly. Exact correspondence with the given d-values‡ is rarely obtained owing to the difficulties of measurement, e.g., locating the true center of the spots, and because λL may change with the

† See C. S. Barrett, *Structure of Metals* (McGraw-Hill, Book Co., New York, 2nd Ed., 1953), p. 632.

‡ These may be obtained from the standard A.S.T.M. card index file.

position of the photographic plate, specimen, etc. With experience the patterns may often be indexed merely by inspection. Probably the most convenient method is to make a model of the reciprocal lattice of the material being examined, including only the allowed reflections. Orienting the model into various positions to compare with the diffraction pattern will enable indexing to be carried out directly if the points on the model are also previously indexed.

2. Determination of Slip Planes and Foil Thickness

Now that the pattern is indexed, then, after correcting for the angle of rotation, it is possible to obtain the direction of the slip traces by comparing the micrograph and diffraction pattern. In the case of Figs. A.1 and A.2 the rotation is 7° (see Fig. A.3). As can be seen from Figs. A.1, A.2, A.3, the traces *A* and *B* do not lie parallel to any direction given by the diffraction pattern; they are both symmetrical about [001]. Now the slip systems for aluminum are {111} ⟨110⟩. Of the four possible slip planes $(\bar{1}\bar{1}1)$, $(1\bar{1}1)$ make angles of 43°, and $(\bar{1}\bar{1}\bar{1})$, $(1\bar{1}\bar{1})$ make angles of 69° to the foil surface, i.e., $(0\bar{3}1)$. Examination of Fig. A.1 shows that the traces *A* and *B* make about the same projected widths, i.e., both traces occur on planes making the same angle to the foil surface.

To determine which planes are operating, the stereographic projection must be used, bringing the $(0\bar{3}1)$ face parallel to the projection plane. This can be done by using a standard (001) projection, whereby the stereogram is rotated so as to bring the $[0\bar{3}1]$ pole to the center. This involves a west to east rotation of 71.5°. When this is done, the poles of the other principal planes are plotted by a similar rotation using a piece of tracing paper over a Wulff net, and this gives the results shown in Fig. A.4. The great circles corresponding to the four {111} slip planes are then drawn at 90° to each pole, as shown at *ab*, *cd*. A comparison of Figs. A.3 and A.4 shows that the trace *A* must be due to slip on $(1\bar{1}1)$, and the trace *B* must occur on $(\bar{1}\bar{1}1)$; this result is consistent with the angle γ made by the two traces. Actually the angle γ between the traces shown in Fig. A.1 is less than the angle γ drawn in Fig. A.4. As was pointed out earlier this discrepancy is due to the fact that the foil is not exactly in $[0\bar{3}1]$ orientation and possibly because of buckling. As an exercise the reader can deduce exactly how far the foil is from $[0\bar{3}1]$ by rotating the stereogram of Fig. A.4 until the angle γ on this diagram is the same as the angle γ of Fig. A.1. It should be clear that the traces *A*, *B* cannot be due to slip on $(1\bar{1}\bar{1})$ and $(\bar{1}\bar{1}\bar{1})$ since the angle between the traces of these planes would be too large to be consistent with that in Fig. A.1.

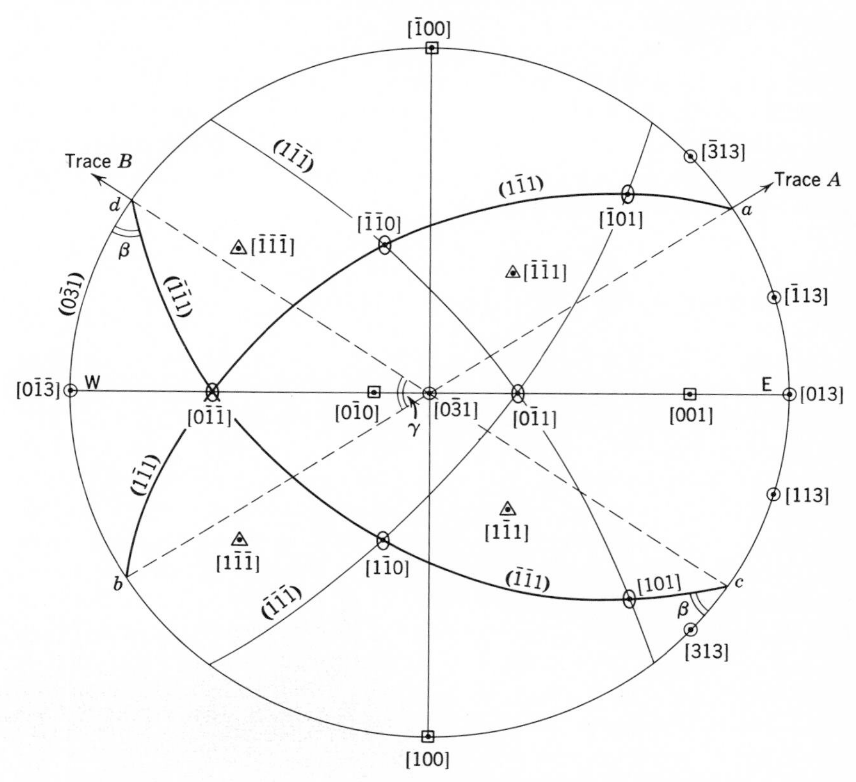

FIG. A.4. Stereographic projection of the [0$\bar{3}$1] orientation showing that the slip planes are (1$\bar{1}$1) for trace *A* and ($\bar{1}\bar{1}$1) for trace *B*. The angle β is the angle between ($\bar{1}\bar{1}$1) and the foil surface (=43°).

An alternative method for determining the slip planes is that illustrated in Fig. A.5, where the (0$\bar{3}$1) plane is projected onto a standard [001] stereogram. The intersection of the great circle (0$\bar{3}$1) with the (1$\bar{1}$1) and ($\bar{1}\bar{1}$1) great circles will give the correct angle (γ Fig. A.1) between these planes in projection onto (0$\bar{3}$1).

It is not always possible to determine the slip systems with this unambiguity. For example, a trace of [0$\bar{1}\bar{1}$] in a foil of orientation [0$\bar{1}$1] could only have two possible slip planes, namely, ($\bar{1}\bar{1}$1) or (1$\bar{1}$1) both at 35° to (0$\bar{1}$1) as shown in Fig. A.6, since the other two slip planes ($\bar{1}\bar{1}\bar{1}$) and (1$\bar{1}\bar{1}$) are perpendicular to the foil surface and no traces on these planes would be observed. Now, since it is not possible to determine which is the top or bottom surface of the foil except in certain circumstances (see section

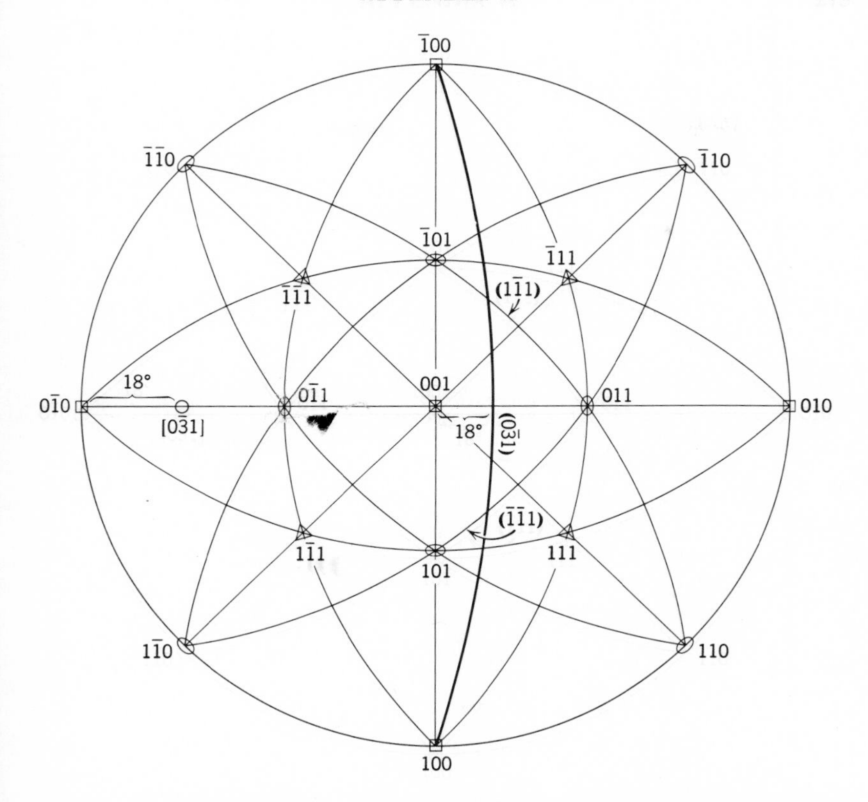

FIG. A.5. Alternative method to that of Fig. A.4 for determining the slip plane traces. The $(0\bar{3}1)$ plane is shown on a standard [001] projection. The angle between $(1\bar{1}1)$ and $(\bar{1}\bar{1}1)$ along the great circle $(0\bar{3}1)$ corresponds to the angle γ between the traces in Fig. A.1.

5.2.14), one cannot distinguish between $(\bar{1}\bar{1}1)$ and $(1\bar{1}1)$ for this orientation.

By plotting the poles of other planes on the stereogram, a complete crystallographic analysis of the foil can be made. This kind of approach is particularly valuable for foil orientations of high indices where the geometry is not as easy to visualize as in the case of cube or other low-index planes.

It is now possible to calculate the thickness of the foil. The projected width x of the slip plane corresponding to trace B is related to the thickness t of the foil by $t = x \tan \alpha$, where α is the angle subtended at the surface, i.e., the angle between $(\bar{1}\bar{1}1)$ or $(1\bar{1}1)$ and $(0\bar{3}1)$. From the micrograph shown in Fig. A.1 the average value of x for the B slip trace near region C

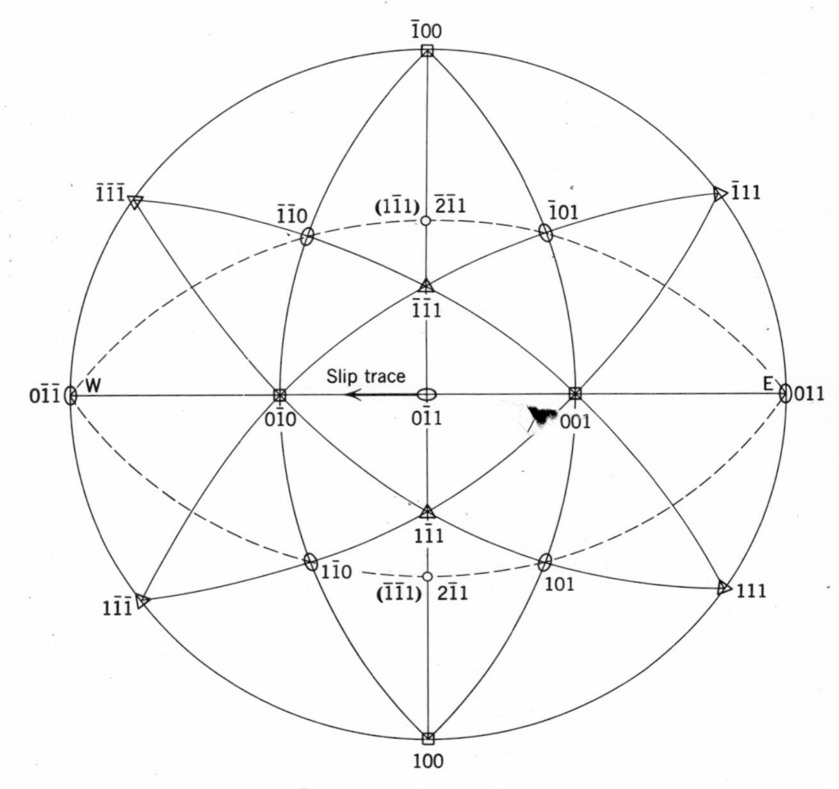

FIG. A.6. Stereographic projection of $[0\bar{1}1]$ orientation for cubic crystals. Both $(1\bar{1}1)$ and $(\bar{1}\bar{1}1)$ have the common trace [011].

is 7 mm at a magnification of 29,000 times. Thus, with $\alpha = 43°$, the thickness of the foil near C is

$$t = \frac{7 \times 10^7}{2.9 \times 10^4} \times 0.936 = 2260 \text{ Å}$$

This value also checks with the projection y of the slip trace A in the same region. This result is consistent with our conclusion that trace A occurs on $(1\bar{1}1)$ and not $(\bar{1}\bar{1}\bar{1})$ since a $(\bar{1}\bar{1}\bar{1})$ trace would indicate a foil thickness of ~5400 Å near region C, and this is too thick considering the clarity of the contrast in Fig. A.1. Once the thickness has been determined it is then possible to obtain accurate quantitative data from micrographs (e.g., density of defects, volume fraction of precipitates, etc.).

The thickness of the film can also be obtained from the number of thickness extinction fringes which occur at stacking faults, inclined boundaries, and at the edges of foils. If the extinction distance t_0 is calculated for the particular orientation involved, the thickness of the foil at the nth fringe is $t_0 n$. However, this is only accurate when the deviation x from the Bragg angle is known, since the number of fringes depends both on x and t (section 2.3.3).

Appendix B

Physical Constants and Conversion Factors

Physical Constants

Velocity of light c	$= 2.9978 \times 10^{10}$ cm/sec
Electronic charge e	$= 4.8025 \times 10^{-10}$ esu or 1.6020×10^{-20} emu
Specific electronic charge e/m_0	$= 5.27 \times 10^{17}$ esu/gm or 1.7592×10^{7} emu/gm
Rest mass of electron m_0	$= 9.1066 \times 10^{-28}$ gm
Plancks constant h	$= 6.624 \times 10^{-27}$ erg/sec
Avogadro's number N	$= 6.0228 \times 10^{23}$/mole
Boltzmann's constant k	$= 1.3805 \times 10^{-16}$ erg/deg
Gas constant R	$= 8.314 \times 10^{7}$ erg/° K/mole

Conversion Factors

Electrical:	1° K	$= 8.6166 \times 10^{-5}$ ev
	1 coulomb	$= c/10$ or 3×10^{-9} esu of charge
	1 ampere	$= c/10$ or 3×10^{-9} esu of current
	1 volt	$= 1/300$ esu of potential $= 10^{8}$ emu
Energy of 1 electron volt (1 ev)		$= 1.602 \times 10^{-12}$ erg $\equiv 1.163 \times 10^{4}$ deg K

Lengths:	1 angstrom (Å)	$= 10^{-8}$ cm $= 10^{-4}$ micron (μ)
	1 in.	$= 2.54$ cm
	1 mil	$= 0.001$ in. $= 25.4\ \mu$
	1 radian	$= 57.296$ deg

Appendix C

The Literature of Electron Microscopy

Bibliography

The N.Y.S.E.M. International Bibliography of Electron Microscopy (New York Society of Electron Microscopists) published quarterly in card form since 1950. 1950 to 1955 issues are now available in book form (from 2 East 63rd Street, New York 21).

Conference Reports

Proc. Conference on Electron Microscopy, Delft, Holland, 1949 (Hoogland, Delft, 1950).

Proc. 3rd International Conference on Electron Microscopy, London, England, 1954 (Roy. Mic. Soc., 1956).

Proc. Conference on Electron Microscopy, Stockholm, Sweden, 1956 (Almqvist and Wiksell, Stockholm, 1957).

Proc. 1st Asian and Oceanic Conference on Electron Microscopy, Tokyo, Japan, 1957 (Electrotechnical laboratory, Tokyo, 1957).

Proc. 4th International Conference on Electron Microscopy, Berlin, 1958 (Springer-Verlag, 1960).

Proc. Conference on Electron Microscopy, Delft, Holland, 1960 (De Nederlandse Vereniging voor Electronenmicroscopie 1961).

American Society for Testing Materials S.T.P. 155 (1954), S.T.P. 245 (1958), S.T.P. 262 (1959).

Proc. E.M.S.A. are published annually in *Journal of Applied Physics*.

Growth and Perfection of Crystals, Proc. Cooperstown Conference, New York, 1958 (John Wiley and Sons, New York).

Structure and Properties of Thin Films, Proc. Bolting Landing Conference, New York, 1959 (John Wiley and Sons, New York).

The Impact of Transmission Electron Microscopy on Theories of the Strength of Crystals. *Proc. 1st International Materials Conference*, University of California, Berkeley, 1961. (Ed. G. Thomas and J. Washburn, to be published by Interscience Publishers, Inc., New York.)

Textbooks

V. K. Zworykin et al., *Electron Optics and the Electron Microscope* (John Wiley and Sons, New York, 1945).

R. W. G. Wykoff, *Electron Microscopy* (Interscience Publishers, New York), 1949.

V. E. Cosslett, *Practical Electron Microscopy* (Butterworths, London), 1951.

C. E. Hall, *Introduction to Electron Microscopy* (McGraw-Hill Book Co., New York), 1953.

R. B. Fischer, *Applied Electron Microscopy* (Indiana University Press), 1954.

Periodicals

Most of the technical journals in the field of interest should be consulted (e.g., see references to Chapter 5). The *Japanese Journal of Electron Microscopy* is devoted solely to this subject.

Appendix D

General Outline for the Establishment of a Laboratory for Electron Microscopy

A laboratory for electron microscopy requires about four rooms: a microscope room, a preparation room, a dark room, and also a smaller room for accommodating the power supply equipment and rotary pumps. In addition to this, it is recommended that a workroom be made available near the microscope for keeping files, spare parts, etc., and as a space for cleaning the microscope parts.

The Microscope Room

The microscope room, which must be dry and free from dust, should be at least 9 ft high and have an area at least 10 by 10 sq ft, in order to provide good working conditions. When possible, the microscope room should be provided with a window for ventilation purposes and also to allow occasional work in daylight (e.g., during cleaning). It is also more comfortable for the operator if the room is air-conditioned; this also keeps dust to a minimum.

A room of such dimensions will accommodate the following accessories in addition to the electron microscope: three swivel stools for the microscopist and two observers; a table with drawers; a table lamp; a wash basin with drainpipe and a waterpipe with built-in water filter for controlling the flow of water for cooling the microscope. In the work room, space should be allowed for a caster-mounted tool table; a wooden or glass cabinet for accessories and spare parts; a desk and chair; a filing cabinet; a small fireproof wall cabinet for highly combustible cleaning solvents; a plate storage cabinet and a small shelf for storing filled plate holders, relatively small accessories, tools for handling specimen holders, cleaning rags made of lens paper, poplin, or linen, and chamois leathers for wiping the parts of the microscope.

Behind the microscopist, two light bulbs should be fixed to the ceiling. Each bulb should be assigned a separate switch. The first light, having a powerful bulb, should have a wall switch about 16 in. higher than normal to avoid the possibility of its being switched on inadvertently, whereas the second, less powerful with only about 15 watts, may be operated from the microscope.

The floor, walls, and ceiling must be without joins and so designed that they will not harbor particles of dust or dirt which might enter the microscope. It is to be recommended, therefore, that the floor be covered with linoleum and the walls coated with paint which is free from dust and absorbs moisture. For work in the dark it has been found agreeable to use all white furniture. This is also conducive to cleanliness.

It must be remembered, when designing a microscope room, that the doors through which the microscope will have to pass must be large enough. If possible the entrance should be provided with a light sluice. If, for reasons of space, it is not possible to have a light sluice, measures should be taken to insure that there is not too much light at the door. Windows, if any, should therefore be blacked out.

The setting up of the microscope does not normally call for shock-mounting. However, as the microscope and the mains (power) supply cabinet may weigh about 1200 lb each, it is necessary to make sure that the floor is able to take the load.

It is advisable to install all the cables, waterpipes, and the exhaust pipe for the forepump beneath the floor. Steel tubing should preferably be used. It is less advisable to build in a duct (cross section approximately $15\frac{1}{2}$ sq in.) since this would have to be covered over, thereby interrupting the continuity of the floor with additional joins.

Cabling connections should be made according to the cable layout plan supplied with the microscope. The greatest care should be observed in connecting up the transfer contacts of the electric power leads.

The pumping system, the power supply to the lenses, and the high potential are controlled from the microscope, although the forepumps and the power cabinet are usually installed in a small adjoining room.

A waterpipe and flow-off basin for two streams of water are required for cooling the diffusion pumps and usually the lenses. The basin should be located as close to the floor as possible. A filter with stop-cock should be provided in order to insure that the cooling water fed to the microscope is as clean as possible.

The Preparation Room

As the preparation of specimens is an important factor in any laboratory for electron microscopy and frequently takes up a considerable amount of time, particular attention should be paid to the equipment in this room. In addition to furniture and special equipment, enough work places should be provided so that assistant microscopists may work on several specimens at the same time, thus getting the maximum use out of the microscope during working hours. On these grounds it is recommended that the room should be as large as possible; even the use of two rooms should not be considered excessive.

The equipment placed in the preparation room will depend primarily on the type of work to be done. An exhaust hood installed next to the sink should contain electric lighting, water and gas taps, a compressed air nozzle, dc and ac electric power sockets, and windows on each side and in the front. A cabinet for storing bottles of acid and organic solvents, etc., may be installed below the ventilation system and in direct communication with the ventilation shaft. At the side of the exhaust hood a space should be left for storing gas cylinders and other such items. Other necessary items are a switchboard, a small work bench complete with tools and drawers, power sockets for electric lamps, etc., at least two glass or wooden cabinets, shelves for glassware, and several work tables and chairs.

Beside the work tables, to be used primarily for preparation, there should be a large bench of the type found in chemical laboratories and fitted with electric power sockets, gas, water, steam, compressed air, and suction pipes.

The smaller work tables should, whenever possible, be arranged facing the light and covered with dark green plastic (e.g., formica), on which should be placed an acid proof plate of glass.

Besides the usual laboratory equipment, the following items† should also be kept: an ample selection of microscope grids (100, 200 mesh, slot types, etc.); platinum crucible with lid for annealing platinum apertures; chamois leather; nylon and linen cloths and cocktail sticks for cleaning work; various tweezers; Petri dishes; lacquer and solvent; dissecting scalpels; scissors; needles and wax pencils.

† In the United States most of the materials necessary for electron microscopy are obtained from E. J. Fullam, Inc., Schenectady, New York.

In addition to these, the preparation room must also be equipped with the following units: vacuum evaporator; carbon deposition apparatus; vacuum-freeze drying chamber; optical microscope with epi and epi-trans illumination for testing the various apertures and for an initial examination of preparations; a stereomicroscope with epi illumination; microtome adjustable for cuts finer than $\frac{1}{10}\,\mu$; hair dryer; refrigerator; fume cupboard; furnaces; viewing screens for plate observation and measuring diffraction patterns; vacuum desiccators for storing specimens; voltage dividing circuits for electropolishing, and ancilliary equipment.

Of the collection of chemicals that are required, the following list comprises those most frequently needed for cleaning apertures: hydrofluoric acid, sulfuric acid, hydrochloric acid, nitric acid, ammonium hydroxide, and sodium bisulfite. Other chemicals are needed for preparation of specimens, e.g., electropolishing solutions, replicating materials, etc.

The Dark Room

The dark room must be equipped with particular care. Above all, the room must not be too small. There should be a clear separation between wet work and dry work. A labyrinthine passage should be installed between the dark room and the adjoining room as a matter of course. Since all the results from electron microscopy depend on good photographic reproduction, every effort should be made to provide high quality materials for developing and enlarging plates.

Many laboratories use a microdensitometer of the type originally designed for the analysis of spectrographs, but which may also be used to great advantage for inspecting plates that are still wet. Such practice will render much enlargement work unnecessary by allowing the prior selection of sections that are particularly suitable for enlargement.

The Smaller Room

The smaller room should have a large enough area to accommodate the power supply and HT cabinet or the stabilizer set.

In order to minimize interference from magnetic fields, the power supply cabinet must be set up at least 4 ft away from the microscope, although in no case further than 30 ft.

A suitable place for the forepump is often provided in the microscope stand. In certain circumstances, however, it may be installed in the small room instead, since this reduces the amount of noise and vibration in the microscope room.

A vacuum desiccating unit should also be provided in order to prepump the loaded cassettes (i.e., plate holders). During this operation most of the moisture is expelled by pumping on ballast. This procedure will reduce the time taken for the microscope to out-gas freshly loaded plates, and insures maximum efficiency of operation.

Appendix E

Table I. Atomic scattering amplitudes $f(\theta)$ for electrons[a] †

Values in ångström units

Element	Z	Ref[b]	sin θ/λ (Å^{-1})											
			0.00	0.05[c]	0.10	0.15	0.20	0.25	0.30	0.35	0.40	0.50	0.60	0.70
H	1	E	0.529	0.508	0.453	0.382	0.311	0.249	0.199	0.160	0.130	0.089	0.064	0.04
Li	3	HF	3.31	2.78	1.88	1.17	0.75	0.53	0.40	0.31	0.26	0.19	0.14	0.11
Be	4	HF	3.09	2.81	2.23	1.63	1.16	0.83	0.61	0.47	0.37	0.25	0.19	0.14
B	5	HF	2.82	2.62	2.23	1.78	1.37	1.04	0.80	0.62	0.49	0.33	0.24	0.18
C	6	HF	2.45	2.26	2.06	1.74	1.43	1.15	0.92	0.74	0.60	0.41	0.30	0.22
N	7	HF	2.20	2.09	1.91	1.68	1.44	1.20	1.00	0.83	0.69	0.48	0.35	0.27
O	8	HF	2.01	1.95	1.80	1.61	1.42	1.22	1.04	0.88	0.75	0.54	0.40	0.31
F	9	HF	2.12	2.01	1.90	1.71	1.50	1.29	1.11	0.95	0.81	0.60	0.45	0.35
Ne	10	HF	1.85	1.80	1.69	1.55	1.40	1.24	1.09	0.95	0.83	0.63	0.49	0.38
Na	11	HF	4.89	4.21	2.97	2.11	1.59	1.29	1.09	0.95	0.83	0.64	0.51	0.40
Mg	12	HF	5.01	4.60	3.59	2.63	1.94	1.50	1.21	1.01	0.87	0.67	0.53	0.43
Ar	18	HF	4.71	4.40	4.07	3.56	3.03	2.52	2.07	1.71	1.42	1.00	0.74	0.57
Ca	20	HF	10.46	8.71	6.40	4.54	3.40	2.69	2.20	1.84	1.55	1.12	0.84	0.65
Mn	25	TFD	6.2	5.93	5.34	4.49	3.66	2.97	2.43	2.04	1.73	1.29	0.99	0.79
Fe	26	TFD	6.4	6.13	5.48	4.62	3.76	3.05	2.51	2.10	1.79	1.33	1.03	0.82
Co	27	TFD	6.5	6.32	5.62	4.73	3.87	3.14	2.58	2.16	1.84	1.37	1.06	0.84
Ni	28	TFD	6.7	6.41	5.74	4.85	3.97	3.22	2.65	2.23	1.89	1.41	1.09	0.87
Cu	29	TFD	6.8	6.61	5.89	4.97	4.06	3.30	2.72	2.29	1.95	1.45	1.13	0.90
Zn	30	H	7.4	6.70	5.67	4.61	3.75	3.11	2.63	2.26	1.97	1.53	1.21	0.97
		TFD	7.0	6.70	6.03	5.08	4.16	3.38	2.79	2.35	2.00	1.49	1.16	0.92
Ca	31	H	9.1	8.42	6.49	4.95	3.90	3.19	2.68	2.29	1.99	1.54	1.23	0.99
		TFD	7.2	6.89	6.15	5.20	4.25	3.46	2.86	2.41	2.05	1.53	1.19	0.95
Ge	32	H	10.4	9.76	7.40	5.37	4.13	3.33	2.75	2.33	2.01	1.56	1.24	1.01
		TFD	7.3	7.08	6.29	5.32	4.35	3.54	2.93	2.46	2.10	1.57	1.22	0.97

Element	Z	Ref[b]	sin θ/λ (Å^{-1})											
			0.00	0.05[c]	0.10	0.15	0.20	0.25	0.30	0.35	0.40	0.50	0.60	0.70
As	33	H	9.2	8.52	7.04	5.52	4.33	3.46	2.85	2.39	2.05	1.58	1.26	1.02
		TFD	7.5	7.18	6.41	5.43	4.44	3.62	2.99	2.52	2.15	1.61	1.25	1.00
Se	34	TFD	7.6	7.37	6.56	5.53	4.54	3.70	3.06	2.58	2.20	1.65	1.28	1.02
Br	35	TFD	7.8	7.47	6.68	5.63	4.63	3.78	3.13	2.64	2.25	1.69	1.32	1.05
Kr	36	TFD	7.9	7.56	6.80	5.74	4.71	3.85	3.19	2.69	2.31	1.73	1.35	1.08
Rb	37	TFD	8.0	7.75	6.92	5.85	4.80	3.93	3.26	2.75	2.35	1.77	1.38	1.10
Sr	38	TFD	8.2	7.85	7.04	5.96	4.89	4.00	3.32	2.80	2.40	1.80	1.41	1.13
Y	39	TFD	8.3	8.04	7.16	6.06	4.98	4.07	3.38	2.86	2.45	1.84	1.44	1.15
Zr	40	TFD	8.5	8.14	7.28	6.16	5.06	4.15	3.45	2.91	2.50	1.88	1.47	1.17
Nb	41	TFD	8.6	8.23	7.40	6.27	5.15	4.22	3.51	2.97	2.54	1.92	1.50	1.20
Mo	42	TFD	8.7	8.42	7.52	6.36	5.24	4.29	3.57	3.02	2.59	1.95	1.53	1.22
Tc	43	TFD	8.9	8.52	7.63	6.47	5.31	4.36	3.63	3.08	2.64	1.99	1.56	1.25
Ru	44	TFD	9.0	8.62	7.75	6.56	5.40	4.43	3.69	3.13	2.68	2.03	1.58	1.27
Rh	45	TFD	9.1	8.81	7.85	6.66	5.48	4.50	3.75	3.18	2.73	2.06	1.61	1.30
Pd	46	TFD	9.3	8.90	7.97	6.75	5.56	4.57	3.81	3.23	2.77	2.10	1.64	1.32
Ag	47	TFD	9.4	9.00	8.07	6.85	5.64	4.64	3.87	3.28	2.82	2.13	1.67	1.34
Cd	48	TFD	9.5	9.19	8.19	6.95	5.72	4.71	3.93	3.34	2.86	2.17	1.71	1.37
In	49	TFD	9.6	9.29	8.31	7.03	5.80	4.78	3.99	3.39	2.91	2.20	1.73	1.39
Sn	50	TFD	9.8	9.38	8.40	7.13	5.88	4.84	4.05	3.44	2.95	2.24	1.76	1.41
Sb	51	TFD	9.9	9.48	8.50	7.22	5.95	4.91	4.10	3.49	3.00	2.27	1.79	1.44
Te	52	TFD	10.0	9.57	8.62	7.31	6.03	4.97	4.16	3.54	3.04	2.31	1.81	1.46
I	53	TFD	10.1	9.77	8.71	7.39	6.11	5.04	4.22	3.59	3.08	2.34	1.84	1.48
Xe	54	TFD	10.2	9.86	8.81	7.49	6.19	5.10	4.27	3.64	3.13	2.38	1.87	1.51
Cs	55	TFD	10.4	9.96	8.93	7.57	6.26	5.17	4.33	3.68	3.17	2.41	1.90	1.53
Ba	56	TFD	10.5	10.05	9.02	7.66	6.34	5.23	4.39	3.73	3.21	2.45	1.93	1.55
La	57	TFD	10.6	10.15	9.12	7.75	6.40	5.30	4.44	3.78	3.26	2.48	1.95	1.57

Ce	58	TFD	10.7	10.24	9.21	7.84	6.49	5.36	4.50	3.83	3.30	2.51	1.98	1.60
Pr	59	TFD	10.8	10.44	9.31	7.92	6.56	5.42	4.55	3.88	3.34	2.55	2.01	1.62
Nd	60	TFD	10.9	10.53	9.41	8.01	6.63	5.48	4.60	3.93	3.38	2.58	2.03	1.64
Pm	61	TFD	11.0	10.63	9.53	8.10	6.70	5.55	4.66	3.97	3.43	2.61	2.06	1.66
Sm	62	TFD	11.1	10.72	9.62	8.17	6.77	5.61	4.71	4.02	3.47	2.65	2.09	1.69
Eu	63	TFD	11.2	10.82	9.72	8.25	6.85	5.67	4.77	4.07	3.51	2.68	2.11	1.71
Gd	64	TFD	11.4	10.91	9.79	8.34	6.91	5.73	4.82	4.11	3.55	2.71	2.14	1.73
Tb	65	TFD	11.5	11.01	9.88	8.42	6.98	5.79	4.87	4.16	3.59	2.74	2.17	1.75
Dy	66	TFD	11.6	11.11	9.98	8.50	7.05	5.85	4.92	4.20	3.63	2.78	2.19	1.77
Ho	67	TFD	11.7	11.20	10.08	8.58	7.12	5.91	4.98	4.25	3.67	2.81	2.22	1.80
Er	68	TFD	11.8	11.30	10.17	8.66	7.19	5.97	5.03	4.30	3.71	2.84	2.25	1.82
T	69	TFD	11.9	11.49	10.27	8.74	7.26	6.03	5.08	4.34	3.75	2.87	2.27	1.84
Yb	70	TFD	12.0	11.58	10.36	8.82	7.33	6.09	5.13	4.39	3.79	2.91	2.30	1.86
Lu	71	TFD	12.1	11.68	10.44	8.90	7.40	6.15	5.18	4.43	3.83	2.94	2.32	1.88
Hf	72	TFD	12.2	11.78	10.53	8.98	7.46	6.20	5.23	4.48	3.87	2.97	2.35	1.90
Ta	73	TFD	12.3	11.87	10.63	9.05	7.53	6.26	5.28	4.52	3.91	3.00	2.38	1.93
W	74	TFD	12.4	11.97	10.72	9.13	7.59	6.32	5.33	4.56	3.95	3.03	2.40	1.95
Re	75	TFD	12.5	12.06	10.79	9.21	7.66	6.38	5.38	4.61	3.99	3.06	2.43	1.97
Os	76	TFD	12.6	12.16	10.89	9.29	7.72	6.43	5.43	4.65	4.03	3.19	2.45	1.99
Ir	77	TFD	12.7	12.25	10.96	9.36	7.79	6.49	5.48	4.70	4.06	3.12	2.48	2.01
Pt	78	TFD	12.8	12.35	11.06	9.44	7.86	6.54	5.53	4.74	4.19	3.16	2.50	2.03
Au	79	TFD	12.9	12.45	11.13	9.51	7.92	6.60	5.58	4.78	4.14	3.19	2.53	2.05
Hg	80	TFD	13.0	12.54	11.23	9.58	7.98	6.66	5.63	4.83	4.18	3.22	2.55	2.07

(a) These values are based on the rest-mass of the electron. For diffraction studies using electrons of velocity u these values should be multiplied by $[1 - (u/c)^2]^{-1/2}$.

(b) E: Exact wave function. HF: Hartree–Fock wave function. H: Hartree (nonexchange) wave function. TFD: Thomas–Fermi–Dirac potential function.

(c) The second decimal places are not significant in this column.

† Courtesy of Dr. J. A. Ibers and *Acta Crystallographica* (see also Table II).

Table II

The following tabulations are meant to replace the tabulations for the same atoms in Table I. These calculations are based on new Hartree–Fock wave functions and are reproduced here by the courtesy of Dr. J. A. Ibers. These are very recent results (to September 1961) and are more up-to-date than those given in International Tables for *X-ray Crystallography* (vol. 3).

Element	Z	$\sin\theta/\lambda(\text{Å}^{-1})$											
		0.00	0.05	0.10	0.15	0.20	0.25	0.30	0.35	0.40	0.50	0.60	0.70
He	2	(0.445)	0.431	0.403	0.368	0.328	0.288	0.250	0.216	0.188	0.142	0.109	0.086
Ne	10	(1.66)	1.59	1.53	1.43	1.30	1.17	1.04	0.92	0.80	0.62	0.48	0.38
Al	13	(6.1)	5.36	4.24	3.13	2.30	1.73	1.36	1.11	0.93	0.70	0.55	0.45
Si	14	(6.0)	5.26	4.40	3.41	2.59	1.97	1.54	1.23	1.02	0.74	0.58	0.47
P	15	(5.4)	5.07	4.38	3.55	2.79	2.17	1.70	1.36	1.12	0.80	0.61	0.49
S	16	(5.2)	4.88	4.36	3.63	2.93	2.33	1.85	1.49	1.22	0.86	0.65	0.51
Cl	17	(5.0)	4.69	4.24	3.62	3.00	2.44	1.98	1.61	1.32	0.93	0.69	0.54
K	19	(9.0)	(7.0)	5.43	(4.10)	3.15	(2.60)	2.14	(1.90)	1.49	1.07	0.79	0.61
Sc	21	(9.7	8.35	6.30	4.63	3.50	2.75	2.29	1.92	1.62	1.18	0.89	0.69)
Ti	22	8.9	7.95	6.20	4.63	3.55	2.84	2.34	1.97	1.67	1.23	0.93	0.72

V	23	8.4	7.60	6.06	4.60	3.57	2.88	2.39	2.02	1.72	1.28	0.97	0.76
Cr	24	(8.0	7.26	5.86	4.55	3.58	2.89	2.42	2.06	1.76	1.32	1.01	0.80)
Mn	25	7.7	7.00	5.72	4.48	3.55	2.91	2.44	2.08	1.79	1.36	1.04	0.83
Fe	26	7.4	6.70	5.55	4.41	3.54	2.91	2.45	2.11	1.82	1.39	1.08	0.86
Co	27	7.1	6.41	5.41	4.34	3.51	2.91	2.46	2.12	1.84	1.42	1.11	0.89
Ni	28	6.8	6.22	5.27	4.27	3.48	2.90	2.47	2.13	1.86	1.46	1.14	0.92
Cu	29	(6.5	6.00	5.11	4.19	3.44	2.88	2.46	2.13	1.87	1.47	1.16	0.95)
Zn	30	(6.5)	5.48	5.00	4.14	3.42	2.88	2.47	2.14	1.88	1.48	1.10	0.96
Ga	31	(7.5)	6.70	5.62	4.51	3.64	3.00	2.53	2.18	1.91	1.50	1.20	0.98
Ge	32	(7.8)	6.89	5.93	4.81	3.87	3.16	2.63	2.24	1.94	1.51	1.22	0.99
As	33	(7.8)	6.99	6.05	5.01	4.07	3.32	2.74	2.31	1.99	1.54	1.23	1.01
Se	34	(7.7)	6.99	6.15	5.18	4.24	3.47	2.86	2.40	2.05	1.57	1.23	1.02
Br	35	(7.3)	6.80	6.15	5.25	4.37	3.60	2.97	2.49	2.12	1.60	1.27	1.04
Ag	47	(8.8)	8.24	7.47	6.51	5.58	4.75	4.05	3.46	2.97	2.22	1.70	1.35
W	74	(14)	—	11.80	—	7.43	—	5.16	—	3.85	2.99	2.39	1.96
Hg	80	(13.3)	12.26	10.82	9.18	7.70	6.48	5.50	4.72	4.09	3.16	2.51	2.05

Values in parentheses are interpolations or extrapolations.

Index